W9-CRB-867

# Catalytic Activation of Carbon Monoxide

# Catalytic Activation of Carbon Monoxide

**Peter C. Ford,** Editor
*University of California*

Based on a symposium sponsored by the Division of Inorganic Chemistry at the Second Chemical Congress of the North American Continent (180th ACS National Meeting), Las Vegas, Nevada, August 25–27, 1980.

ACS SYMPOSIUM SERIES 152

AMERICAN CHEMICAL SOCIETY
WASHINGTON, D. C. 1981

6483-1164
sep/ae
chem

XD81
1084
CHEM

Library of Congress CIP Data

Catalytic activation of carbon monoxide.
(ACS symposium series, ISSN 0097-6156; 152)

Includes bibliographies and index.

1. Catalysis—Congresses. 2. Carbon monoxide—Congresses. 3. Chemistry, Organic—Synthesis—Congresses.

I. Ford, Peter C. II. American Chemical Society. Division of Inorganic Chemistry. III. Chemical Congress of the North American Continent (2nd: 1980: Las Vegas, Nev.). IV. Series.

TP156.C35C39 661'.8 81-1885
ISBN 0-8412-0620-1 AACR2
ACSMC8 152 1-358 1981

Copyright © 1981

American Chemical Society

All Rights Reserved. The appearance of the code at the bottom of the first page of each article in this volume indicates the copyright owner's consent that reprographic copies of the article may be made for personal or internal use or for the personal or internal use of specific clients. This consent is given on the condition, however, that the copier pay the stated per copy fee through the Copyright Clearance Center, Inc. for copying beyond that permitted by Sections 107 or 108 of the U.S. Copyright Law. This consent does not extend to copying or transmission by any means—graphic or electronic—for any other purpose, such as for general distribution, for advertising or promotional purposes, for creating new collective work, for resale, or for information storage and retrieval systems.

The citation of trade names and/or names of manufacturers in this publication is not to be construed as an endorsement or as approval by ACS of the commercial products or services referenced herein; nor should the mere reference herein to any drawing, specification, chemical process, or other data be regarded as a license or as a conveyance of any right or permission, to the holder, reader, or any other person or corporation, to manufacture, reproduce, use, or sell any patented invention or copyrighted work that may in any way be related thereto.

PRINTED IN THE UNITED STATES OF AMERICA

5/28/81 SD DS

TP156
C35
C39
CHEM

ACS Symposium Series

**M. Joan Comstock,** *Series Editor*

*Advisory Board*

David L. Allara
Kenneth B. Bischoff
Donald D. Dollberg
Robert E. Feeney
Jack Halpern
Brian M. Harney
W. Jeffrey Howe
James D. Idol, Jr.
James P. Lodge
Marvin Margoshes
Leon Petrakis
Theodore Provder
F. Sherwood Rowland
Dennis Schuetzle
Davis L. Temple, Jr.
Gunter Zweig

C6293

# FOREWORD

The ACS Symposium Series was founded in 1974 to provide a medium for publishing symposia quickly in book form. The format of the Series parallels that of the continuing Advances in Chemistry Series except that in order to save time the papers are not typeset but are reproduced as they are submitted by the authors in camera-ready form. Papers are reviewed under the supervision of the Editors with the assistance of the Series Advisory Board and are selected to maintain the integrity of the symposia; however, verbatim reproductions of previously published papers are not accepted. Both reviews and reports of research are acceptable since symposia may embrace both types of presentation.

# CONTENTS

# PREFACE

The catalytic activation of carbon monoxide is a research area currently receiving major attention from academic, industrial, and government laboratories. There has been a long standing interest in this area; however, the new attention obviously is stimulated by concerns with the present and future costs and availability of petroleum as a feedstock for the production of hydrocarbon fuels and of organic chemicals. One logical alternative source to be considered is "synthesis gas," mixtures of carbon monoxide and hydrogen that can be produced from coal and other carbonaceous materials.

Potential applications of synthesis gas include conversion to liquid fuels via the Fischer–Tropsch reaction, production of hydrogen via the shift reaction (for ammonia manufacture and for the direct liquifaction of coal) and the production of methanol, ethylene glycol, and other oxygenated organic chemicals. Efficient catalysts for such processes need to be designed and their fundamental reaction mechanism chemistry understood. The symposium was organized to focus on these questions. The major emphasis was directed toward homogeneous catalysis; however, several authors addressed the question of characterizing catalysis pathways on surfaces. The chapters included in this volume comprise the major part of the papers presented and are organized in the order of presentation. The symposium was sponsored by the Inorganic Division of the American Chemical Society and also received some financial support from the Chevron Research Company, for which the Editor is grateful.

PETER C. FORD
Santa Barbara, California
November 30, 1980

1

# Activation of Carbon Monoxide by Carbon and Oxygen Coordination

## Lewis Acid and Proton Induced Reduction of Carbon Monoxide

D. F. SHRIVER

Department of Chemistry, Northwestern University, Evanston, IL 60201

In its free state, carbon monoxide is highly resistant to attack by hydrogen and a variety of other common reducing agents. The reactivity of coordinated CO is much greater than that of the free molecule and metal surfaces are in general even more effective than simple coordination compounds in promoting CO reduction. One great challenge to the inorganic chemist is to make the connection between chemistry which occurs on the surfaces of metals and the more readily studied reactions of discrete molecular organometallics. One possible mode of CO activation which has been invoked in heterogenous catalysis is C and O bonding to a surface. This bifunctional activation of CO may lead to CO cleavage and eventual incorporation of a surface carbide into organic products, or to direct incorporation of the C and O coordinated CO into an organic group (1-3). Bifunctional activation also is thought to be important in molecular systems (4,5), but it is fair to say that the evidence for, and understanding of this phenomenon has been very rudimentary. In this paper we present the results of studies at Northwestern which were designed to provide clear-cut evidence for bifunctional CO activation in molecular systems and to provide information on the important

0097-6156/81/0152-0001$05.00/0
© 1981 American Chemical Society

chemical variables in these reactions. We first describe Lewis acid promotion of the alkyl migration (CO insertion) reaction, including recent results on the combination of this acid promoted alkyl migration reaction with CO reduction. This repetitive sequence of CO insertion and carbonyl reduction provides a means of building hydrocarbon chains under mild conditions. Finally, proton induced CO reduction will be described, and the most recent mechanistic information on this reaction will be presented. As a prelude to these discussions, we outline two fundamental reactions of coordinated CO.

## Electrophilic and Nucleophilic Attack of Coordinated CO

The attack on coordinated carbon monoxide by nucleophiles was first extensively developed in synthetic organometallic chemistry by E. O. Fischer and his students (6); as discussed by others in this volume, this reaction provides one route to the reduction of coordinated CO and to catalysis of the water gas shift reaction. Those carbonyl groups which are susceptible to attack by nucleophiles are electron deficient, as judged by their high CO stretcing frequencies (7).

By contrast, metal carbonyls having low CO stretcing frequencies are susceptible to attack of the CO oxygen atom by electrophiles such as $Al(CH_3)_3$, $AlBr_3$, or $BF_3$. This chemical evidence and a variety of physical evidence indicate that a low CO stretching frequency corresponds to high electron density on the CO ligand (8). Carbonyl groups in this category include bridging carbonyls, terminal carbonyls in metal carbonyl anions, and terminal carbonyls in donor substituted metal carbonyls, structures 1a through 1c. The attack on bridging carbonyls by electrophilic

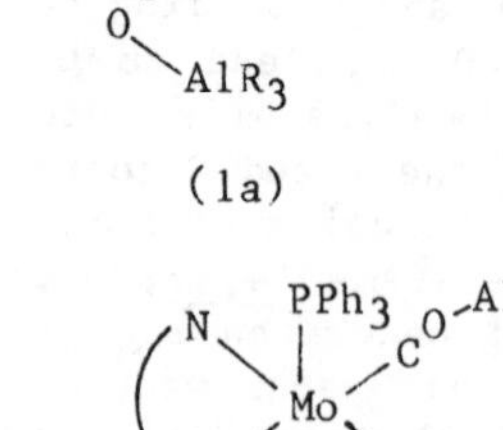

(1a)

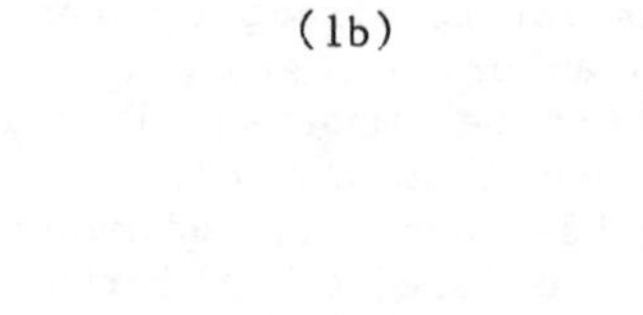

(1b)

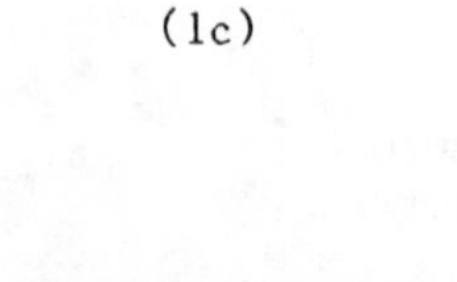

(1c)

reagents is a common feature of the chemistry of polynuclear carbonyls, and it may lead to a variety of CO rearrangements (8,9). One striking physical effect of Lewis acid addition to the oxygen end of CO is a very large reduction in the CO stretching frequency, which implies a large decrease in CO bond order, Figure 1 (10). This phenomenon will be discussed in more detail at the end of the paper, and for the present it will suffice to point out that the addition of a Lewis acid to the carbonyl oxygen favors carbene-like resonance structures, which arise from the polarization of the π system, equation 2.

$$L_nM{-}C{\equiv}O + AlX_3 \longrightarrow L_nM{=}C{=}O{\smallsetminus}AlX_3 \qquad (2)$$

The very large perturbing influence of C and O bonding on the CO bond order led us to explore the influence of Lewis acid and proton acid promoted reactions of metal carbonyl complexes.

## Acid Promoted CO Insertion

Owing in part to its great commercial importance, the CO insertion reaction is perhaps the most thoroughly studied metal carbonyl reaction other than substition (11-13). As shown in equation 3a, the currently

$$L_n\overset{R}{\overset{|}{M}}{-}CO \underset{k_{-1}}{\overset{k_1}{\rightleftharpoons}} L_nM{-}C\overset{\diagup R}{\underset{\diagdown\!\!\diagdown O}{}} \qquad K = \frac{k_1}{k_{-1}} \qquad (3a)$$

accepted mechanism for this reaction is the migration of the alkyl group onto a coordinated CO, to yield a coordinatively unsaturated metal acyl intermediate (which perhaps may be solvent stabilized). This intermediate is then attacked by an incoming ligand to produce a stable acyl complex, equation 3b. When a

$$L_nM{-}C\overset{\diagup R}{\underset{\diagdown\!\!\diagdown O}{}} + L' \underset{k_{-2}}{\overset{k_2}{\rightleftharpoons}} L_nL'M{-}C\overset{\diagup R}{\underset{\diagdown\!\!\diagdown O}{}} \qquad (3b)$$

stable product is formed, $k_{-2} \ll k_2$, the kinetic expression takes the form given in equation 4, with two limiting conditions, equations 5a and 5b.

$$\text{rate} = \frac{k_1k_2[L'][L_nMR(CO)]}{k_{-1} + k_2[L']} \qquad (4)$$

$$\text{rate} = Kk_2[L'][L_nMR(CO)] \text{ when } k_2[L'] \ll k_{-1} \qquad (5a)$$

$$\text{rate} = k_1[L_nMR(CO)] \text{ when } k_2[L'] \gg k_{-1} \qquad (5b)$$

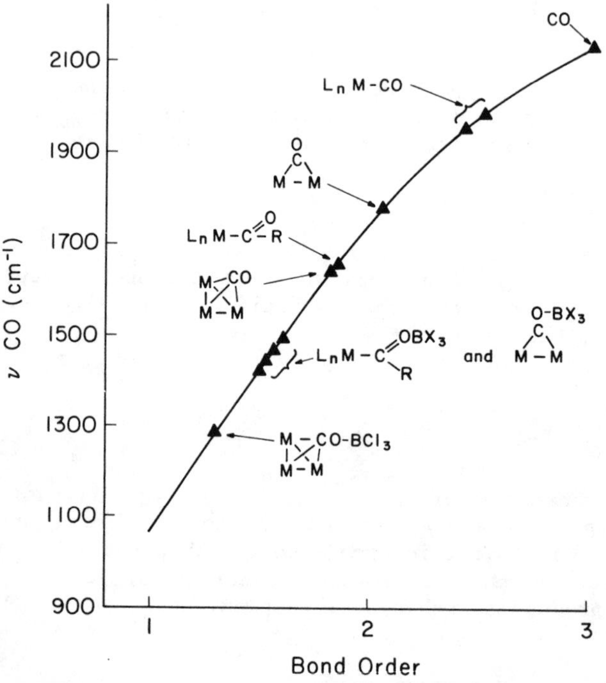


*Figure 1. Bond order vs. CO stretching frequency; the curve was determined from data on organic compounds, data for organometallic compounds have been entered on the curve based on observed CO stretching frequencies.*

The simple second-order kinetics of equation 5a apply for $Mn(CO)_5(CH_3)$ when L=CO at subatmospheric pressures. It is under this set of conditions that we have studied the Lewis acid promoted CO insertion reaction, see Figure 2.

Prior to our studies it was recognized that ion pairing with anionic metal carbonyls could promote CO insertion and related reactions (14-16). Both kinetic and non-kinetic evidence suggests the importance of ion pairs in these types of reactions (14,17). For example, a small cation was found to greatly accelerate the CO insertion reaction relative to the same reaction with a large cation, equation 6 (14).

$$M^+[RFe(CO)_4^-] + L' \xrightarrow{THF} M^+[L'Fe(CO)_3(CRO)] \qquad (6)$$

$$\text{rate } (M^+ = (Ph_3P)_2N^+) : \text{rate } (M^+ = Li^+) \quad 1:10^3$$

Our work on the bifunctional activation of CO insertion was prompted by the thought that strong molecular Lewis acids should be more effective and more general than simple cations. It already had been observed that molecular Lewis acids would promote a molecular Fischer-Tropsch type reaction (5), and that iron diene complexes can be converted to polycyclic ketones by the action of aluminum halides, equation 7,(18), but information on the course of these reactions was sketchy.

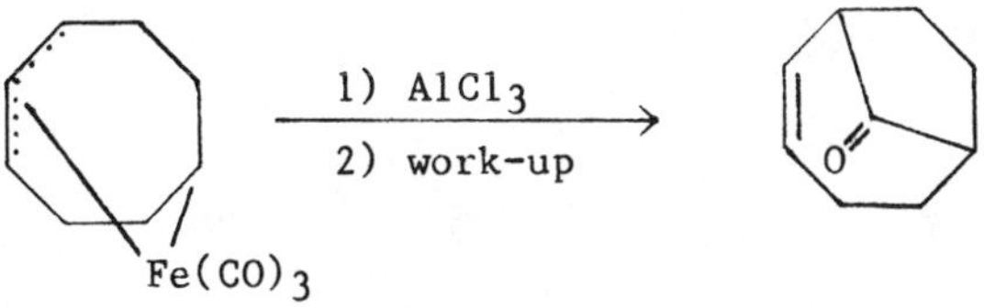


(7)

In the first studies performed at Northwestern, Steven Strauss found that $AlBr_3$ brought about a moderate increase in the rate of CO uptake by $Mn(CO)_5(CH_3)$. Susan Butts then discovered that the reaction occurs in two steps. The first is a very rapid CO insertion to yield a cyclic product, equation 8a, which is followed by the much slower uptake of CO, equation 8b.

(8a)

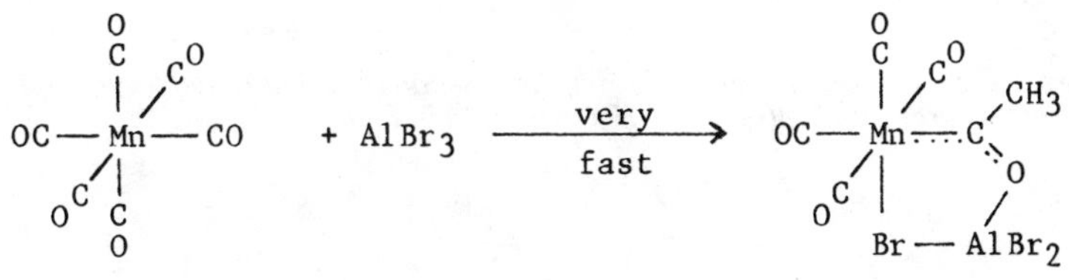

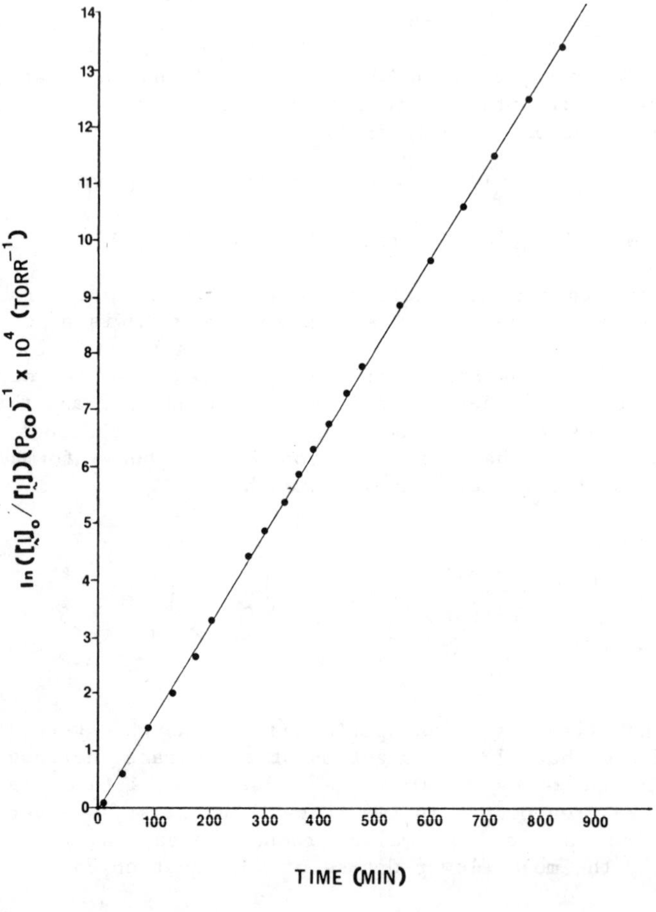


Journal of the American Chemical Society

*Figure 2. Second-order kinetic plot for the reaction of CO at subatmospheric pressures with $Mn(CO)_4(CCH_3OAlBrBr_2)$ in toluene solution* (19)

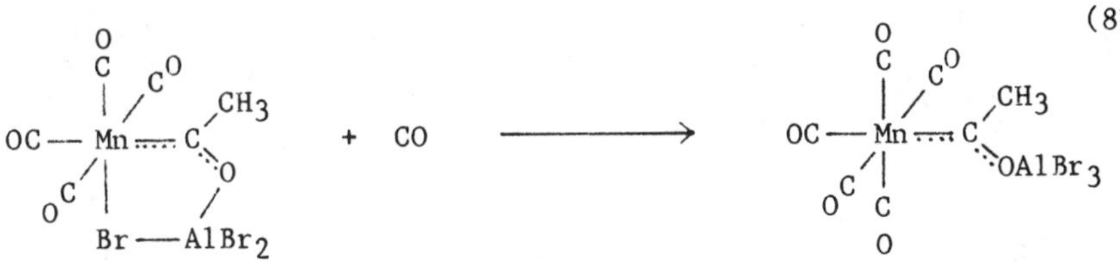

(8b)

The structure of the cyclic reaction product of equation 8a has been determined by Dr Elizabeth Holt at the University of Georgia, and a similar cyclic structure is indicated by spectroscopic data for other metal systems (19).  Reaction 8a is so rapid that the rate has eluded measurement by conventional kinetic methods, but based on the immediate color change upon mixing we estimate that the rate is at least $10^3$ greater than the forward rate in reaction 3a.  Thus the role of the Lewis acid is not simply to capture the coordinatively unsaturated intermediate, $Mn(CO)_4(COR)$, but rather to promote the alkyl migration process. Attack by the Lewis acid must occur prior to or very early in the sequence of alkyl migration.   One possible variation on this general picture is prior equilibrium complex formation between $Mn(CO)_5R$ and $AlBr_3$, eq. 9a, followed by alkyl migration, eq. 9b. We have no experimental evidence on whether or not the reaction

$$RMn(CO)_5 + AlBr_3 \rightleftharpoons RMn(CO)_4(COAlBr_3) \qquad (9a)$$

$$RMn(CO)_5(COAlBr_3) \longrightarrow Mn(CO)_4(\overline{CROAlBrBr_2}) \qquad (9b)$$

rate is increased by simultaneous bromide attack on Mn, but the theoretical studies of Berke and Hoffmann suggest that nucleophilic attack on the central metal is not likely to assist the alkyl migration reaction (13).  There is no direct evidence for the equilibrium complex formation depicted in equation 9a; indeed the high CO stretching frequencies of $Mn(CO)_5(CH_3)$ are in a range for which stable complexes have not been observed between carbonyls and Lewis acids such as $AlBr_3$.  However, the frequency-basicity correlation does not exclude the possibility of minute but kinetically important amounts of the complex being formed.   It also is possible that a pre-equilibrium does not exist but instead alkyl migration occurs simultaneously with Lewis acid attack, equation 10.  Whatever the finer details of this step, there is no doubt of the great acceleration of alkyl migration by molecular Lewis acids.

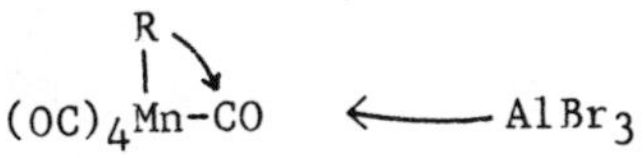

$$(OC)_4\overset{\overset{\displaystyle R}{|}}{Mn}\text{-}CO \longleftarrow AlBr_3 \qquad (10)$$

The second slower step in the over-all reaction, eq. 8b, obeys second order kinetics, eq. 11, and Figure 2.   The relative rate increases on going from Br to Cl to F bridging groups, Table I (19). In view of these data, and of the general occurence of a dissociative pathway for substitution reactions on octahedral 18 electron complexes, the most likely rate determining process in these reactions is Mn-halide bond breaking.

$$rate = k[CO][\overline{Mn(CO)_4(CROAlBr}Br_2)] \quad (11)$$

The CO insertion process also can be promoted by proton acids (20). The only compound to be studied in detail is $Mn(CO)_5(CH_3)$, for which very weak acids such as acetic acid bring

Table I. Initial Rate of CO Uptake Under Uniform Conditions

| Complex | Relative Initial Rate |
|---|---|
| $Mn(CH_3)(CO)_5$ | 1 |
| $\overline{Mn(C(OAlBr}Br_2)CH_3)(CO)_4$ | 4 |
| $\overline{Mn(C(OAlCl}Cl_2)CH_3)(CO)_4$ | 23 |
| $\overline{Mn(C(OBF}F_2)CH_3)(CO)_4$ | 43 |

about little rate enhancement, acids of intermediate strength cause appreciable rate enhancement, and strong acids such as HBr bring about a competitive cleavage of the $Mn-CH_3$ bond, Table II.

Table II. HX Promoted CO Insertion

$$Mn(CO)_5(CH_3) + CO \xrightarrow{HX} Mn(CO)_5(\overset{O}{C}CH_3)$$

| HX | Relative Rate |
|---|---|
| HBr | ($CH_4$ evolved) |
| $HOOCCF_3$ | >9 (some $CH_4$) |
| $HOOCCCl_2H$ | 7 |
| $HOOCCClH_2$ | 2 |
| none | 1 |

Kinetic studies by Butts and Richmond indicate that both the monomer and dimer of dichloroacetic acid promote the reaction in an aromatic solvent, equations 12a and 12b, (20).

$$Mn(CO)_5(CH_3) + CO \xrightarrow[\phi Me \text{ soln.}]{HA} Mn(CO)_5(CCH_3O) \qquad (12a)$$

$$rate = k\{[HA] + \alpha[H_2A_2]\}[Mn(CO)_5(CH_3)][CO] \qquad (12b)$$

where HA = $Cl_2HCOOH$ and $\alpha \approx 0.2$

A very exciting recent development in joint research with Professor Burwell is the observation by Correa, Nakamura and Stimson, that metal oxide surfaces also promote CO insertion (21). In this study it was possible to characterize the nature of the reaction products formed by the interaction of alkylmetal carbonyls with metal oxide surfaces, which had been activated to produce surface acid and base sites. The reaction was followed by two methods, in the first instance the evolution of CO, $H_2$, $CH_4$ and other light molecules was measured in the course of the interaction of the organometallic molecule with the surface. Secondly the nature of the reaction product on the surface was deduced from Fourier transform infrared spectroscopy of the surface species. Fortunately we have available molecular analogues of the surface species so the structural inferences from infrared spectroscopy are quite strong. For example a sample of γ-alumina was heated to 900°C in high purity helium, cooled, and then exposed to $Mn(CO)_5(CH_3)$. The resulting infrared spectrum, when compared with the molecular species, indicates that a surface acetyl is formed with a cyclic structure analogous to those seen in the molecular Lewis acid promoted reactions, equation 13.

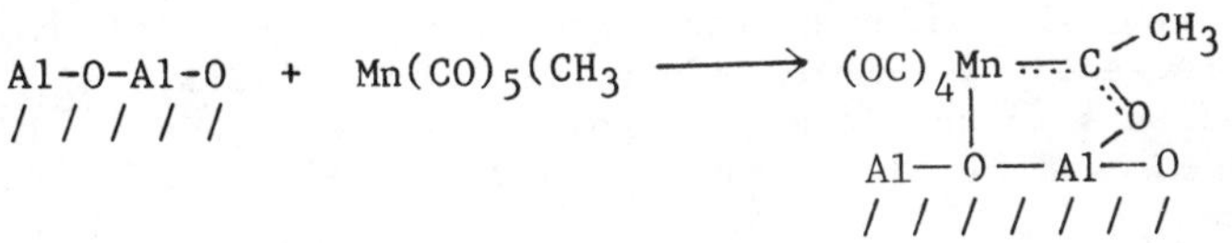

$$Al-O-Al-O + Mn(CO)_5(CH_3) \longrightarrow (OC)_4Mn \cdots C\underset{\underset{\underset{/ / / / / / /}{Al-O-Al-O}}{O}}{\overset{CH_3}{\diagup}} \qquad (13)$$

This insertion reaction is very fast, as judged by the immediate color change upon exposure of the alumina surface to the organometallic, and in keeping with these interpretations no appreciable amounts of gaseous products are evolved on the time scale of the measurements. One implication of this observation is that the metal oxides, which are frequently employed to support conventional metallic heterogenous catalysts, may play an active role in promoting CO reduction by the interaction of surface Lewis acid sites with the oxygen end of CO. Indeed support effects are well documented for the conversion of CO to hydrocarbons (22-23), and therefore we speculate that interactions, such as those suggested in equation 14, may be important. The usual

(14)

explanation is that The $Al_2O_3$ surface withdraws electron density from the metal particle, thereby changing its reactivity. We do not believe that this is a chemically reasonable explanation, because $Al_2O_3$ lacks low lying conduction bands which might accept electron density. More localized interactions, such as those between the surface atoms and the metal particle should be ineffective in greatly altering the charge on the metal particle, because the presence on the surface of both electron acceptors, $Al^{+3}$, and donors, $O^{-2}$, should yield mutually compensating effects.

## Addition of $MX_3$ Across CO Multiple Bonds

The organic analogues of the reactions to be discussed here are the borane reductions of aldehydes and ketones and the addition of metal alkyls across ketonic carbonyls, equation 15. In contrast to the ease of these organic reactions, qualitative data which has accumulated in our laboratory over the last decade demonstrates that the carbonyl group in organometallics is fairly resistant to addition across CO. For example, many stable adducts of organometallic carbonyls with aluminum alkyls are known, eq. 1c, but under similar conditions a ketone will quickly react by addition of the aluminum alkyl across the CO bond. A similar reactivity pattern is seen with boron halides.

There is good evidence that addition across ketonic CO groups is preceeded by simple adduct formation (24-26), and it is thought that this adduct formation polarizes the carbonyl, making the carbon susceptible to attack by the $R^-$, $H^-$, or $X^-$ nucleophiles, equation 15.

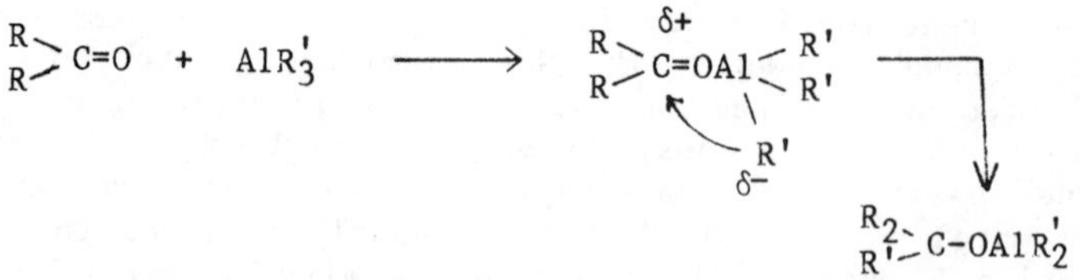


(15)

The resistance of metal carbonyls to addition across the CO bond may reflect the influence of the adjacent electron rich metal center, which can delocalize electron density onto the car-

bonyl carbon.  The electron density shift from the metal to the
carbonyl carbon will thereby partially offset the polarization of
the carbonyl by the Lewis acid, and thus moderate the reactivity
of the carbonyl carbon toward nucleophiles.  Vibrational spectroscopic evidence for this electron delocalization upon adduct formation has been cited in an earlier section.

The depressed reactivity of the CO bond in metal carbonyls
relative to organic carbonyls is not apparent in the case of $BH_3$
and $AlH_3$.  For example, Masters and coworkers have observed that
$H_3B \cdot THF$ reduces metal acyl compounds to the corresponding alkyls,
eq. 16.  Although no mechanistic studies have been reported, it

$$L_nM-\overset{\overset{O}{\parallel}}{C}-CH_3 + H_3B \cdot THF \longrightarrow L_nM-CH_2CH_3 \qquad (16)$$

was proposed that this reduction is preceeded by the formation of
a $BH_3$ adduct at the acyl oxygen (4).  Recently, $AlH_3$ reductions
of metal carbonyls to produce hydrocarbons have been reported as
well (27,28).

Investigations in our laboratory by Rebecca Stimson have
demonstrated that it is possible to combine the borane reduction
of a metal acyl with the Lewis acid promoted CO insertion reaction which has been discussed earlier in this paper (29).  In
this reaction, which is presumed to proceed by equation 17, the

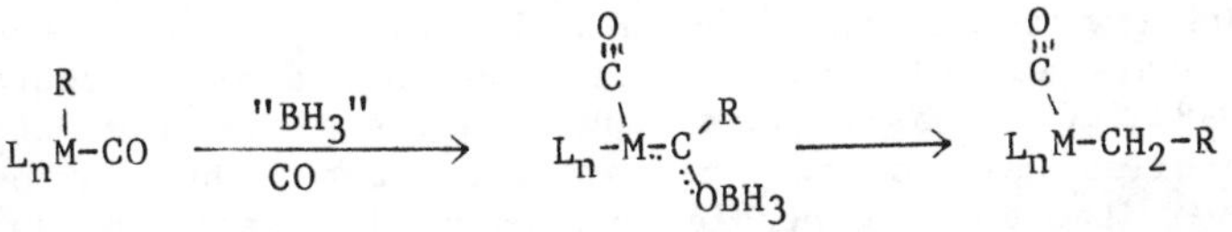

$$\overset{R}{\underset{L_nM-CO}{|}} \xrightarrow[CO]{"BH_3"} L_n-\overset{\overset{O}{\underset{\parallel}{C}}}{M\doteq C}\overset{R}{\underset{OBH_3}{\diagdown}} \longrightarrow L_nM-CH_2-R \qquad (17)$$

$BH_3$ acts both as a reducing agent for the acyl carbonyl and as a
promoting agent for subsequent CO insertion into the metal-alkyl
bond.  As yet the process has been carried as far as $C_4H_9$, with
$Mn(CO)_5(CH_3)$, CO, and $H_3B \cdot THF$ as reactants.

## CO Conversion to Methylidynes

As previously illustrated in Figure 1, the CO bond order is
greatly reduced by the addition of an acceptor to the carbonyl
oxygens of a metal carbonyl.  The CO stretching frequencies and
structures of these adducts bear close resemblances to those in
compounds which are regarded as methylene (carbene) and methylidyne (carbyne) complexes.  Comparisons between some of these
analogues are given in 18a and 18b.  (For the sake of clarity,

$$L_nM \doteq C\overset{O-AlBr_3}{\underset{CH_3}{\diagdown}} \qquad vs \qquad L_nM \doteq C\overset{O-CH_3}{\underset{CH_3}{\diagdown}} \qquad (18a)$$

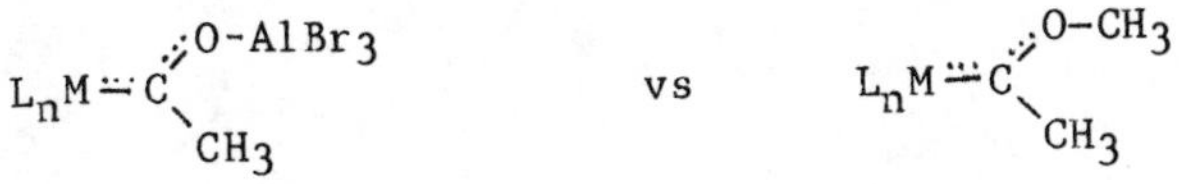

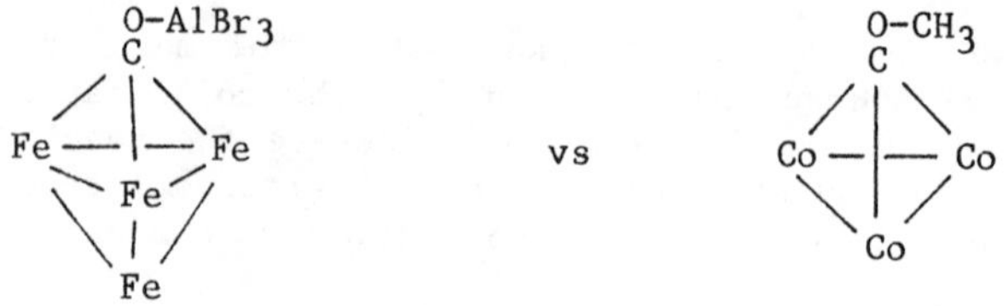

(18b)

terminal CO ligands will be omitted from the structural representations of the polynuclear carbonyls throughout the rest of the text.) The analogy with the methylidenes is strengthened by the recent discovery that bridging carbonyls in anionic polynuclear carbonyl anions are susceptible to alkylation and protonation, equations 19a and 19b (30-35). These reactions contrast

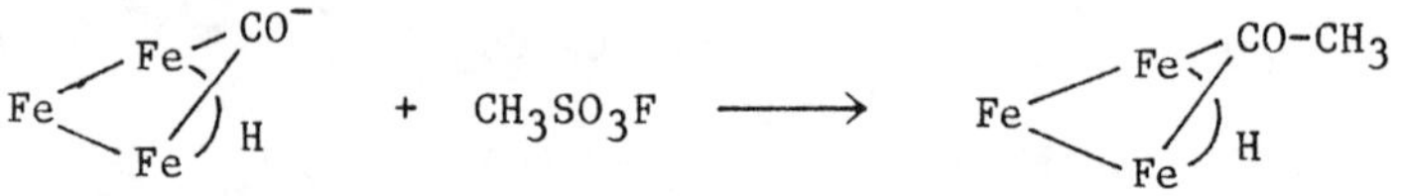

(19a)

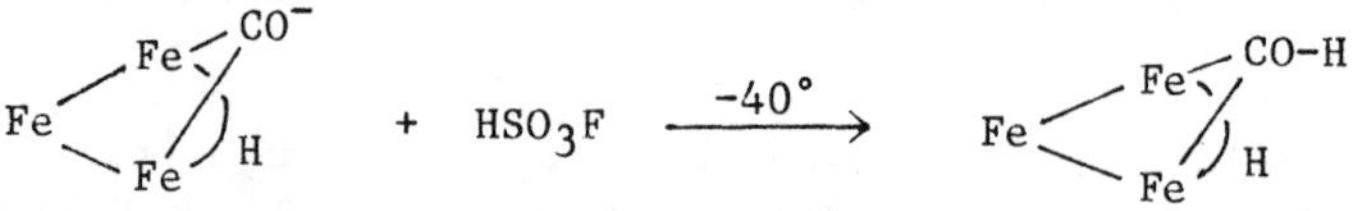

(19b)

markedly with the usual products obtained from mononuclear carbonyl anions, where $CH^+$ or $H^+$electrophyles attack metal centers (36). Keister has shown that in the case of $HRu_3(CO)_{11}^-$ the protonation at low temperatures occurs initially at a carbonyl oxygen, and this is followed by migration to the metal center, producing $H_2Ru_3(CO)_{11}$ (37). This result suggests that the attack by electrophiles may often initially go onto the carbonyl ligands, even though the stable and only observable products indicate that the metal center is the ultimate nucleophile.

Examples of O-alkylation have also been demonstrated for triruthenium and triosmium anionic clusters, as well as the tetrairon cluster $Fe_4(CO)_{13}^{2-}$ (31-33). This reaction has considerable promise as an entry into many different methylidyne complexes through the replacement of the $OR^-$ group, eq. 20 (38).

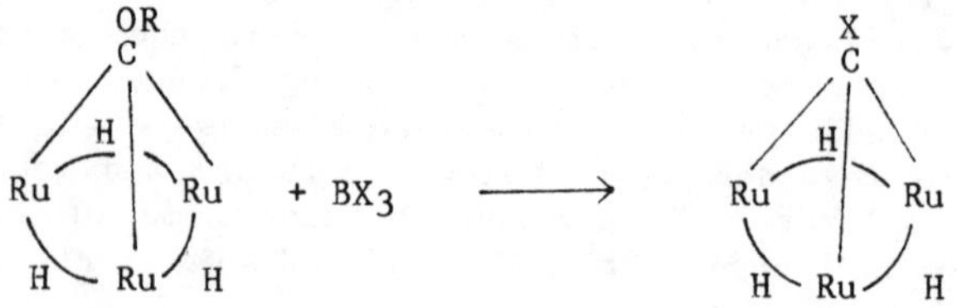

(20)

Protonation of the carbonyl oxygen is as yet only recognized in a couple of cases (34,35). Even though O-alkylation and perhaps O-protonation open up useful synthetic paths in metal carbonyl chemistry, their main interest for the purposes of this survey is their intermediacy in the further reduction of CO.

## Proton Induced Reduction of CO

The observation of O-protonation with the attendant formal reduction of the carbonyl carbon suggested to us that further protonation steps might lead to methane or methanol formation. In this process the necessary electrons for the reduction would be provided by the metal cluster, as indicated schematically in equation 21. After considerable experimentation with reactants

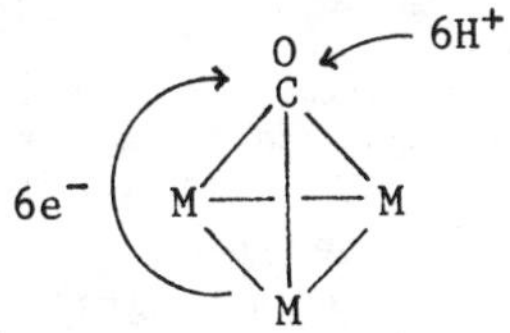


$$\longrightarrow M_3^{6+} + CH_4 + H_2O \qquad (21)$$

and reaction conditions, Kenton Whitmire demonstrated the first reaction of this type, eq. 22 (29). A yield of about 0.5 $CH_4$ per

$$Fe_4(CO)_{13}^{2-} \xrightarrow[\sim 3\text{ days, room t}]{\text{neat } H_2SO_3CF_3} CH_4 + Fe^{2+} + CO + H_2 + \text{Fe clusters} \qquad (22)$$

cluster was obtained, and isotopic experiments demonstrated that the carbon in the $CH_4$ originated from a carbonyl carbon and not from some adventitious source. It was also shown that the hydrogen in the methane is derived directly from the proton, and not via intermediate $H_2$. Furthermore, the quantity of $Fe^{2+}$ produced is sufficient to account for the equivalents of electrons necessary to yield $CH_4$ and $H_2$ (39). This reaction clearly fits our original concept of proton induced reduction.

With the general nature of reaction 22 established, we are now concentrating on a more detailed understanding of the mechanism. The first step undoubtedly is the known monoprotonation and rearangement of the tetrahedral iron cluster, $Fe_4(CO)_{13}^{2-}$ to yield a butterfly arrangement of the iron atoms in the product $HFe_4(CO)_{13}^-$ (40). This interesting reaction, which produces a unique $\eta^2$-CO ligand, is shown schematically in equation 23. The next step is likely to be the protonation of the unique bridging CO ligand to yield 24a, which is analogous to the known compound 24b (41). In keeping with this interpretation, Whitmire has shown by isotope tracer studies that the unique $\eta^2$-CO in 24b

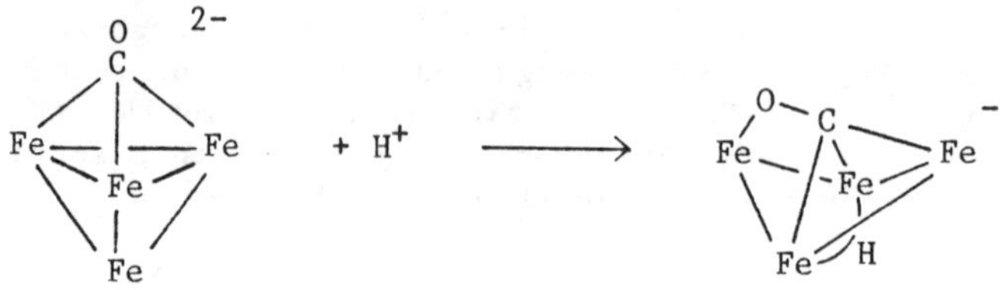

(23)

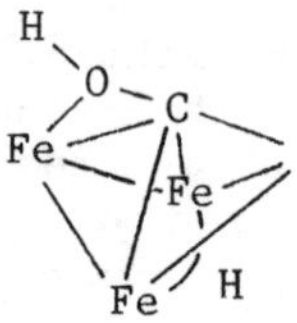

(24a)

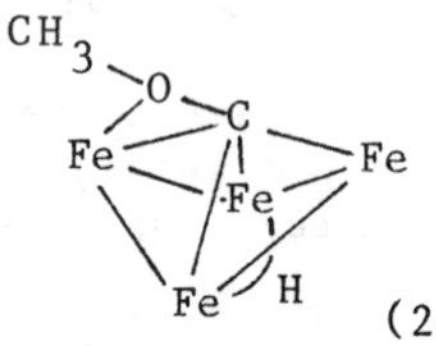

(24b)

undergoes proton induced reduction to yield $CH_4$. The exact route by which this transformation to methane occurs is still not fully understood, however the reaction mixture which has been quenched before extensive reaction has occurred, contains significant quantites of the previously known methyne (42,43), illustrated in equation 25. Whitmire has shown that this methyne in the

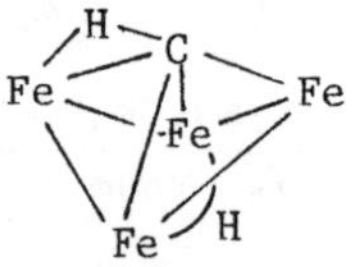

(25)

presence of neat $HSO_3CF_3$ yields some $CH_4$. Therefore, we are inclined to write the sequence of reactions shown in equation 26 to describe the proton induced reduction of CO in $Fe_4(CO)_{13}^{2-}$.

The above observations strongly indicate that O-protonation is an important step in this particular reaction for the reduction of coordinated CO. Recent studies in our laboratory provide other examples of proton induced reduction in metal cluster systems, and an example of proton induced CO reduction has recently been reported by Atwood (44). It thus appears that protons as well as Lewis acids are effective in the bifunctional activation of coordinated CO.

## Principles of Bifunctional CO Activation

The foregoing examples clearly demonstrate that the attachment of an electron acceptor to the oxygen of coordinated CO activates this molecule toward a variety of reduction reactions.

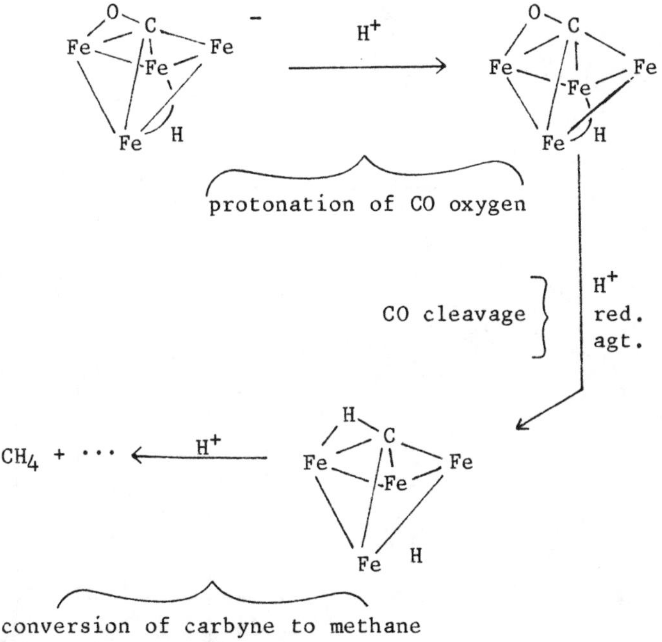


(26)

This bifunctional activation can be ascribed to a variety of factors (10,13,14): (1) lowering of the CO bond order to more nearly approximate that of the products, (2) stabilization of electron migration toward oxygen, which typically occurs during reduction, (3) lowering of a specific unoccupied MO which facilitates the formation of the transition state, and (4) polarization of the CO bond, making the carbon susceptible to attack by nucleophiles. These are all the simple consequence of transforming CO into a more polar group by the electron acceptor. This perturbation can be illustrated by comparison of the electronic structure of CO with that of the more polar isoelectronic molecule $CF^+$. Because of the very simple relationships which exist between the MO's of heteronuclear diatomic molecules, one can immediately make several predictions about the relative properties of these ligands. The highest occupied $\sigma$ orbital will be lowered in energy so $CF^+$ will be a poorer $\sigma$ donor, the energy of the $\pi^*$ LUMO will be lowered and the amplitude on C will be increased thereby increasing the $\pi$ acceptor character of the ligand (eq. 27), the

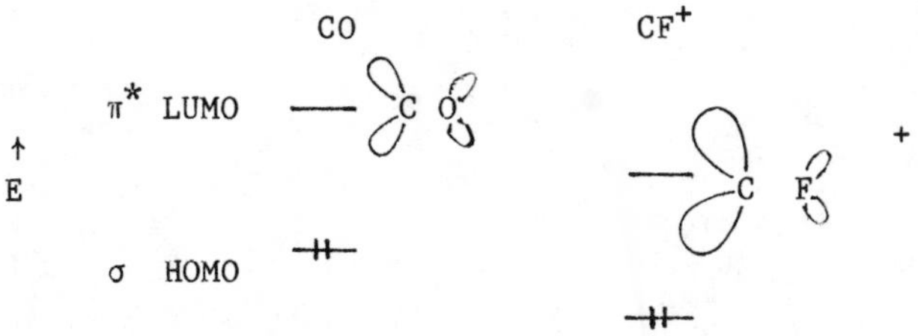


(27)

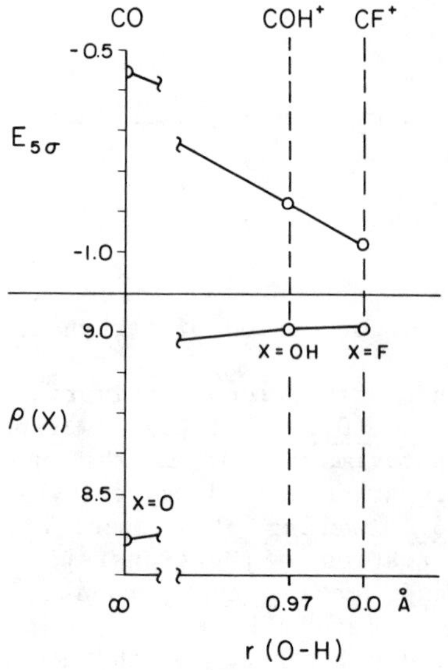


*Figure 3. Upper frame: Energy in Hartrees of the 5 σ orbital for CO, $COH^+$, and $CF^+$. Lower frame: Electron density on O, OH, and F.*

$\pi$ overlap population will be reduced between C and X, and there will be a drift of charge from C to X making the C more susceptible to attack by electrophiles (in the actual complex this last perturbation will be attenuated by the metal to C $\pi$ bonding). Some of these changes and the similarity of $CF^+$ to an O protonated CO are shown in Figure 3, where the results of *ab initio* MO calculations (45), are presented in graphical form.

Acknowledgments.

Our research on CO activation is supported by the NSF. In addition to members of my group and colleagues who are mentioned in the text, I appreciate the contributions of former co-workers Dr. Hamdallah Hodali, who discovered the O-protonation of CO, and Dr. Norvell Nelson who discovered the Lewis acid attack of CO. Work on the proton induced reduction of CO has been aided by the exchange of information on iron butterfly compounds with Professor Earl Mutterties, Dr. John Bradly and Dr. Jack Williams.

Literature Cited

1. Denny, P. J.; Whan, D. A. "Catalysis"; Vol. 2, Specialist Periodical Reports, The Chemical Society, London, 1978, p. 46.
2. Blyholder, G.; Emmet, P.H. Jr *Phys. Chem.*, 1959, *63*, 962.
3. Blyholder, G.; Goodsel, A. J. *J. Catal*, 1971, *23*, 374.
4. van Dorn, J. A.; Masters, C.; Volger, H. C. *J. Organometal. Chem.*, 1976, *105*, 245.
5. Demitras, G. C.; Muetterties, E. L. *J. Am. Chem. Soc.*, 1977, *99*, 2976.
6. Fischer, E. O.; Schubert, U. *J. Organometal. Chem.*, 1975, *100*, 59.
7. Darensbourg, M. Y.; Cerder, H. L.; Darensbourg, D. J.; Hasday, C. *J. Am. Chem. Soc.*, 1973, *95*, 5919.
8. Shriver, D. F. *J. Organometal. Chem.*, 1975, *94*, 259.
9. Holt, E.; Whitmire, K.; Shriver, D. F. *Chem. Comm.*, in press.
10. Stimson, R. E.; Shriver, D. F. *Inorg. Chem.*, 1980, *19*, 1141.
11. Calderazzo, F. *Angew Chem.*, *Inte. Ed. Engl.*, 1977, *16*, 299.
12. Wojcicki, A. *Adv. Organometal. Chem.*, 1973, *11*, 87.
13. Berke, H.; Hoffmann, R. *J. Am. Chem. Soc.*, 1978, *100*, 7224.
14. Collman, J. P.; Finke, R. G.; Cawse, J.; Brauman, J. I. *J. Am.Chem. Soc.*, 1978, *100*, 4766.
15. Calderzaao, F.; Noack, K. *Coord. Chem. Rev.*, 1966, *1*, 118.
16. Nitay, M.; Priester, W.; Rosenblum, M. *J. Am. Chem. Soc.*, 1978, *100*, 3620.
17. Darensbourg, M. Y.; Darensbourg, D. J.; Burns, D.; Drew, D. A. *J. Am. Chem. Soc.*, 1976, *98*, 3127.
18. Johnson, B. F. G.; Lewis, J.; D. J. Thompson, D. J.; Heil, B. *J. Chem. Soc. Dalton Trans.*, 1975, 567; Karl in, K. D.; Johnson, B. F. G.; Lewis, J. *J. Organometal. Chem.*, 1978, *160*, C21.

19. Butts, S. B.; Holt, E. M.; Strauss, S. H.; Alcock, N. W.; Stimson, R. E.; Shriver, D. F. J. Am. Chem. Soc., 1979, 101, 5864; ibid., 1980, 102, 5093.
20. Butts, S. B.; Richmond, T.; Shriver, D. F. Inorg. Chem., in press.
21. Correa, F.; Nakamura, R.; Stimson, R. E.; Burwell, R. L. Shriver, D. F. J. Am. Chem. Soc., 1980, 102, 5112.
22. Vannice, M. A. J. Catal., 1975, 40, 129.
23. Vannice, M. A. J. Catal., 1975, 37, 462.
24. Gutsche, C. D. "The Chemistry of Carbonyl Compounds", Prentice-Hall, Englewood Cliffs, N. J., 1967 ; p. 74ff.
25. Neumann, H. M.; Laemmle, J.; Ashby, E. C. J. Am. Chem. Soc., 1973, 95, 2596.
26. Brown, H. C. "Organic Syntheses via Boranes", Wiley-Interscience, New York, N.Y., 1975.
27. Masters, C.; Van der Woude, C.; van Doorn, J. A. J. Am. Chem. Soc., 1979, 101, 1633.
28. Atwood, Jim D., private comunication, 1980.
29. Stimson, R. E., unpublished observations, Northwestern University, 1980.
30. Shriver, D. F.; Lehman, D.; Strope, D. J. Am. Chem. Soc., 1975, 97, 1594.
31. Hodali, H. A; Shriver, D. F. Inorg. Chem., 1979, 18, 1236.
32. Johnson, B. F. G.; Lewis, J.; Orpen, A. G.; Suss, G. J. Organometal. Chem., 1979, 173, 187.
33. Gavens, P.D.; Mays, M. J. J. Organometal. Chem., 1978, 162, 389.
34. Hodali, H. A.; Shriver, D. F.; Ammlung, C. A. J. Am. Chem. Soc., 1978, 100, 5239.
35. Fachinetti, G. J. Chem. Soc. Chem. Commun., 1979, 379.
36. King, R. B. Acc. Chem. Res., 1970, 3, 417.
37. Keister, J. B. J. Organometal. Chem., 1980, 190, C36.
38. Keister, J. B. J. Chem. Soc. Chem. Commun., 1979, 214; and private communication, 1980.
39. Whitmire, K.; Shriver, D. F. J. Am. Chem. Soc., 1980, 102, 1456.
40. Manassero, M.; Sansoni, M.; Longoni, G. J. Chem. Soc. Chem. Commun., 1976, 919.
41. Whitmire, K.; Shriver, D. F.; Holt, E. M. J. Chem. Soc. Chem. Commun., 1980, in press.
42. Muetterties, E. L.; Tachikawa, M. J. Am. Chem. Soc., 1980, 102, 4541.
43. Beno, M. A.; Williams, J. M.; Tachikawa, M.; Muetterties, E. L. J. Am. Chem. Soc., 1980, 102, 4542.
44. Wong, A.; Harris, M.; Atwood, J. D. J. Am. Chem. Soc., 1980, 102, 4529.
45. Summers, N. L.; Tyrrell, J. J. Am. Chem. Soc., 1971, 99, 3960.

RECEIVED December 8, 1980.

# 2

# Experimental and Theoretical Studies of Mechanisms in the Homogeneous Catalytic Activation of Carbon Monoxide

H. M. FEDER, J. W. RATHKE, M. J. CHEN, and L. A. CURTISS

Chemical Engineering Division, Argonne National Laboratory,
9700 South Cass Avenue, Argonne, IL 60439

At the outset it is important to clarify the scope of this discussion by the elimination of areas which will not be considered. When one notes that the term "activation of carbon monoxide" may mean a process as little perturbative of the C-O bond as its end-on attachment to a metal atom in carbonyls, or as strongly perturbative as its dissociation to atoms on a metal surface, the need for limits becomes obvious. In this discussion we will consider only the activation of carbon monoxide in the sense that isolable products are formed by the addition of hydrogen to the molecule without complete rupture of all carbon-oxygen bonds, *i.e.* oxygenates are formed.

Within this context carbon monoxide is not the inert molecule so frequently depicted on the basis of its formal triple bond and the remarkable similarity of its physical properties to those of the isoelectronic molecule dinitrogen. (Indeed, if it were, atmospheric carbon monoxide would present no hazard!) It is, in fact, a fairly readily activated molecule; the industrial process for the production of methyl formate (1) is well known, but it is less widely appreciated that this process is an example of a homogeneous, selective, base-catalyzed, activation of carbon monoxide which has for its net chemistry

$$MeOH + CO \xrightleftharpoons{[B]} MeO_2CH \; [Me = CH_3] \qquad (1)$$

the insertion of CO into an O-H bond. It should also be noted that the same reactants yield an isomer of methyl formate, acetic

0097-6156/81/0152-0019$05.00/0
© 1981 American Chemical Society

acid, also with good selectivity, when the catalyst solution contains a carboxylic acid, water, methyl iodide, and the transition metal ion, $[Rh(CO)_2I_2]^-$. The mechanism of this elegant (and industrially important (2)) reaction is fairly well understood (3). A further important point made by these two examples should be noted. It is that selectivity for the formation of only one (or few) out of the many thermodynamically possible products from a given set of reactants is often the most sought-after characteristic of a catalyst system. By the same token, analysis of the factors which determine the distribution of products from a given catalyst system can lead to desirable modifications of its behavior.

## HYDROGENATION OF CO

When our work in this area began (1978) it was fashionable to assume that the activation of CO toward hydrogenation by molecular catalysts was the exclusive province of multiple transition metal centers ("clusters"). This belief was based mainly on two observations. (a) The heterogeneous Fischer-Tropsch process for the production of catenated hydrocarbons and oxygenates was known to be of the catalysis type known as "demanding" (4). This was generally interpreted as meaning that metal atoms in specific geometric arrangements were required to guide the formation of the transition state. (b) Then current studies were disclosing the formation of carbon monoxide hydrogenation products -- methane (5), ethane (6), methanol and various polyhydroxylic compounds (7) -- by clusters which appeared intact subsequent to the catalyzed reaction.

There were, however, countervailing indications the significance of which only became apparent in retrospect. (a) Methanol and ethylene glycol had been reported by Gresham (8) to be among the products of CO hydrogenation by soluble cobalt catalysts under conditions that are now known to give mononuclear cobalt species almost exclusively. (b) Ziesecke (9), also using cobalt catalysts under similar conditions, reported that methanol and ethanol occurred among the products of the homogeneously catalyzed homologation of n-propanol to n-butanol.

Our own research was not based on scepticism concerning the essential role of clusters nor on the obscure significance of 25-year old observations. Rather, it was based on the idea that organotransition metal compounds capable of thermal generation of radicals or radical pairs (10,11) might afford a different entry into the problem of CO hydrogenation with mononuclear catalyst complexes. From this perspective, the immediate success (12) of the attempted reactions with $HCo(CO)_4$ or $HMn(CO)_5$ made it possible for the first time to study the reductive transformations of CO with readily characterizable catalysts by the conventional techniques of physical organic chemistry. We (12,13) and others (14, 15,16,17) have now obtained evidence concerning the reaction pathways for the formation of oxygenates by examining kinetic orders, activation parameters, product distributions, solvent effects, *in-situ* IR spectra, isotopic labeling, and kinetic isotope effects. The remainder of this section will be devoted to the conclusions which may be drawn from the evidence.

A. Nuclearity. For Co (12,13), Mn (12), and Ru (14) based catalytic systems a mononuclear species has been established in each case as playing an essential role in homogeneous hydrogenation to oxygenates. In catalytic systems (15) based on precursor complexes containing Fe, Ni, Rh, Pd, Os, Ir, or Pt dissolved in organic solvents, kinetic orders have not been measured, nor have equilibria among the oligomers been determined, so that nuclearity of the catalytically active species is unknown. (For Pd (15,17) and Pt (15) it is not even certain that homogeneous catalysis occurred; for Ru carbonyls the problem of avoiding the confusion caused by partial decomposition to metal has been discussed (14)). King *et al.* (17) have examined Co-based systems at 180-200°/200 atm/*p*-dioxane in some detail. Precursor compounds added as multinuclear species, $Co_2(CO)_8$, $(\mu_3\text{-MeC})Co_3(CO)_9$ (I), $Co_2(CO)_6[\mu_2\text{-CPh}]_2$ (II), and $Co_4(CO)_{10}(\mu_4\text{-PPh})_2$ (III), [Ph=$C_6H_5$], which IR spectroscopy revealed as forming $HCo(CO)_4$ under reaction conditions, catalyzed the formation of oxygenates. They reported that cobalt added as the compounds $Me_3SnCo(CO)_4$ (IV) and $Co_2(CO)_2[\mu_2\text{-MeN}(PF_2)_2]_3$ (V), formed carbonyl-free species under the same reaction conditions and were inactive for the formation of oxygenates. It appears reasonable to conclude that under these conditions only $HCo(CO)_4$ is the active catalyst for homogeneous hydrogenation to oxygenates. This is not to deny that cluster catalysis may be important under other conditions. For example III catalyzes hydroformylation at 130°/60 atm, may be recovered intact from reaction mixtures (18a), and has been shown (18b) to exhibit selectivity different from its fragmentation products. In this case the evidence for cluster catalysis is good.

B. Activation of Hydrogen. Activation of hydrogen is clearly as important for the catalytic processes under consideration as the activation of CO. For Co and Mn the hydrogen activation mechanism is the equilibrium:

$$H_2 + M_2(CO)_{2x} \rightleftharpoons 2HM(CO)_x \quad [(M,x) = (Co,4),(Mn,5)] \tag{2}$$

For ruthenium the hydrogen activation mechanism is not clearly established, although King (17) reports observing carbonyl bands in the infrared specta of catalytically active solutions in *p*-dioxane which are suggestive of minor amounts of $H_2Ru_4(CO)_{13}$ and $H_4Ru_4(CO)_{12}$ along with the major species $Ru(CO)_5$. In the aqueous alkaline water-gas shift system described by Pettit *et al.* (19) the appearance of minor amounts of methanol among the products may or may not involve the activation of molecular hydrogen. The well-known hydride-forming reaction (3a) is almost certainly involved in the

$$Fe(CO)_5 + 2OH^- \rightarrow HFe(CO)_4^- + HCO_3^- \tag{3a}$$

initial stages; the analog of (2), namely (3b),

$$H_2 + Fe_2(CO)_8^{2-} \rightleftharpoons 2HFe(CO)_4^- \quad (3b)$$

may be involved as well.

C. Product Distributions (13). A distinguishing characteristic of all the catalytic systems under consideration is the occurrence of methanol and methyl formate among the products, their ratio being a function of the transition metal and the experimental conditions. Polyhydroxylic products, principally ethylene glycol, also occur with the $d^9$ metals but have not been identified with certainty among products with the $d^7$, $d^8$, or $d^{10}$ metals. Reactions which are clearly secondary abound in these CO hydrogenation systems; they contributute greatly to the low specificity which plagues them. Among the secondary reactions which can occur are:

(a) alcohol homologation, the formation of ethanol from methanol, n-propanol from ethanol, *etc.*, *via* the formation of metal-alkyls, metal-acyls, and aldehydes;

(b) transesterification, the formation of formate esters of various alcohols by equilibration with methyl formate;

(c) methanolysis, the reaction of accumulated methanol with an intermediate metal-acyl to produce methyl esters of higher acids, particularly methyl acetate;

(d) alkane, particularly methane, formation from cleavage of an intermediate alkyl;

(e) aldolization, *via* condensations of intermediate aldehydes;

(f) $CO_2$ formation, induced by the accumulation of water by-product and its reaction with formate esters to produce formic acid which is readily decomposed catalytically.

It should be noted at this point that primary and secondary reaction products can be distinguished not only by kinetic data (13) but also by suppression of the secondary reactions. *E.g.*, substitution of 2,2,2-trifluoroethanol for *p*-dioxane as solvent for $HCo(CO)_4$ suppresses homologation and methane formation; addition of a phosphine to give the less acidic catalyst $HCo(CO)_3PR_3$ has the same effect, as has the substitution of the less acidic catalyst $HMn(CO)_5$.

With respect to quantitative results, Rathke and Feder have shown (13) that when account is taken of the secondary reactions in the cobalt-catalyzed system the fractions of primary products $MeOH(f_1)$, $MeO_2CH(f_2)$, and $(CH_2OH)_2(f_3)$ can be rationalized as follows:

$$f_2/(f_1 + f_3) \propto P_{CO}^a \, P_H^o \; (1 \leq a \leq 2) \quad (4a)$$

$$(f_3/f_1) \propto P_H^c \, P_{CO}^d \; (c,d \sim 1) \quad (4b)$$

The studies of Keim *et al.* (15a) and Fahey (16), which were carried out to much higher pressures (2500 atm. *vs* 300 atm.) but with constant gas compositions, generally confirm these relations for $HCo(CO)_4$ in that they show the methanol fraction decreases with increasing pressure, and the ethylene glycol/methanol ratio increases rapidly with increasing pressure; but precise relations cannot be discerned because secondary reaction products are not taken into account. Fahey's study (16) of the rhodium-catalyzed reaction shows a selectivity unaffected by total pressure at constant gas composition.

D. Activity. Reliable values for relative activities of various homogeneous transition metal complexes to give oxygenates do not exist, although qualitative ordering (15a) has been done. Such orderings should, however, be suspect, especially when the comparisons are done at a single specified temperature and pressure, because of variation in the stabilities of the carbonyls, variation of the activation energies for total hydrogenation products and for specified products, variable solvent effects, ligand effects, *etc.* These phenomena may account for appraent discrepancies between reports of good activity (12) for $Mn_2(CO)_{10}$ at 240° and 300 atm. and relative inactivity (16) for the same compound at 225° and 2500 atm. They may also account for the discrepant reports that ruthenium carbonyl has no significant activity for ethylene glycol formation (14) and has slight activity (15a), because the temperature at which these studies were done were 260° and 230°, respectively. (In work with cobalt (13) it was noted that selectivity toward formation of ethylene glycol decreases with increasing temperature.) It would be wrong to attach too much significance to relatively minor discrepancies in the absence of certain knowledge that reactors were scrupulously cleaned between runs, *etc.*

The most that can be said for relative activities at this stage is that to a good approximation in each period $d^9 > d^8 > d^{10}$, and in each group 4d>3d>5d. Only for cobalt has sufficient rate law data been obtained (12,13) to allow future comparisons of absolute activities with respect to CO hydrogenation to oxygenates.

$$\frac{d[P]}{dt} = k^{(2)} [HCo(CO)_4]P_H \; ; \; ([P] = \Sigma \text{ Concn. products}) \quad (P_H = \text{partial pressure hydrogen}) \qquad (5a)$$

$$\log_{10}[k^{(2)}(\text{atm.}^{-1}\text{sec}^{-1})] = A - 41{,}000/\Theta; (\Theta = 2.303\ RT) \qquad (5b)$$

$$A = 12.40\ (p\text{-dioxane}),\ 13.02\ (2,2,2\text{-trifluoroethanol}). \qquad (5c)$$

Note that the pre-exponential factors indicate only small entropies of activation in the Eyring form of the rate equations. This is a significant observation which indicates that the decrease of entropy associated with the incorporation of a hydrogen molecule at or prior to the transition state must be compensated for by a dissociation or decrease of coordination number.

Fahey (16) finds the rate of reaction in the rhodium-based system to be proportional in the (3.3 ± 0.5) power of the total gas composition; he concludes that pressure-induced declustering to species of varying activities is involved.

E. Solvent Effects. Significant solvent effects have been noted; *e.g.* Rh and Fe-based complexes are more active by about a decade in the polar aprotic solvnet N-methylpyrrolidinone ($\varepsilon_{25} \cong 32$) than in toluene (15a). Likewise, as noted above, the $HCo(CO)_4$ based system in 2,2,2-trifluoroethanol ($\varepsilon_{25}=26.7$) is about four times as active as in *p*-dioxane ($\varepsilon_{25}=2.2$) and about 20 times as active as in benzene ($\varepsilon_{25}=1.8$). It seems unlikely that such relatively small changes can be ascribed to development of large charge separations in the transition state. We attribute the observed increases with solvent polarity to stabilization of an important polar intermediate by a general dipolar mechanism. Carbonyl insertions (20), for example, are of this type. In this same connection it should be mentioned that although anionic clusters of variable nuclearity are regarded as the catalytic species in certain reactions, *e.g.* the Rh carbonyl-based ethylene glycol process (7), the formation of ion pairs (the counter-ions deriving from so-called promotors) may in fact be crucial with respect to activity and selectivity.

Specific solvent effects, such as may be ascribed to tight binding of certain solvent molecules in the coordination sphere of a complex, appears not be have been encountered in primary hydrogenation of CO. The specific solvent effect of 2,2,2-trifluoroethanol which strongly inhibits the secondary reaction of methanol homologation in the cobalt system has been ascribed to strong hydrogen bond donation of the type $CF_3CH_2OH \cdots O(H)Me$; such complexes effectively prevent further protonation of the alcohols.

Another specific effect of some consequence arises from the use of aqueous organic solvent mixtures. For the cobalt system, dry solvents induce CO hydrogenations which are stoichiometric for water (13a) (within 2% material balance) according to the reaction:

$$aCO + (a+b-c)H_2 = C_aH_{2b}O_c + (a-c)H_2O \qquad (6)$$

The production of $CO_2$ is negligible. When large excesses of water are added new secondary reactions occur and selectivities change. As already noted, these include hydrolysis of formates followed by formation of carbon dioxide, enhancement of ethylene glycol formation, and strong acceleration of alcohol homologation and of its secondary consequences, *e.g.* methane formation. The cause of the increase in methanol homologation rate is, in our opinion, a solvation-enhanced increase in the ability of $HCo(CO)_4$ to protonate alcohols. The effect of water on ethylene glycol production is probably quite real (21) but the cause is less evident.

F. Kinetic Isotope Effect. One of the more important observations regarding the cobalt-based homogeneous CO hydrogenation system is that the substitution of gaseous deuterium for gaseous hydrogen causes an inverse (*i.e.*, $k_H < k_D$) kinetic isotope effect, $k_D^{(2)}/k_H^{(2)} = 0.73$ at 180° in *p*-dioxane (13b). Because

$k_H \neq k_D$ it is immediately evident that hydrogen is transferred in or before the transition state and not, say, an electron followed by a fast proton transfer. Because inverse kinetic isotope effects are somewhat infrequently observed in homogeneous catalyses the rules which govern mechanisms leading to such results have been extensively discussed in the literature (23). In brief, for a significant inverse kinetic isotope effect to be observed in thermal reactions it is necessary, according to transition-state theory, that the transition state position be strongly assymmetric with respect to the reactant state immediately preceding and the product state immediately following. Accordingly, this reaction should be far from thermoneutral; in the present case, the large activation endothermicity guarantees that the transition state will be close to the immediately following product state. Also, it is necessary that in the sum of steps leading to the transition state, including pre-equilibria, there occurs a net transfer of hydrogen atoms from lower vibrational frequency bonds to higher vibrational frequency bonds.

For the cobalt-based system the molecularity of the transition state indicated by the reaction order is $H_3CoC_4O_4$ and the reactants are $H_2$ and $HCo(CO)_4$. Thus, two hydrogen atoms start with values of $\nu \sim 3200$ cm$^{-1}$ and one with $\nu \sim 1830$ cm$^{-1}$. If in the transition state the strong H-H bond is not yet completely broken, then we should expect to find the H atom originally attached to cobalt bound to carbon or oxygen ($\nu \sim 2900$-$3400$ cm$^{-1}$) in the transition state.

## MECHANISM

The mechanism we have proposed (13) for the cobalt-based catalytic cycle is shown (slightly modified) in Chart I. For the reasons described immediately above, the transition state is presumed to occur between the left- and right-hand sides of eq. 8.

Fahey (16), Keim *et al.* (15a), and ourselves (13) appear to have independently arrived at the equilibrium formation of the formyl complex 1 as a key step. Some of the arguments advanced are given in Ref. 13b. Other principles used in the construction of the scheme include: hydrogenolyses, indicated by $\xrightarrow{[H]}$, are irreversible; branching occurs *via* alternative hydrogen transfers to either carbon or oxygen of an attached aldehydic group; CO "insertions" into cobalt-carbon bonds are reversible, while those into cobalt-oxygen bonds are not. By use of the usual assumption of steady state concentrations of all cyclic intermediates the experimental forms of the rate and selectivity equations (4) and (5) are readily recovered.

$$HCo(CO)_4 \rightleftharpoons \underset{\underline{1}}{(CO)_3Co(CHO)} \quad (7)$$

$$\underline{1} + H_2 \rightleftharpoons \underset{\underline{2}}{(CO)_3(H)_2Co(CHO)} \quad (8)$$

$$\underline{2} \longrightarrow \underset{\underline{3}}{(CO)_3(H)Co(CH_2O)} \rightleftharpoons CH_2O + HCo(CO)_3 \qquad (9)$$

$$\underline{3} \rightleftharpoons \underset{\underline{4}}{(CO)_3Co(-OCH_3)} \qquad (10a)$$

$$3 \rightleftharpoons \underset{\underline{5}}{(CO)_3Co(-CO_2CH)} \qquad (10b)$$

$$\underline{4} + 2\ CO \rightarrow \rightarrow \underset{\underline{6}}{(CO)_4Co(-CO_2CH_3)} \qquad (11)$$

$$\underline{6} \xrightarrow{[H]} HCO_2CH_3 + [Co] \qquad (12)$$

$$\underline{5} \xrightarrow{[H]} CH_3OH + [Co] \qquad (13)$$

$$\underline{5} + CO \rightleftharpoons \underset{\underline{7}}{(CO)_3Co(-COCH_2OH)} \qquad (14)$$

$$\underline{7} + H_2 \longrightarrow \underset{\underline{8}}{(CO)_3(H)Co(CHO - CH_2OH)} \qquad (15)$$

$$\underline{8} \xrightarrow{[H]} (CH_2OH)_2 + [Co] \qquad (16)$$

CHART I. Outline Mechanism of Primary Reactions

A number of other mechanistic possibilities were examined and rejected in the course of this work. (a) The original suggestion (12) of a rate-determining step involving formation of radical pairs

$$HCo(CO)_4 + CO \longrightarrow HCO + Co(CO)_4 \qquad (17)$$

was rejected when the expected first-order dependence on $P_{CO}$ did not materialize. The involvement of $Co(CO)_4$ radicals (24) in the kinetics of $2HCo(CO)_4 \rightarrow H_2 + Co_2(CO)_8$, and, presumably, the reverse step has been recently demonstrated, as have one-electron reactions of trityl radical with $HCo(CO)_4$ and of benzophenone ketyl with $CO_2(CO)_8$. Thus, in Chart I each hydrogenolysis, indicated by $\xrightarrow{[H]}$ is a rapid reaction which may involve attack by $Co(CO)_4$ radicals, scission of cobalt-carbon or cobalt-oxygen bonds, hydrogen abstractions from $HCo(CO)_4$, and $Co(CO)_4$ radical pairings. (b) A second possibility, the equilibrium formation of low concentrations of formaldehyde, followed by a rate-determining, product-forming step which is inverse first-order in $P_{CO}$, has been considered (13b); it was rejected on the basis that catalyzed decomposition of formaldehyde to $H_2$ + CO was not observed. Nevertheless, the coincidence of products from the catalyzed reactions of formaldehyde and of synthesis gas (13,16) points strongly to the involvement of a common intermediate. Fahey (16) suggests that intermediate 3 dissociates formaldehyde; he finds supportive evidence in the rhodium-based system by observation of minor yields of 1,3-dioxolane, the ethylene glycol trapped acetal of formaldehyde. For reasons to be discussed later, we believe the formation of free formaldehyde is not on the principal reaction pathway. (c) We have also rejected two aspects of the reaction mechanism proposed by Keim, Berger, and Schlupp (15a): (i) the production of formates *via* alcoholysis of a formyl-cobalt bond, and (ii) the production of ethylene glycol *via* the cooperation of two cobalt centers. Neither of these proposals accords with the observed kinetic orders and the time invariant ratios of primary products.

## Results of Molecular-Orbital Theory Investigation

Although the intermediates indicated in Chart I have adequate precedents in organometallic chemistry and have been discussed elsewhere, it was thought that additional insight would be afforded by applying current methods of theoretical chemistry. In particular, it was hoped that calculations within current capabilities would help establish approximate structures and relative electronic energies for various intermediates. The reader is referred to recent studies of this type for the hydroformylation reaction (25) and for organometallic migration reactions (26). The molecular orbital methods which have been employed for such studies include extended Hückel theory (EHT), CNDO, and *ab initio* LCAO-SCF.

The method employed for an initial effort to explore a complex reaction sequence should be fast and inexpensive, but sufficiently reliable to enable meaningful deductions concerning structures and relative energies to be made. The results can then be usefully employed as starting points for more elaborate and accurate methods for detailing the energy hypersurfaces. For this study we have chosen to use Anderson's modification (27) of extended Hückel theory (MEHT). MEHT differs from EHT by the inclusion of a pair-wise correction for atom-atom repulsions. Pensak and McKinney (28) [PM], using this method, have recently reported a systematic study of first-row transition metal carbonyl complexes for which experimental bond distances and angles were reliably reproduced, along with key bond dissociation energies. We have chosen to perform our initial M.O. calculations also with MEHT, but have made two modifications to the [PM] parameter choices: (i) The set of coulomb integrals (diagonal elements), $H_{ii}$, for Co, H, C, and O were replaced by the set of values used by Hoffman (26a). (ii) The Wolfsberg-Helmholz parameter, K, used for the calculation of off-diagonal elements $H_{ij}$ *via* the relation

$$H_{ij} = \frac{1}{2} S_{ij} K_{ij}(H_{ii} + H_{jj})\exp(-o.13\ R) \qquad (18)$$

(R = internuclear distance; $S_{ij}$ = overlap integral)

is given the fixed value 2.25 by [PM]. Such a treatment fails to give good bond energies for small common molecules which may be involved in reactions; thus there is an inherent distortion of energy changes in reactions. The modifications we employ assigns the following values: (a) $K_{H,H} = K_{H,Co} = 1.75$. This value produces a bond length and a dissociation energy for $H_2$ in good agreement with experiment, and a bond energy for $H\text{-}Co(CO)_4$ dissociation at its experimental bond length in good agreement with experiment. (b) $K_{H,C} = 1.88$. This value reproduces experimental bond dissociation energies in $H_2CO$. (c) $K_{H,O} = 2.05$. This value reproduces experimental bond dissociation energies in $H_2O$. (d) $K_{C,O} = 2.15$. It appears necessary to distinguish carbon-oxygen bonds with formal order two from carbon-oxygen bonds in carbonyl ligands, for which K = 2.23 is employed. The [PM] parameter is used for all other bonds.

As a result of these changes the [PM] lengths for M-C and C-O bonds are essentially undisturbed. We believe this minimum parameterization is both necessary and sufficient to give a reasonable approximation to the energy hypersurface.

The calculations were done in two stages. The first stage calculations were done with the constant value K = 2.25. Bond lengths and bond angles were optimized. The bond angles were then fixed, and those bond lengths which were close to their reference values were fixed at their reference values. The reference values are as follows: Co-H, 1.56Å; Co-C, 1.82Å; Co-$C_{ax}$[$HCo(CO)_4$], 1.76Å; C-O (carbonyl), 1.14Å; C-O (aldehyde), 1.20Å; C-H, 1.10Å. In the second stage the parameterized values of K discussed in (ii) were employed and the remaining geometric variables optimized to obtain the molecular electronic energy values. By this technique the following results were obtained.

(a) <u>$HCo(CO)_4$</u>. In the $C_{3v}$ form the H-Co-$C_{eq}$ angle optimized at 86°; the experimental value is 80.3°. The molecular electronic energy, -923.16e.v., was taken as a reference point for subsequent calculations.

(b) <u>$Co(CO)_3(CHO)$</u>. The 16-electron formyl cobalt tricarbonyl, postulated as participating in an unfavorable equilibrium with $HCo(CO)_4$, was optimized in the symmetries shown in Fig. 1. Structure B optimized in a surprising way: the angles $\Theta_1$ and $\Theta_2$ closed to 0° and 5°, respectively, while β opened to 171°. The result is a nearly square planar structure, shown in Fig. 2. It should be noted that the near planarity of this structure may be sensitive to the values chosen for the coulomb integrals. The calculated relative internal energy (0 Kelvins) of this form is +9 kcal mole. (Uncertainties of not less than 10 kcal/mole should be assigned to this and subsequent relative values.) Our previous estimate (<u>13b</u>), based on thermochemical considerations,

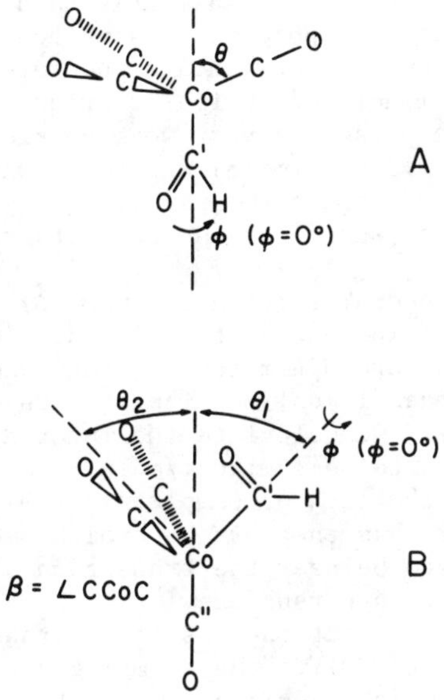


*Figure 1. Structures A and B of the $Co(CHO)(CO)_3$ complexes*

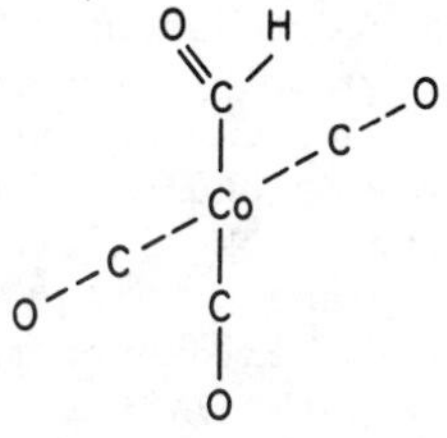

*Figure 2. The square planar structure that is obtained for the ground state of $Co(CHO)(CO)_3$*

was +16 kcal/mole at 200°C. The approximate $C_{3v}$ form, A, optimized at $\Theta = 86°$ and a relative energy of +39 kcal/mole; it is unlikely to be the ground state of the pair. The possibility of a dihapto formyl ($\eta^2$-CHO) interaction was investigated for both structures A and B, but in both cases it was less stable. Evidently, the oxophilicity of earlier transition metals is an important factor in stabilizing the $\eta^2$-CHO configuration.

(c) $H_2\cdots(Co(CO)_3(CHO)$. In common with other $d^8$ square planar complexes $Co(CO)_3(CHO)$ should lend itself to oxidative additive to yield an octahedral molecule; the approach of a molecule of $H_2$ should lead to the cis-addition product. Stage one calculations were carried out for parallel and perpendicular orientations of $H_2$ with respect to the base plane. A very weak complex (stabilization energy approximately 0.5 kcal/mole) was found with $H_2$ located about 3Å from the cobalt center in the parallel orientation. The perpendicular orientation was close in energy to the parallel orientation.

(d) $Co(H)_2(CO)_3(CHO)$. Two octahedral structures, *mer* (A) and *fac* (B) shown in Fig. 3, were considered for the cis-addition product. The bond lengths displayed were optimized at stage one; the energies were found to be less than 1 kcal/mole apart. Using the reference bond lengths and angles, we calculate the internal energy of B relative to $HCo(CO)_4 + H_2$ to be +52 kcal/mole. Considering the uncertainty limits and the lack of complete optimization, we remark that this structure has the correct stoichiometry and approximate relative energy to be near the transition state (at +41 kcal/mole) on the energy hypersurface.

(e) $(H)Co(CO)_3CH_2O)$. We next looked at the result of transferring a hydrogen atom in $(H)_2Co(CO)_3(CHO)$ to the formyl group. The two obvious candidate complexes which result contain either a formaldehyde grouping or a hydroxycarbene grouping. Geometry optimization at stage one was done with the formaldehyde H-C-H angle and bond lengths fixed, and the other variables shown on Fig. 4 optimized. On optimization the angle $\alpha$ opened to about 175° and $\beta$ became 87°, *i.e.* the $HCo(CO)_3$ fragment is nearly a square planar structure. The angles $\gamma$ and $\delta$ are 88° and 76°, respectively, and are not complementary, *i.e.* the formaldehyde molecule is no longer exactly planar, the hydrogens being tilted back ∿16°, but its approximate plane is parallel to the basal plane. Variation of the rotation angle, $\phi$, produced no potential well deeper than 0.5 kcal/mole, *i.e.* rotation in the plane is essentially unhindered. The point and angle of intersection between the X-axis and the C-O bond of the formaldehyde fragment were allowed to vary. The optimized location is close to the carbon atom and the distance, R, is 2.34Å. The relative energy of this complex at stage two is +40 kcal/mole, downhill (exoergic) from the octahedral complex at +52 kcal/mole. The calculated energy of dissociation of the formaldehyde complex to $HCo(CO)_3$ plus formaldehyde is +6 kcal/mole. The thermochemically estimated value (13b) was +10 kcal/mole.

The distance of the formaldehyde oxygen from the cobalt

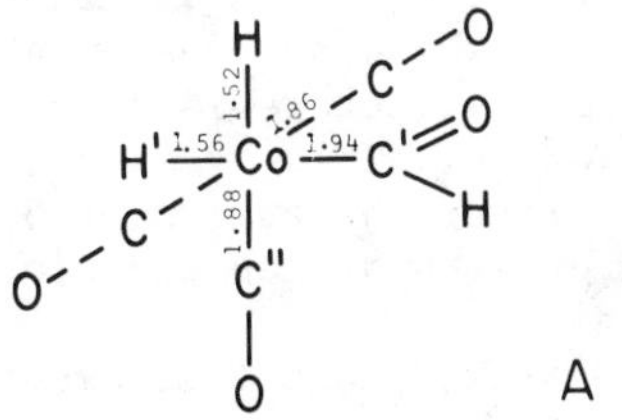


A

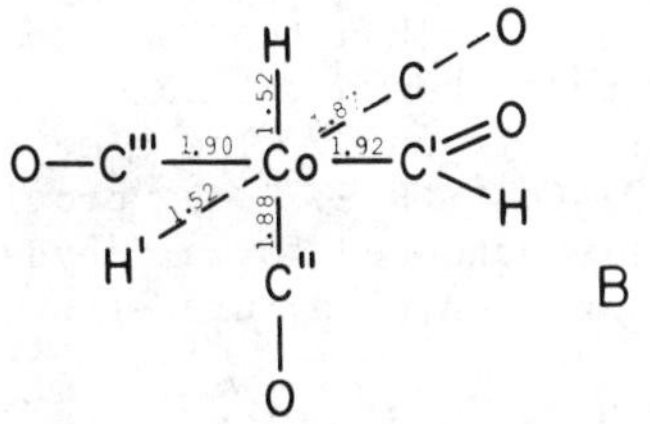


B

*Figure 3. Structures A and B of the $(H)_2Co(CHO)(CO)_3$ complex. The numbers are optimized bond lengths (in Å) at stage one. The carbonyl ligands were held fixed at 1.09Å, the optimized value from the stage one calculations on $HCo(CO)_4$.*

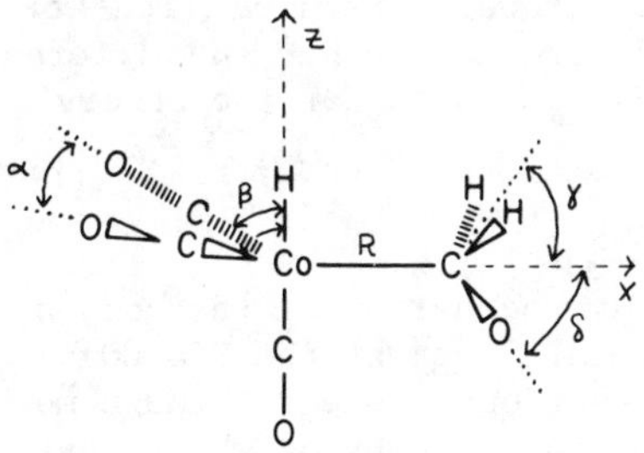


*Figure 4. Optimized structure of $Co(H)(CO)_3(CH_2O)$*

center is 2.85Å which indicates that there is little dihapto character in the interaction of $HCo(CO)_3$ with $H_2CO$. This result is supported by a recent study of the X-ray crystal structure of some acyl derivatives of ruthenium by Roper *et al.* (30). They found the mean difference between the oxygen-metal bond length and carbon-metal bond length to be 0.52Å; this should be compared to a difference of 0.51Å in our case. We also find a structure, having a more nonplanar $HCo(CO)_3$ fragment, which has the oxygen of $H_2CO$ closer to the cobalt center. This structure is slightly less stable than the complex described above. These results, showing that cobalt can form a bond to either the carbon or oxygen of $H_2CO$, indicate that the next internal hydrogen transfer could go either to the carbon or oxygen.

The hydroxycarbene isomer $(H)Co(CO)_3(CHOH)$ was also examined. It yielded a complex with molecular electronic energy more than 60 kcal/mole higher on the energy scale. The hydroxycarbene complex is not likely to play a significant role in the catalytic cycle. It is of some interest to inquire why the 18e hydroxycarbene complex $(H)(CO)_3Co(=CHOH)$ is less stable than the 16e isomer $(H)(CO)_3Co(CH_2O)$. The results suggest that the formation of the carbonyl double bond makes the critical difference. The electronically delocalized structure $(H)(CO)_3Co^{+\delta}-CH_2-O^{-\delta}$ may provide some extra stabilization for the formally unbonded formaldehyde moiety. The resonance form is dipolar and could be further stabilized by polar solvents.

(f) $(CO)_3Co(CH_2OH)$ and $(CO)_3Co(OCH_3)$. These molecules have been examined only briefly by the method outlined, the hydroxymethyl and methoxy groups being held fixed in the methanol experimental geometry. They form quite energetically from the formaldehyde complex by internal hydrogen transfer to either formaldehyde oxygen or carbon and rearrangement. It would require calculations of great delicacy and accuracy to determine whether the difference between the two transitional pathways accounts for the observed branching ratio.

## CONCLUSIONS

Detailed investigation of the hydrogenation of the carbon monoxide molecule, as homogeneously catalyzed by the $HCo(CO)_4$/$Co_2(CO)_8$ system, reveals that the reactions proceed through mononuclear transition states and intermediates, many of which have established precedents. The major pathway requires neither radical intermediates nor free formaldehyde. The observed rate laws, product distributions, kinetic isotope effects, solvent effects, and thermochemical parameters are accounted for by the proposed mechanistic scheme. Significant support of the proposed scheme at every crucial step is provided by a new type of semi-empirical molecular-orbital calculation which is parameterized *via* known bond-dissociation energies. The results may serve as a starting point for more detailed calculations. Generalization to other transition-metal catalyzed systems is not yet possible.

ACKNOWLEDGEMENTS

This work was supported by the Office of Chemical Sciences, Division of Basic Energy Sciences, U.S. Department of Energy. The authors gratefully acknowledge stimulating and useful discussions with A. B. Anderson, J. S. Bradley, J. K. Burdett, R. Hoffmann, and J. Halpern, but they are not responsible for any errors of opinion or fact herein.

LITERATURE CITED

1. Kirk-Othmer "Encyclopedia of Chemical Technology" 3rd Ed. John Wiley and Sons, New York, 1978, 4, 780.
2. J. F. Roth, J. H. Craddock, A. Herschman, and F. E. Paulik, Chem. Tech, 1971, 600.
3. D. Forster, Ann. N.Y. Acad. Sci., 1977, 295, 79.
4. M. Boudart, Proc. Robert A. Welch Found. Conf. Chem. Res., 1971, 14, 299.
5. M. G. Thomas, B. F. Beier, and E. L. Muetterties, J. Am. Chem. Soc., 1976, 98, 1296.
6. G. C. Demitras and E. L. Muetterties, J. Am. Chem. Soc., 1977, 99, 2796.
7. R. L. Pruett, Ann. N.Y. Acad. Sci., 1977, 295, 239, and patents referenced therein.
8. W. F. Gresham (to E. I. Dupont) Brit. Patent 655237 (1951), CA, 46, 7115h (1951); W. F. Gresham and C. F. Schweitzer, U.S. Patent 2,534,018; W. F. Gresham, U.S. Patent 2,636,046 (1953).
9. K. H. Ziesecke, Brennstoff-Chem., 1952, 33, 385; CA, 1955, 49, 6870.
10. H. M. Feder and J. Halpern, J. Am. Chem. Soc., 1975, 97, 7186.
11. R. L. Sweaney and J. Halpern, J. Am. Chem. Soc., 1977, 99, 8335.
12. J. W. Rathke and H. M. Feder, J. Am. Chem. Soc., 1978, 100, 3623.
13. (a) J. W. Rathke and H. M. Feder, in "Catalysis in Organic Syntheses - 1980", W. R. Moser, Ed. Academic Press, N.Y., In press; (b) J. W. Rathke and H. M. Feder, Ann. N.Y. Acad. Sci., 1980, 333, 45.
14. J. S. Bradley, J. Am. Chem. Soc., 1979, 101, 7419; private communication.
15. (a) W. Keim, M. Berger, and J. Schlupp, J. Catalysis, 1980, 61, 359; (b) A. Deluzarche, R. Fonseca, G. Jenner, and A. Kiennemann, Erdol u. Kohle, 1979, 32, 313.
16. D. R. Fahey, Preprints of the Petroleum Div., ACS, 1980, 25, submitted for publication.

17. R. B. King, A. D. King, Jr., and K. Tanaka, submitted for publication.
18. (a) R. C. Ryan, C. V. Pittman, Jr., and J. P. O'Connor, J. Am. Chem. Soc., 1977, 99, 1986; (b) C. V. Pittman, Jr., G. M. Wilemon, W. D. Wilson, and R. C. Ryan, Angew. Chem. Int. Ed. Engl., 1980, 19, 478.
19. R. Pettit, C. Mauldin, T. Cole, and H. Kang, Ann. N.Y. Acad. Sci., 1977, 295, 151.
20. (a) R. J. Mawby, F. Basolo, and R. G. Pearson, J. Am. Chem. Soc., 1964, 86, 3994; (b) F. Calderazzo and F. A. Cotton, Inorg. Chem., 1962, 1, 30.
21. T. Onoda (to Mitsubishi Chem. Ind.), Jap. Kokai, 76,128,903 (1976).
22. See, *e.g.*, B. J. Klingler, K. Mochida, and J. K. Kochi, J. Am. Chem. Soc., 1979, 101, 6626.
23. "Isotopes in Organic Chemistry, 2, Isotopes in Hydrogen Transfer Process", E. Buncel and C. C. Lee, Eds., Elsevier Scientific Publishing Co., Amsterdam and New York, (1976).
24. (a) R. W. Wegman and T. L. Brown, J. Am. Chem. Soc., 1980, 102, 2494; (b) F. Ungvary and L. Marko, J. Organometallic Chem., 1980, 193, 383.
25. (a) A. Dedieu, Inorg. Chem., 1980, 19, 375; (b) J. P. Grima, F. Choplin, and G. Kaufmann, J. Organometallic Chem., 1977, 129, 221; (c) V. Bellagamba, R. Ercoli, A. Gamba, and G. B. Suffritti, ibid., 1980, 190, 381.
26. (a) H. Berke and R. Hoffmann, J. Am. Chem. Soc., 1978, 100, 366; (b) M. Ruiz, A. Flores-Riveras, and O. Novarro, J. Catalysis, 1980, 64, 1.
27. A. B. Anderson, J. Chem. Phys., 1975, 62, 1187 and personal communication.
28. D. A. Pensak and R. J. McKinney, Inorg. Chem., 1979, 18, 3407.
29. E. A. McNeill and F. R. Scholer, J. Am. Chem. Soc., 1977, 99, 6293.
30. W. R. Roper, G. E. Taylor, J. M. Waters, and L. J. Wright, J. Organometallic Chem., 1979, 182, C46.

RECEIVED January 12, 1981.

# 3

# Heterobimetallic Carbon Monoxide Hydrogenation

## Hydrogen Transfer to Coordinated Acyls: The Molecular Structure of $(C_5H_5)_2Re[(C_5H_5)_2ZrCH_3](OCHCH_3)$

JOHN A. MARSELLA, JOHN C. HUFFMAN, and KENNETH G. CAULTON

Department of Chemistry and Molecular Structure Center, Indiana University, Bloomington, IN 47405

There is mounting evidence that the intramolecular transformation of a hydridocarbonyl, M(H)CO, into a $^1\eta$-formyl, MC(O)H, is not thermodynamically favorable (1). This has directed attention towards reactions which provide exceptional stability to the formyl species. The oxophilic character of the early transition metals may provide such stabilization in the form of dihapto binding (I). This unusual donor behavior was first

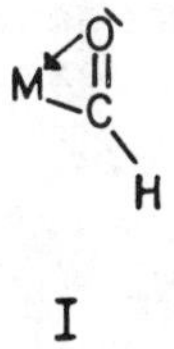


I

0097-6156/81/0152-0035$05.00/0
© 1981 American Chemical Society

demonstrated crystallographically by Floriani, et. al. (2,3) for homologous acetyl complexes of titanium and zirconium.

One approach to promoting the kinetics of hydrogen transfer to bound carbon monoxide is based on maximizing the difference in polarity of the carbon (eg. δ+) and hydrogen (eg. δ-) involved (4). This strategy leads naturally to a bimolecular approach, based upon MCO and M′H. The additional degree of freedom which follows from employing two different transition metals is noteworthy as an alternative to cluster activation or catalysis.

In view of the fact that early transition metal alkyls insert CO under very mild conditions (2,3), we chose to examine the reactions of electron-rich metal hydrides (5) with the resultant dihapto acyl complexes. Such acyls obviously benefit from reduction of the CO bond order from three (in C≡O) to two. More significantly, the dihapto binding mode will significantly enhance the electrophilic character of the acyl carbon.

In the course of this work, we found that addition of $Cp_2ReH$ (6) to $Cp_2Zr[C(O)Me]Me$ (2) yielded a product whose spectroscopic properties were in accord with the stoichiometry $Cp_2Re[Cp_2ZrMe](OCHMe)$ (7). The presence of a chiral carbon produced a slight inequivalence (0.0016 ppm) in the cyclopentadienyl ring protons attached to zirconium. These results do not distinguish between structures II and III, both being

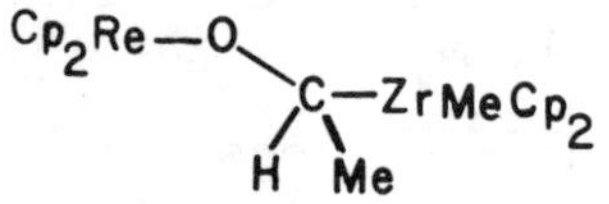


II

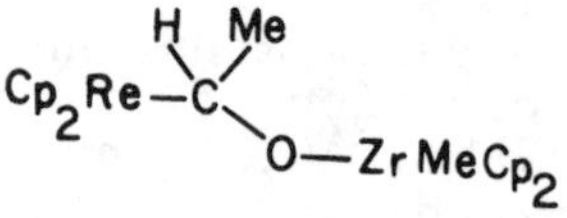


III

reasonable in view of the high oxophilicity of rhenium and zirconium. Moreover, in view of the fact that $[Cp_2ZrClAlEt_3]_2CH_2CH_2$ exhibits a structure (IV)

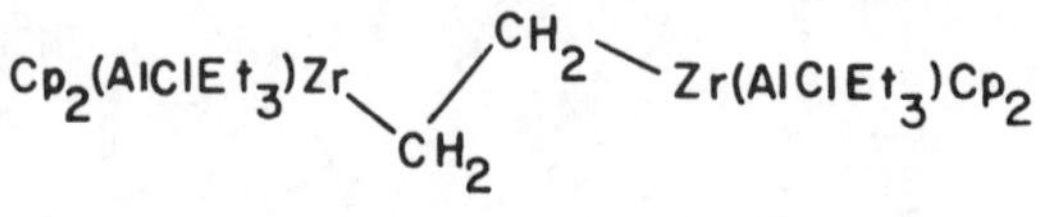


IV

with remarkably acute $ZrC_{\alpha}C_{\beta}$ angles (76°) (8), structure V, with a bridging acetaldehyde ligand, also

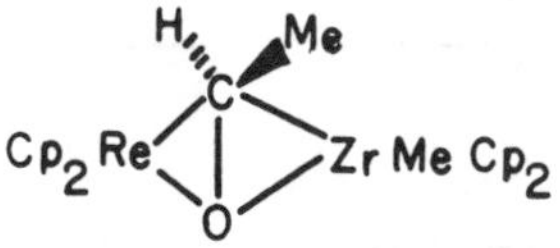


V

merits consideration. A structure of this type may represent the transition state in the fluxional process displayed by $(Cp_2ZrCl)_2OCH_2$ (9). We now report a solution to this structural problem by means of crystallographic methods.

## Experimental

Synthesis. Crystals of $Cp_2Re[Cp_2ZrMe](OCHMe)$ were grown from a toluene/hexane solution (*ca.* 2:1) in the following manner. Equimolar amounts of $Cp_2Zr[C(O)Me]$-Me and $Cp_2ReH$ were dissolved in a minimum of toluene. Within several hours, the solution had taken on the dark orange color of the dimeric product. Hexane was added and the resulting solution was allowed to stand for several days at room temperature until bright orange crystals formed. The compound is sensitive to both oxygen and moisture.

Crystallography. The crystal was transferred to the goniostat using inert atmosphere techniques. Crystal data and parameters of the data collection (at -173°, 5° $\leq 2\theta \leq$ 45°) are shown in Table I. A data set collected on a parallelopiped of dimensions 0.09 x 0.18 x 0.35 mm yielded the molecular structure with little difficulty using direct methods and Fourier techniques. Full matrix refinement using isotropic thermal parameters converged to R = 0.17. Attempts to use anisotropic thermal parameters, both with and without an absorption correction, yielded non-positive-definite thermal parameters for over half of the atoms and the residual remained at *ca.* 0.15.

Data was then collected on a smaller crystal. The residuals improved, but several non-hydrogen anisotropic thermal parameters converged to non-positive-definite values. There was no evidence for

Table I. Crystal Data for $(C_5H_5)_2ZrCH_3(OCHCH_3)Re(C_5H_5)_2$

| | | |
|---|---|---|
| Formula | | $C_{23}H_{27}OZrRe$ |
| Color | | yellow |
| Crystal Dimensions (mm) | | 0.032 x 0.019 x 0.064 |
| Space Group | | P $2_1$/a |
| Cell Dimensions (at -173°C; 28 reflections) | | |
| | a = | 20.762(13) Å |
| | b = | 7.843(5) |
| | c = | 12.724(8) |
| | β = | 72.28(2)° |
| Z (Molecules/cell) | | 4 |
| Cell Volume | | 1973.74 |
| Calculated Density (gm/cm$^3$) | | 2.009 |
| Wavelength | | 0.71069 Å |
| Molecular Weight | | 596.89 |
| Linear Absorption Coefficient | | 67.4 |
| Min. Absorption = | | 0.64 |
| Max. Absorption = | | 0.79 |
| Total Number of Reflections collected | | 3449 |
| Number of unique intensities | | 2598 |
| Number with F > 0.0 | | 2302 |
| Number with F > σ (F) | | 2141 |
| Number with F > 2.33 σ (F) | | 1894 |
| Final Residuals | | |
| R(F) | | .087 |
| Rw(F) | | .068 |
| Goodness of fit for the last cycle | | 1.33 |
| Maximum Δ/σ for last cycle | | .05 |

decomposition during the data collection (four standard reflections varied randomly within ± 0.8 σ), nor was there evidence of disorder or solvent molecules in the crystal lattice. Consequently, we report here the results from this second crystal but using isotropic thermal parameters for all atoms except Zr and Re. These data are corrected for absorption. In the final least squares cycles, all hydrogen atoms whose positions are fixed by assumed $sp^2$ or $sp^3$ hybridization were included in fixed positions with C-H = 0.95 Å and B = 3.0 Å$^2$; methyl hydrogens were not included.

The results of the X-ray study are contained in Tables II-IV. Anisotropic B's and a table of observed and calculated structure factors are available (10). The molecular structure is shown in Figures 1 and 2.

The cyclopentadienyl ring carbons deviate by less than 0.6 σ from their respective least squares planes. The average C-C distances in the four rings are identical within experimental error. Metal-to-ring midpoint lines intersect the ring planes at angles of 87.4° and 87.7° (Re) and 86.8° and 88.3° (Zr). The shortest intramolecular nonbonded contacts are from C(24) to C(3), 2.75 Å, and to C(9), 2.83 Å. The shortest distances to oxygen are from C(3) and C(15) (both 2.95 Å). All inter-ring carbon-carbon distances exceed 3 Å. Intermolecular C···H contacts (calculated with all C-H distances fixed at 1.08 Å) exceed 2.6 Å while intermolecular H···H contacts exceed 2.2 Å.

## Results and Discussion

Overall Structure. The results indicate that structure III is correct and that the reaction is a geminal addition of the Re-H bond to the acetyl carbon. The cyclopentadienyl rings on the same metal center are in the semi-staggered configuration typically found for bent metallocene structures (11), while rings on different metals assume a cog-like arrangement (Figure 3a). The rings are arranged so that all four ring centroids fall approximately in one plane; deviations of these centroids are ± 0.1 Å from their least squares plane. The arrangement minimizes end-to-end interactions of cyclopentadienyl rings with both methyl groups, as can be seen in Figure 3. These space filling models clearly show that both methyl groups are located in cavities formed by the canted rings at the opposite end of the molecule. This arrangement of cyclopentadienyl rings contrasts with that observed (12) in $Cp_2W{=}C(H)OZr(H)Cp^*_2$ ($Cp^* = C_5Me_5$). In this carbene complex, both the minimal steric requirements

Table II. Fractional Coordinates for $(Cp)_2(Me)ZrOCHCH_3Re(Cp)_2$[a,b]

| | $10^4X$ | $10^4Y$ | $10^4Z$ | $10B(Å^2)$ |
|---|---|---|---|---|
| Re | 8828(0) | 4288(1) | 6798(1) | 14 |
| Zr | 6469(1) | 5769(3) | 8129(2) | 13 |
| C(3) | 8125(12) | 2186(35) | 6577(21) | 21(5) |
| C(4) | 8250(13) | 1904(37) | 7565(23) | 26(6) |
| C(5) | 8951(12) | 1616(33) | 7306(21) | 16(5) |
| C(6) | 9247(12) | 1792(33) | 6192(21) | 15(5) |
| C(7) | 8740(13) | 2144(34) | 5679(21) | 19(5) |
| C(8) | 9338(11) | 6862(30) | 6347(19) | 9(4) |
| C(9) | 8842(12) | 7010(33) | 7476(21) | 16(5) |
| C(10) | 9030(12) | 5870(37) | 8156(20) | 22(5) |
| C(11) | 9643(12) | 5033(32) | 7519(20) | 17(5) |
| C(12) | 9824(12) | 5601(40) | 6392(21) | 25(5) |
| C(13) | 6228(14) | 8105(37) | 9564(23) | 27(6) |
| C(14) | 6865(14) | 8391(39) | 8943(24) | 30(6) |
| C(15) | 6859(14) | 8829(37) | 7892(23) | 29(6) |
| C(16) | 6202(13) | 8735(35) | 7856(21) | 23(6) |
| C(17) | 5790(12) | 8322(34) | 8913(21) | 19(5) |
| C(18) | 5995(12) | 4293(39) | 6759(20) | 21(5) |
| C(19) | 5466(12) | 5279(33) | 7412(21) | 20(5) |
| C(20) | 5273(13) | 4626(35) | 8482(21) | 23(5) |
| C(21) | 5655(12) | 3208(32) | 8530(20) | 15(5) |
| C(22) | 6132(12) | 2990(33) | 7479(20) | 17(5) |
| O(23) | 7363(8) | 5414(23) | 7083(14) | 22(3) |
| C(24) | 7989(11) | 5639(36) | 6329(19) | 17(4) |
| C(25) | 7976(14) | 5304(38) | 5104(24) | 34(6) |
| C(26) | 6733(12) | 4401(39) | 9566(21) | 26(5) |

[a]The isotropic thermal parameter listed for those atoms refined anisotropically is the isotropic equivalent.

[b]Numbers in parenthesis in this and all following tables refer to the error in the least significant digits.

Table III. Bond Distances (Å)

| | | | | | |
|---|---|---|---|---|---|
| Re | M(1)[a] | 1.90 | Zr | M(3)[b] | 2.21 |
| Re | M(2)[a] | 1.90 | Zr | M(4) | 2.24 |
| Re | C(3) | 2.28(3) | Zr | C(13) | 2.53(3) |
| Re | C(4) | 2.28(3) | Zr | C(14) | 2.55(3) |
| Re | C(5) | 2.23(3) | Zr | C(15) | 2.52(3) |
| Re | C(6) | 2.18(3) | Zr | C(16) | 2.44(3) |
| Re | C(7) | 2.25(3) | Zr | C(17) | 2.48(3) |
| | av. | 2.24(4) | | av. | 2.50(4)[c] |
| Re | C(8) | 2.27(2) | Zr | C(18) | 2.53(3) |
| Re | C(9) | 2.31(3) | Zr | C(19) | 2.54(2) |
| Re | C(10) | 2.27(3) | Zr | C(20) | 2.55(3) |
| Re | C(11) | 2.23(2) | Zr | C(21) | 2.57(2) |
| Re | C(12) | 2.23(3) | Zr | C(22) | 2.50(3) |
| | av. | 2.26(3) | | av. | 2.54(3) |
| Re | C(24) | 2.27(2) | Zr | C(26) | 2.32(3) |
| | | | Zr | O | 1.95(2) |
| | | | O | C(24) | 1.37(3) |
| | | | C(24) | C(25) | 1.59(4) |
| C(3) | C(4) | 1.38(4) | C(13) | C(14) | 1.34(4) |
| C(3) | C(7) | 1.43(3) | C(13) | C(17) | 1.41(4) |
| C(4) | C(5) | 1.41(3) | C(14) | C(15) | 1.38(4) |
| C(5) | C(6) | 1.37(3) | C(15) | C(16) | 1.38(4) |
| C(6) | C(7) | 1.43(3) | C(16) | C(17) | 1.40(4) |
| | av. | 1.40(3) | | av. | 1.38(3) |
| C(8) | C(9) | 1.50(3) | C(18) | C(19) | 1.39(3) |
| C(8) | C(12) | 1.43(3) | C(18) | C(22) | 1.46(4) |
| C(9) | C(10) | 1.38(3) | C(19) | C(20) | 1.39(3) |
| C(10) | C(11) | 1.44(3) | C(20) | C(21) | 1.38(3) |
| C(11) | C(12) | 1.44(3) | C(21) | C(22) | 1.41(3) |
| | av. | 1.44(4) | | av. | 1.41(3) |

[a] M(1) and M(2) are the midpoints of the $C_5H_5$ rings bound to Re.

[b] M(3) is the midpoint of the ring C(13) thru C(17).

[c] Esd's on average values are calculated using the scatter formula $\sigma(av) = [\Sigma(d_i-\overline{d})^2/(N-1)]^{1/2}$ where $d_i$ is one of N individual values and $\overline{d}$ is their average.

Table IV. Bond Angles (deg).

| | | | | |
|---|---|---|---|---|
| Re | C(24) | O | 113.3(2) | |
| Re | C(24) | C(25) | 115.1(2) | |
| O | C(24) | C(25) | 111.7(2) | |
| O | Zr | C(26) | 93.9(8) | |
| Zr | O | C(24) | 164.3(2) | |
| M(1) | Re | M(2) | 150.5 | |
| M(3) | Zr | M(4) | 129.6 | |
| M(1) | Re | C(24) | 104.9 | |
| M(2) | Re | C(24) | 104.6 | |
| M(3) | Zr | O | 107.9 | |
| M(4) | Zr | O | 110.8 | |
| M(3) | Zr | C(26) | 103.3 | |
| M(4) | Zr | C(26) | 104.9 | |
| C(4) | C(3) | C(7) | 110.8(2) | |
| C(3) | C(4) | C(5) | 106.3(2) | |
| C(4) | C(5) | C(6) | 109.4(2) | |
| C(5) | C(6) | C(7) | 109.4(2) | |
| C(3) | C(7) | C(6) | 104.0(2) | |
| | | | 108.0(3) | av. |
| C(9) | C(8) | C(12) | 107.7(2) | |
| C(8) | C(9) | C(10) | 108.0(2) | |
| C(9) | C(10) | C(11) | 108.2(2) | |
| C(10) | C(11) | C(12) | 109.8(2) | |
| C(8) | C(12) | C(11) | 106.3(2) | |
| | | | 108.0(1) | av. |
| C(14) | C(13) | C(17) | 109.3(3) | |
| C(13) | C(14) | C(15) | 108.4(3) | |
| C(14) | C(15) | C(16) | 108.6(3) | |
| C(15) | C(16) | C(17) | 107.6(2) | |
| C(13) | C(17) | C(16) | 106.1(2) | |
| | | | 108.0(1) | av. |
| C(19) | C(18) | C(22) | 106.5(2) | |
| C(18) | C(19) | C(20) | 108.6(2) | |
| C(19) | C(20) | C(21) | 109.9(2) | |
| C(20) | C(21) | C(22) | 107.7(2) | |
| C(18) | C(22) | C(21) | 107.2(2) | |
| | | | 108.0(1) | av. |

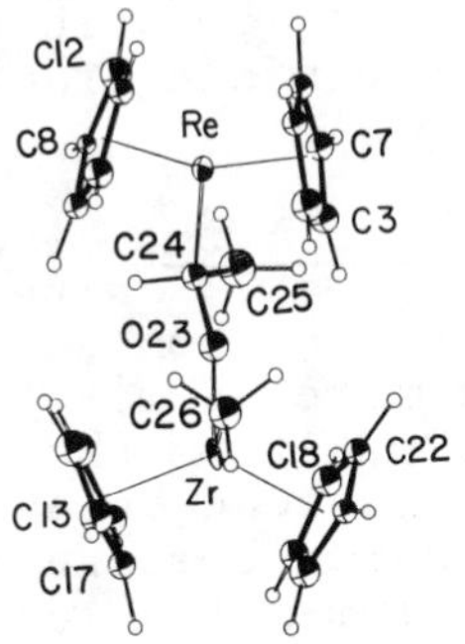


*Figure 1. ORTEP drawing of $Cp_2$-$ReCH(Me)OZrMeCp_2$ showing atom labeling scheme; unlabeled ring carbons follow the numerical sequence determined by the atom labels given.*

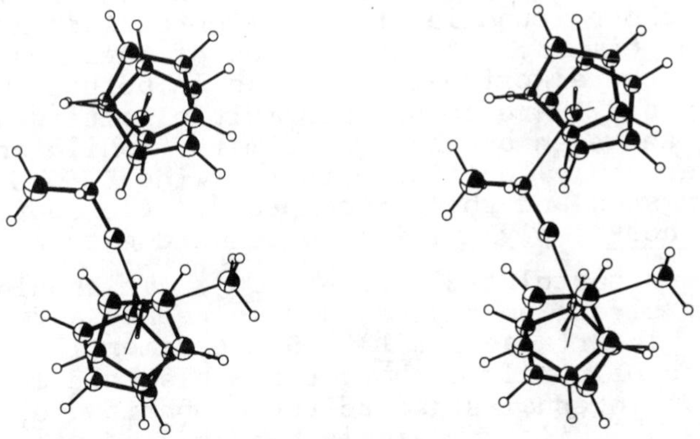

*Figure 2. Stereo view of $Cp_2ReCH(Me)OZrMeCp_2$ approximately perpendicular to the bridging group; the Rh fragment is at the top.*

of the bridge linking the metal atoms and the bulkier nature of the $C_5Me_5$ rings on zirconium dictate that the four rings arrange themselves so as to minimize ring-ring repulsions; that is, the plane containing the Cp centroids and the tungsten atom is almost perpendicular to the plane containing the $C_5Me_5$ centroids and the zirconium atom. The same staggered arrangement is seen in the salt $[Cp_4W_2H_3^+][ClO_4^-]$, due to the short W-W (nonbonded) separation (13). The compound $[Cp^*_2ZrN_2]N_2$ (14) and complexes of the type $[Cp_2MX]_2O$ (11,15,16,17,18) also show staggering of $Cp_2M$ units, but orbital overlap requirements are a major determining factor in these complexes (19).

Structural Features Around the Zr Atom. The arrangement of ligands around zirconium is quite typical of $Cp_2MXY$ structures. The methyl carbon and oxygen atoms form an angle of 93.9(8)° at zirconium. The centers of the Cp rings average 2.23 Å from the zirconium atom and the ring midpoints subtend an angle of 130° at Zr. These values are all quite comparable to those in $Cp_2ZrCl_2$ (11). The zirconium-methyl carbon distance is 2.32(3) Å, identical within experimental error to that in $Cp_2ZrCH_3[C(O)CH_3]$ (2). The interesting feature at the zirconium center is the Zr-O bond. Table V shows a comparison of some parameters of this bond with those found in other oxygen-containing metallocene dimers. An examination of these parameters shows that the alkoxide-like ligand in both $Cp_2W{=}C(H)O{-}ZrMeCp^*_2$ and III are bound with multiplicities approaching those in oxo-bridged dimers. While steric effects certainly contribute to the wide Zr-O-C angles in the tungsten and rhenium compounds, the short Zr-O bond distances imply multiple bonding due to $p_\pi \rightarrow d_\pi$ donation, as postulated earlier (20). It should be noted that zirconium-oxygen "single" bonds average 2.198(9) Å in $Zr(acac)_4$ (21). Since dimer III contains a Zr-C bond of purely σ character, this bond length provides an internal standard for comparison of bond order. Thus, while the single bond radius of oxygen is 0.11 Å shorter than that of carbon, the Zr-O bond in III is 0.37 Å shorter than the $Zr{-}CH_3$ bond; a significant π-component is present in this Zr-OR bond.

In $Cp_2MXY$ complexes (M = Ti, Zr, Hf), the ligands X and Y together donate a total of four electrons to the neutral fragment $Cp_2M$. When Y is a non-π-donor, as $CH_3$ in complex III, all π bonding is provided by X(OR); this leads to maximum contraction of the Zr-O distance. It was noted previously (3) that the Ti-Cl distance in

Table V. Selected Structural Data on Dimeric Cyclopentadienyl Complexes

| Compound | d(M-O)(Å) | ∠M-O-M (deg) | ∠Zr-O-C (deg) | Reference |
|---|---|---|---|---|
| $(Cp_2ZrCl)_2O$ | 1.94(1)<br>1.95(1) | 168.9(8) | --------- | 15 |
| $(Cp_2ZrSPh)_2O$ | 1.964(3)<br>1.968(3) | 165.8(2) | --------- | 16 |
| $Cp_2W{=}C(H)OZr(Me)(C_5Me_5)_2$ | 1.973(10) | -------- | 166.4(7) | 12 |
| $Cp_2ReC(H)(Me)OZr(Me)Cp_2$ | 1.95(2) | -------- | 164.3(17) | this work |
| $(Cp_2HfMe)_2O$ | 1.941(3) | 173.9(3) | --------- | 17 |
| $[(Cp_2NbCl)_2O](BF_4)_2$ | 1.88(1) | 169.3(8) | --------- | 11 |
| $\{[Cp_2Ti(H_2O)]_2O\}(ClO_4)_2$ | 1.829(2) | 175.8(5) | --------- | 18 |

$Cp_2TiCl[\eta^2-C(O)Me]$, is 0.13 Å longer than in $Cp_2TiCl_2$. Intramolecular competition for $\pi$-donation in this complex is dominated by the $\eta^2$-acetyl group, so that the Ti-Cl bond has nearly pure $\sigma$ character. On the other hand, the Zr-alkyl bond lengths in III and in $Cp_2ZrCH_3$-$[\eta^2-C(O)CH_3]$ (2) are identical, as expected. A final example of this competition exists in $Cp_2Ti$(p-nitrobenzoate)$_2$ (22). Here, the two compositionally identical carboxylate ligands do not perform identical donor functions. Instead, one functions as a $\pi$-donor (Ti-O = 1.94 Å and ∠ Ti-O-C = 157°) while the other serves as a $\sigma$-donor (Ti-O = 2.04 Å and ∠ Ti-O-C = 136°).

Structural Features Around the Re Atom. This portion of the molecule consists of a carbon atom bound symmetrically between two tilted Cp rings. None of the points Re, C(24), M(1), M(2) deviates by more than 0.02 Å from their least squares plane (M(i) is the ring centroid). This is the structure predicted for a $d^4$ complex of $Cp_2MX$ stoichiometry (19). The complexes $Cp_2VCl(d^2)$ (23), $Cp_2Ti$(2,6-ditertbutyl-4-methylphenyl) (24), $Cp_2Ti$(2,6-dimethylphenyl)($d^1$) (25), and $Cp_2V$-$\{N_2(SiMe_3)_2\}(d^1)$ (26) also have the symmetric structure exhibited by the $Cp_2ReC$ fragment in III. According to MO calculations (19), the more electron rich the metal center in a "bent" metallocene, the greater the M(1)-M-M(2) angle. Indeed, the M(1)-Re-M(2) angle in III is 150°, while it is in the range 136°-141° in the compounds cited above (19,23,24,25,26). The largest previously reported value for this parameter for metal-substituted metallocenes is 149° for $[Cp_2MoHLi]_4$ (27) (also formally $d^4$). Large angles have also been reported for $Cp_2MoD_2$ (148.2°) (28), $Cp_2W{=}C(H)OZr(H)Cp^*_2$ (145.6°), $[Cp_2W_2H_3]^+$ (148.2°), $[CpMoHCO]^+$ (144.5°) (29) and $Cp_2MoH_2\cdot ZnBr_2\cdot$DMF (143.5°) (30).

When the metal-ring distance is short, it has been postulated that large M(1)-M-M(2) angles are a steric consequence of inter-ring repulsions (27). An examination of Table III shows that the Re-C(Cp) distances are in fact shorter than those to zirconium; they are also shorter than the corresponding distances in several $Cp_2MX$ complexes of titanium and vanadium. However, the Re-C(Cp) distances in $[Cp_2ReBr_2]BF_4$ (11) are comparably short (Re-C = 2.26, ReM = 1.924), but the CpReCp angle in this salt is only 139.5°. Clearly the increase in CpReCp angle is very strongly dependent on electronic effects and on the number of attached ligands, in accord with previous calculations (19). Note that the great disparity (0.28 Å) in the distance from ring carbons to Zr vs. Re in III is not reflected

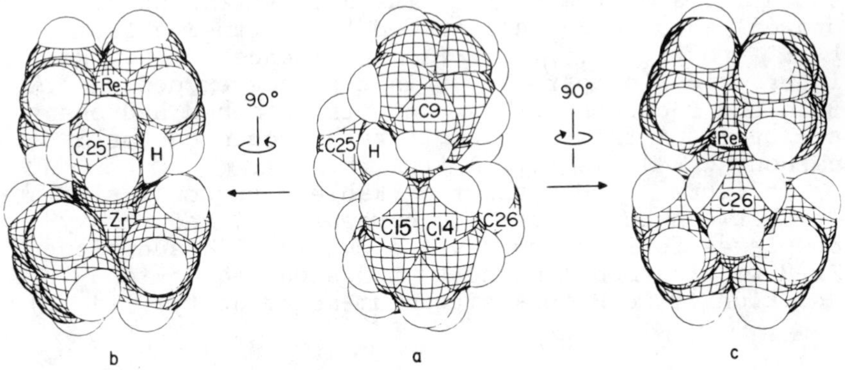


*Figure 3. View of space filling models of $Cp_2ReCH(Me)OZrMeCp_2$: a, molecule oriented as in Figure 2; b, rotated 90° about a vertical axis from* **a;** *c, molecule oriented as in Figure 1; H indicates the hydrogen atom on the tertiary carbon, C(24)*

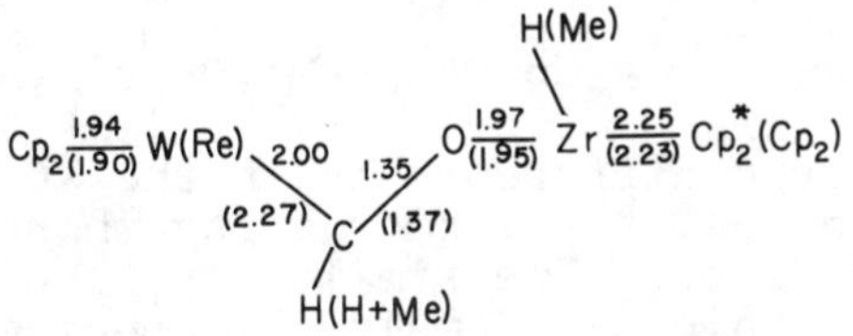


*Figure 4. Structure parameters for $Cp_2WC(H)OZrHCp_2{}^*$; corresponding atoms and parameters in compound* **III** *are shown in parentheses*

in the metal alkyl bond lengths:  $Zr-CH_3 = 2.32(3)$ Å and $Re-CH(Me)O = 2.27(2)$ Å.

The rhenium-$\sigma$ carbon bond distance of $2.27(2)$ Å does not differ statistically from those observed in $CpReMeBr(CO)_2$ $(2.32(4)$ Å$)$ $(\underline{31})$ and $CpRe(\eta^4-C_5H_5Me)Me_2$ $(2.23(3)$ Å$)$ $(\underline{32})$. The distance is shorter $(2.19(1)$ Å$)$ in $Li_2(Re_2Me_8)$ $(\underline{33})$.

Structure determinations of secondary metal alkyl complexes are relatively rare, yet they provide an opportunity to assess interactions of the metal with the $\beta$-atoms of the alkyl. The angles (excluding hydrogen) about $C(24)$ all exceed $109°$, ranging from $111.7°$ to $115.1°$. There is no evidence for any Re$\cdots$O interaction (compare V), this distance exceeding $3$ Å. Both the $\beta$-carbon, $C(25)$, and its attached hydrogens are over $3$ Å from rhenium. The hydrogen on the $\alpha$-carbon, $C(24)$, is $2.76$ Å from rhenium.

Figure 4 shows the remarkable structural similarity between the bimetallic carbene $(\underline{12})$ and alkoxy complexes formed from diverse paths: 1,2 addition of Zr-H to a carbonyl bound to tungsten (eq. 1) and 1,1 addition of Re-H to a zirconium-bound acetyl (eq. 2).

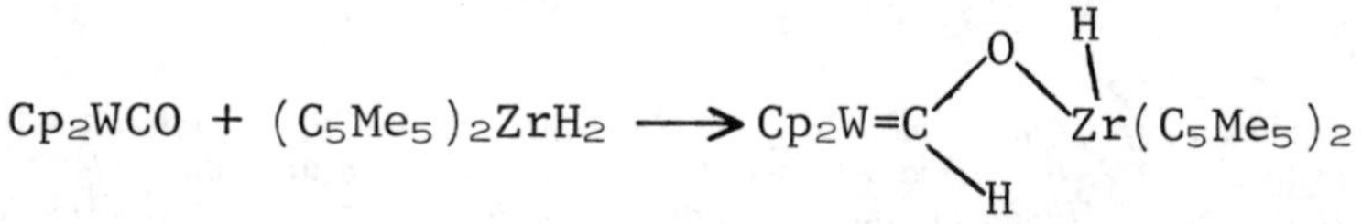

$$Cp_2WCO + (C_5Me_5)_2ZrH_2 \longrightarrow Cp_2W=C \overset{O}{\underset{H}{\diagdown}} \overset{H}{\underset{}{|}} Zr(C_5Me_5)_2 \qquad (1)$$

$$Cp_2ReH + (C_5H_5)_2Zr[C(O)Me]Me \longrightarrow$$

$$\overset{Me}{\underset{|}{}}$$
$$Cp_2Re-CH(Me)O-Zr(C_5H_5)_2 \qquad (2)$$

The major metric difference is in the WC (carbene) and ReC (alkyl) bonds, the former being shorter due to its multiple character. The similarity in CO distances implies negligible O $\rightarrow$ C multiple bonding in the W/Zr compound. This is a consequence of O $\rightarrow$ Zr $\pi$ donation in both compounds (compare the Zr-O bond lengths), which represents the dominant utilization of the oxygen lone pairs.

## Conclusion

This structure determination affirms the spectroscopic indication that the product of the reaction of $Cp_2Zr[C(O)Me]Me$ and $Cp_2ReH$ involves geminal addition of the Re-H bond to the electrophilic acetyl carbon. The unique Zr-O bond in $Cp_2Zr[C(O)Me]Me$ is retained in

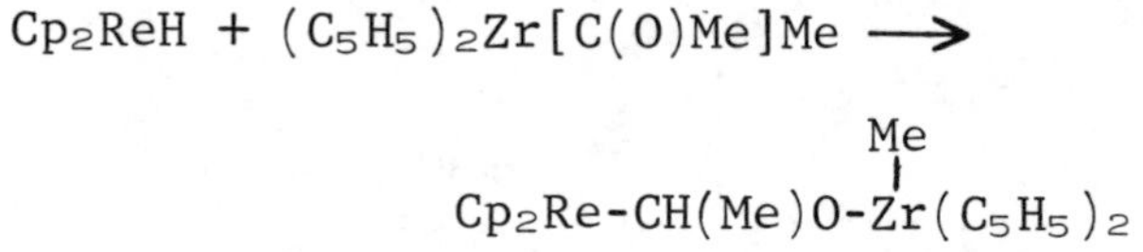

this reaction, suggesting that it contributes to the driving force for this facile reduction of carbon monoxide. To our knowledge, this is the first definitive example of an insertion of an acetyl carbon into an M-H bond and we are continuing our investigation of the importance of such insertions in Fischer-Tropsch syntheses.

## Acknowledgement

This work was supported by NSF Grant No. CHE 77-10059 and by the M. H. Wrubel Computer Center. Gifts of chemicals from Climax Molybdenum Company are gratefully acknowledged.

## Literature Cited

1. Collman, J. P.; Winter, S. R. J. Amer. Chem. Soc., 1973, 95, 4089.

2. Fachinetti, G.; Fochi, G.; Floriani, C. J. Chem. Soc., Dalt. Trans., 1977, 1946.

3. Fachinetti, G.; Floriani, C.; Stoeckli-Evans, H. J. Chem. Soc., Dalt. Trans., 1977, 2297.

4. Marsella, J. A.; Curtis, C. J.; Bercaw, J. E.; Caulton, K. G. J. Am. Chem. Soc., 1980, 102.

5. Labinger, J. A.; Komadina, K. J. Organometal. Chem., 1978, 155, C25.

6. King, R. B., "Organometallic Syntheses," Vol. 1, Academic Press, New York, 1965, p. 80.

7. Marsella, J. A.; Caulton, K. G. J. Am. Chem. Soc., 1980, 102, 1747.

8. Kaminsky, W.; Kopf, J.; Sinn, H.; Vollmer, H.-J. Angew. Chem., Int. Ed. Engl., 1976, 15, 629.

9. Fachinetti, G.; Floriani, C.; Roselli, A.; Pucci, S. J. Chem. Soc., Chem. Commun., 1978, 269.

10. Huffman, J. C., Indiana University Molecular Structure Center, Report No. 7944.

11. Prout, K.; Cameron, T. S.; Forder, R. A.; Critchley, S. R.; Denton, B.; Rees, G. V. Acta Cryst. Sect. B, 1974, 30, 2290.

12. Wolczanski, P. T.; Threlkel, R. S.; Bercaw, J. E. J. Am. Chem. Soc., 1978, 101, 218.

13. Klingler, R. J.; Huffman, J. C.; Kochi, J. K. J. Am. Chem. Soc., 1980, 102, 208.

14. Sanner, R. D.; Manriquez, J. M.; Marsh, R. E.; Bercaw, J. E. J. Am. Chem. Soc., 1976, 98, 8351.

15. Clarke, J. F.; Drew, M.G.B. Acta Cryst. Sect. B, 1974, 30, 2267.

16. Petersen, J. L. J. Organometallic Chem., 1979, 166, 179.

17. Fronczek, F. R.; Baker, E. C.; Sharp, P. R.; Raymond, K. N.; Alt, H. G.; Rausch, M. D. Inorg. Chem., 1976, 15, 2284.

18. Thewalt, U.; Kebbel, B. J. Organometal. Chem., 1978, 150, 59.

19. Lauher, J. W.; Hoffmann, R. J. Am. Chem. Soc., 1976, 98, 1729.

20. Huffman, J. C.; Moloy, K. G.; Marsella, J. A.; Caulton, K. G. J. Am. Chem. Soc., 1980, 102, 3009.

21. Silverton, J. V.; Hoard, J. L. Inorg. Chem., 1963, 2, 243.

22. Kuntsevich, T. S.; Gladkikh, E.; Lebedev, V. A.; Linevag, A.; Bolov, N. V. Sov. Phys.: Crystallogr., 1976, 21, 40.

23. Fieselmann, B. F.; Stucky, G. D. J. Organometal. Chem., 1977, 137, 43.

24. Cetinkaya, B.; Hitchcock, P. B.; Lappert, M. F.; Torroni, S.; Atwood, J. L.; Hunter, W. E.; Zaworotko, M. J. J. Organometal. Chem., 1980, 188, C31.

25. Olthof, G. J.; Van Bolhuis, F. J. Organometal. Chem., 1976, 122, 47.

26. Veith, M. Angew. Chem., Int. Ed. Engl., 1976, 15, 387.

27. Forder, R. A.; Prout, K. Acta Cryst., Sect. B., 1974, 30, 2318.

28. Schultz, A. J.; Stearly, K. L.; Williams, J. M.; Mink, R.; Stucky, G. D. Inorg. Chem., 1977, 16, 3303.

29. Adams, M. A.; Folting, K.; Huffman, J. C.; Caulton, K. G. Inorg. Chem., 1979, 18, 3020.

30. Crotty, D. E.; Corey, E. R.; Anderson, T. J.; Glick, M. D.; Oliver, J. P. Inorg. Chem., 1977, 16, 920.

31. Aleksandrov, G. G.; Struchkov, Yu. T.; Makarov, Yu. V. Zh. Strukt. Khim., 1973, 14, 98.

32. Alcock, N. W. J. Chem. Soc. (A), 1967, 2001.

33. Cotton, F. A.; Grange, L. D.; Mertis, K.; Shive, L. W.; Wilkinson, G. J. Chem. Soc., Dalt. Trans., 1976, 98, 6922.

RECEIVED December 8, 1980.

# 4

# Unusual Carbon Monoxide Activation, Reduction, and Homologation Reactions of 5*f*-Element Organometallics

## The Chemistry of Carbene-Like Dihaptoacyls

PAUL J. FAGAN, ERIC A. MAATTA, and TOBIN J. MARKS

Department of Chemistry, Northwestern University, Evanston, IL 60201

There is currently great interest in understanding chemistry by which metallic reagents transform carbon monoxide, either catalytically or stoichiometrically, into useful organic compounds (1,2,3). In the case of Group VIII transition metal catalysts, vast quantities of acetic acid, alcohols, aldehydes, esters, etc., are currently produced from carbon monoxide, hydrogen, and various organic feedstocks. In regard to mechanism, much of this chemistry is now reasonably well-understood. A key reaction is the migratory insertion of CO (4,5) into a metal-carbon sigma bond (eq.(1)) to produce a metal acyl, A, followed

$$M{-}CH_3 + CO \longrightarrow (CH_3)M \leftarrow CO \longrightarrow M{-}C(CH_3){=}O \qquad (1)$$

A

by scission of the metal-carbon bond via a process such as hydrogenolysis, reductive elimination, olefin insertion, etc. Although this classic picture evolved from "soft," mononuclear transition metal complexes suffices to explain a great deal of carbon monoxide chemistry, it is not clear that it is complete or accurate for understanding processes whereby CO is reduced, deoxygenated, and/or polymerized to form methane, long-chain hydrocarbons, alcohols, and other oxocarbons, especially in cases where heterogeneous catalysts or "hard" metals are involved (6,7,8,9,10).This deficiency of information has led to the search for new modes of carbon monoxide reactivity and to attempts to understand carbon monoxide chemistry in nontraditional environments.

In the past several years, it has become apparent that with proper tuning of ligation, it is possible to prepare organometallic compounds of actinide elements with very high coordinative unsaturation and very high chemical reactivity (11,12,13). In regard to exploring "nonclassical" modes of carbon monoxide acti-

0097-6156/81/0152-0053$06.50/0
© 1981 American Chemical Society

vation, these features, combined with the very large affinity which actinides exhibit for oxygen, offer the possibility of drastically modifying the classical chemistry and of modelling in homogeneous solution, some of the features of heterogeneous CO reduction catalysts (especially those involving actinides) (12,14-16). The prodigious strengths of actinide-oxygen bonds can be appreciated by considering the formation enthalpies of binary oxides (Table I) (17,18). It can also be seen that early

Table I. Representative Formation Enthalpies of Some Binary d- and f-Element Oxides.[a]

| | | | | | |
|---|---|---|---|---|---|
| $TiO_2$ | $VO_2$ | $CrO_2$ | $MnO_2$ | $Fe_3O_4$ | CoO |
| -112 | -93.9 | -69.7 | -62.0 | -64.9 | -56.9 |
| $ZrO_2$ | $NbO_2$ | $MoO_2$ | | $RuO_2$ | RhO |
| -130 | -94.5 | -66.5 | | -27.3 | -19.5 |
| | | $WO_2$ | | $OsO_2$ | $IrO_2$ |
| | | -67.3 | | -35.2 | -18.3 |
| $ThO_2$ | $UO_2$ | | | | |
| -143 | -125 | | | | |

[a] In kcal/mole of O atoms. From references 17 and 18.

transition metals possess similar characteristics, and where mean bond dissociation energy data exist for both classes of metals (e.g., the metal tetrahalides) trends in actinide and early transition metal (e.g., Ti, Zr) parameters are largely parallel (19). The consequences of this oxygen affinity for the making and breaking of metal-to-ligand bonds can be readily assessed in Table II (20,21). Extrapolating to the actinides, it can be surmised that a Th-O bond is stronger than a Th-C bond by ca. 50 kcal/mole.

The purpose of this article is to review recent results on the carbonylation chemistry of actinide-to-carbon sigma bonds, bearing in mind the unique properties of 5f-organometallics cited above. We focus our attention on the properties of bis(pentamethylcyclopentadienyl) actinide acyls. Just as transition metal acyls (A) occupy a pivotal role in classical carbonylation chemistry, it will be seen that many of the unusual

Table II. Mean Bond Dissociation Energy Data for Some Early Transition Metal Complexes[a] and Estimated[b] Values for Thorium and Uranium.

| $MR_n$ | Ti | Zr | Hf | Nb | Ta | Mo | W | Th | U |
|---|---|---|---|---|---|---|---|---|---|
| R | n=4 | 4 | 4 | 5 | 5 | 6 | 6 | 4 | 4 |
| $CH_3$ | 62 | 74 | 79 | | 62 | | 38 | 78[b] | 73[b] |
| $CH_2CMe_3$ | 45 | 54 | 54 | | | | | 53[b] | 50[b] |
| $CH_2Ph$ | 49 | 60 | | | | | | | |
| Cl | 103 | 117 | 119 | 97 | 103 | 73 | 83 | 117[b] | 110[b] |
| $NEt_2$ | 74 | 82 | 88 | | 78 | | 86 | 87[b] | 81[b] |
| OPr-i | 106 | 124 | 128 | 100 | 105 | | | 126[b] | 118[b] |
| F | 140 | 154 | 155 | | | | | 153[a] | 143[a] |

[a] In kcal/mole, from references 19, 20, and 21.

[b] Estimated from the proportionality Ac-R = M-R(Ac-F/M-F), where Ac is the actinide value.

features which actinide elements introduce to CO chemistry manifest themselves in the chemical and physicochemical characteristics of the acyls. It will also be seen that the coordinative unsaturation and oxygen affinity of Th(IV) and U(IV) ions give rise to highly reactive, oxygen-coordinated (dihapto) acyls with marked carbene-like character, and patterns of chemical reactivity never previously observed in solution. We discuss here the synthesis and physicochemical properties of actinide bis(pentamethylcyclopentadienyl) dihaptoacyls, carbon monoxide coupling involving these complexes, rearrangement reactions of the dihaptoacyls, as well as hydride-catalyzed isomerization and hydrogenation reactions.

## Synthesis of Organoactinide Acyls and Properties

As discussed elsewhere (12,16), the carbonylation of bis(pentamethylcyclopentadienyl) thorium and uranium bis(hydrocarbyls), $M[(CH_3)_5C_5]_2R_2$, leads to rapid, irreversible formation of enediolate (B) complexes. Although there is circumstantial

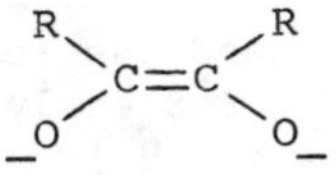

B

evidence (14) that such species could arise from intra- or intermolecular coupling of dihaptoacyl functionalities, the very high reactivity of the organoactinides has so far precluded the observation or isolation of intermediates on the reaction coordinate leading to the enediolate. On the other hand, solutions of actinide bis(pentamethylcyclopentadienyl) chlorohydrocarbyls (11,22,23) and dialkylamidehydrocarbyls (14) absorb an equivalent of carbon monoxide in the course of 0.2-1.5 h at low temperatures to yield actinide acyls. Representative examples are illustrated in eq.(2). Unlike analogous early transition metal systems

$$M[(CH_3)_5C_5]_2(X)R + {}^{*}CO \longrightarrow M[(CH_3)_5C_5]_2(X)({}^{*}COR) \qquad (2)$$

**1a** M=Th, X=Cl, $R=CH_2C(CH_3)_3$, $^{*}C=^{12}C$

**1b** M=Th, X=Cl, $R=CH_2C(CH_3)_3$, $^{*}C=^{13}C$

**2a** M=Th, X=Cl, $R=CH_2C_6H_5$, $^{*}C=^{12}C$

**3a** M=Th, $X=N(CH_3)_2$, $R=CH_3$, $^{*}C=^{12}C$

**4a** M=U, X=Cl, $R=C_6H_5$, $^{*}C=^{12}C$

**4b** M=U, X=Cl, $R=C_6H_5$, $^{*}C=^{13}C$

(24,25,26,27), the insertion in these organoactinides is irreversible. The new compounds were characterized by elemental analysis, infrared and nmr spectroscopy. The C-O stretching frequencies, verified in several cases by $^{13}C$ substitution, provide important information on the metal-acyl bonding. In particular, the energies (Table III) are considerably lower than in classical transition metal acyls ($\nu_{CO}$ = ca. 1630-1680 $cm^{-1}$) and suggest dihaptoacyl ligation (C, D).

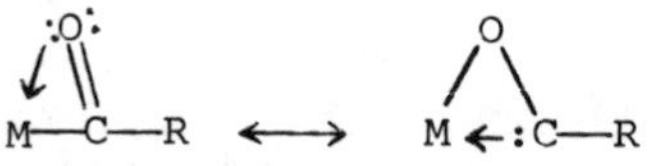


C D

Dihaptoacyl coordination has been previously observed in several transiton metal systems. For later transition metals such as Ru (28) and Mo (29), the metal-oxygen interaction, as judged by metrical parameters and the relatively high $\nu_{CO}$ frequencies, is relatively weak. For early transition metal complexes such as $Ti(C_5H_5)_2(\eta^2\text{-}COCH_3)Cl$ (25) and $Zr(C_5H_5)_2$-$(\eta^2\text{-}COCH_3)CH_3$ (26), the metal-oxygen interaction is significantly stronger, but still does not equal that in the organoactinides. These differences can be demonstrated by several lines of evi-

Table III. Infrared Vibrational Spectroscopic Data for Organoactinide Dihaptoacyls.[a]

| Compound | $\nu_{CO}$ | $\nu_{^{13}CO}$ |
|---|---|---|
| $Th[(CH_3)_5C_5]_2[\eta^2\text{-}COCH_2C(CH_3)_3]Cl$ (1a) | 1469 | |
| $Th[(CH_3)_5C_5]_2[\eta^2\text{-}^{13}COCH_2C(CH_3)_3]Cl$ (1b) | | 1434 |
| $Th[(CH_3)_5C_5]_2[\eta^2\text{-}COCH_2C_6H_5]Cl$ (2a) | 1439 | |
| $Th[(CH_3)_5C_5]_2(\eta^2\text{-}COCH_3)N(CH_3)_2$ (3a) | 1483 | |
| $U[(CH_3)_5C_5]_2(\eta^2\text{-}COC_6H_5)Cl$ (4a) | 1429 | |
| $U[(CH_3)_5C_5]_2(\eta^2\text{-}^{13}COC_6H_5)Cl$ (4b) | | 1404 |

[a]Recorded as Nujol mulls; data expressed in wavenumbers.

dence. First, as already noted, carbonylation is irreversible for the organoactinides studied to date, while it is reversible for analogous transition metal bis(cyclopentadienyl) and bis-(pentamethylcyclopentadienyl) systems. Second, the C-O stretching frequencies of the f-element $\eta^2$-acyls (Table III) are substantially lower than for comparable d-element $\eta^2$-acyls. Examples of the latter are $Zr(C_5H_5)_2[\eta^2\text{-}COCH_2C(CH_3)_3]Cl$ where $\nu_{CO}$ = 1550 $cm^{-1}$ (26), $Zr(C_5H_5)_2(\eta^2\text{-}COCH_3)CH_3$ where $\nu_{CO}$ = 1545 $cm^{-1}$ (25), $Hf(C_5H_5)_2(\eta^2\text{-}COCH_3)CH_3$ where $\nu_{CO}$ = 1550 $cm^{-1}$ (25), and $Zr[(CH_3)_5C_5]_2(\eta^2\text{-}COCH_3)CH_3$ where $\nu_{CO}$ = 1550 $cm^{-1}$ (27). The reduced C-O force constant can be taken as evidence for a major contribution from the carbene-like resonance hybrid D. Further support for the greater importance of this electronic structure in the case of the f-element ions comes from $^{13}C$ nmr spectroscopy. For diamagnetic $Th[(CH_3)_5C_5]_2[\eta^2\text{-}^{13}COCH_2C(CH_3)_3]Cl$, the acyl carbon chemical shift occurs at a remarkably low value of δ 360.2 ($C_6D_6$), reminiscent of carbene complexes (30,31), while for $Zr(C_5H_5)_2[\eta^2\text{-}COCH_2C(CH_3)_3]Cl$, the resonance frequency is at δ 318.7 ($C_6D_6$) (24).

The molecular structure of 1 has been determined by single crystal X-ray diffraction techniques (15) and the result is illustrated in Figure 1. The actinide coordination geometry features the familiar, "bent sandwich" $M[\eta^5\text{-}(CH_3)_5C_5]_2X_2$ structure where X = Cl and $\eta^2\text{-}COCH_2C(CH_3)_3$. Two features of the dihaptoacyl ligation are particularly important. First, the Th-O distance (2.37(2) Å)is 0.07 Å shorter than the Th-C distance (2.44 (2) Å). This ordering is in contrast to the corresponding M-O vs. M-C parameters in $Ti(C_5H_5)_2(\eta^2\text{-}COCH_3)Cl$ (2.19(1) vs 2.07(2) Å) and $Zr(C_5H_5)_2(\eta^2\text{-}COCH_3)CH_3$ (2.290(4) vs. 2.197(6) Å), where the metal-oxygen distance is clearly longer than the metal-carbon distance. Furthermore, the Th-O distance in 1 is only *ca.* 0.17 Å longer than Th-O bond distances in other bis(pentamethylcyclopentadienyl) thorium organometallics containing metal-oxygen bonds, i.e., 2.150(4) Å in the enediolate

$\{Th[(CH_3)_5C_5]_2[\mu\text{-}O_2C_2(CH_3)_2]\}_2$ (16) and 2.27(1) Å in $\{Th[(CH_3)_5C_5]_2[\mu\text{-}CO(CH_2C(CH_3)_3)CO]Cl\}_2$ (15) (vide infra).

The second important structural feature in 1 concerns the orientation of the $\eta^2$-acyl ligand. The C-O vector in the present case points in the direction away from the Cl ligand, while in the aforementioned titanium and zirconium compounds, the C-O vector is oriented toward the non-acyl monohapto ligand (E vs F). The reasons for these differences are not entirely clear.

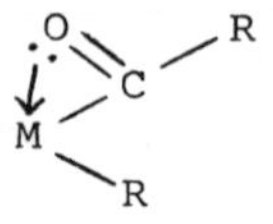


E

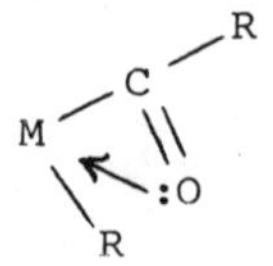


F

Molecular orbital considerations suggest that for a transition metal $M(C_5H_5)_2R_2$ complex, initial CO activation will involve electron flow from the carbon-centered CO $\sigma$ donor orbital into the $M(C_5H_5)_2R_2$ LUMO (32,33). That is, attack will be in the direction perpendicular to the (ring centroid)-M-(ring centroid) plane yielding, after R migration, structure E. Although isomer E has been previously proposed (34) as a fleeting intermediate in $Zr(C_5H_5)_2(p\text{-}C_6H_4CH_3)_2$ carbonylation, the present results are the first case where such a structure has been unambiguously identified. In all probability, the two configurations (E and F) differ little in energy content and can rapidly interconvert. Indeed, for $U[(CH_3)_5C_5]_2(\eta^2\text{-}CONR_2)Cl$ compounds, both isomers are in equilibrium, with $\Delta H = 1.2 \pm 0.1$ (R=$CH_3$), $0.8 \pm 0.3$ (R=$C_2H_5$) kcal/mol; $\Delta S = 8 \pm 1$ (R=$CH_3$), $9 \pm 3$ (R=$C_2H_5$) e.u. Interconversion of the two structures is rapid on the nmr timescale with $\Delta G = 8.9 \pm 0.5$ (R=$CH_3$, -80°C), $8.9 \pm 0.5$ (R=$C_2H_5$, -70°C) kcal/mole (14). In regard to whether the unique spectral characteristics of 1 and the other actinide dihaptoacyls might arise simply from having structure E, the information on $Zr(C_5H_5)_2[\eta^2\text{-}CO(p\text{-}C_6H_4CH_3)](p\text{-}C_6H_4CH_3)$ (34) indicates that structures E and F exhibit similar $\nu_{CO}$ values (1480 and 1505 $cm^{-1}$) and $\delta\,^{13}C$(acyl) shifts (300 and 301 ppm). To summarize, these results indicate that the coordinative unsaturation and oxygen affinity of the actinide environment have perturbed the ligation of inserted carbon monoxide further toward the nonclassical, carbene-like hybrid (D) than has heretofore been achieved. Such a situation provides a unique opportunity to explore new patterns of metal acyl reactivity in solution.

## Carbon Monoxide Oligomerization by Organoactinide Acyls

In the presence of excess CO at room temperature,

$Th[(CH_3)_5C_5]_2[\eta^2\text{-}COCH_2C(CH_3)_3]Cl$ reacts with an additional equivalent of carbon monoxide according to eq.(3).

$$2\ Th[(CH_3)_5C_5]_2[^{*}COCH_2(CH_3)_3]Cl \xrightarrow[25°,\ 12\ h]{^{\ddagger}CO}$$

1a $^{*}C = {}^{12}C$
1b $^{*}C = {}^{13}C$

ca. 50% isolated yield

$$\{Th[(CH_3)_5C_5]_2[^{*}CO(CH_2C(CH_3)_3)\,{}^{\ddagger}CO]Cl\}_2 \qquad (3)$$

5a $^{*}C = {}^{\ddagger}C = {}^{12}C$

5b $^{*}C = {}^{13}C;\ {}^{\ddagger}C = {}^{12}C$

5c $^{*}C = {}^{12}C;\ {}^{\ddagger}C = {}^{13}C$

The compound 5 can be obtained from toluene as dark-violet crystals (starting material 1 is pale-yellow). The empirical formula and molecularity of 5 could be readily established from elemental analysis and cryoscopic molecular weight data, but little other unambiguous structural information was evident in the infrared and nmr data. The crystal structure of 5 was determined by single crystal X-ray diffraction methods (15) and the result is shown in Figure 2. In effect, oligomerization of four carbon monoxide molecules has occurred to produce a dimeric thorium complex of an enedionediolate ligand (G). The

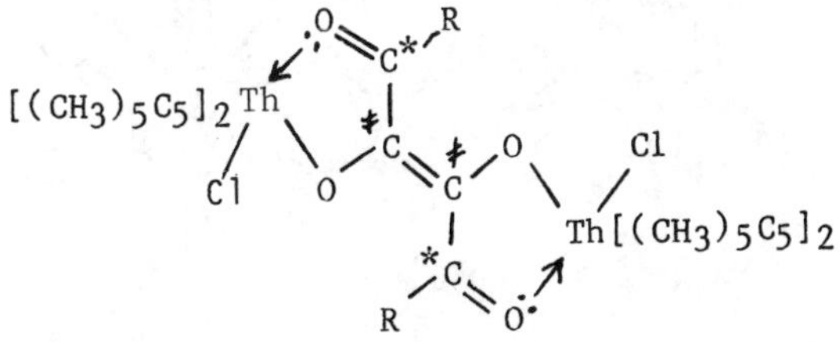


G

metrical parameters are in accord with the resonance structure as drawn. Thus, $C_a$-$O_a$ (Figure 2) (1.26(2) Å) is shorter than $C_b$-$O_b$ (1.34(2) Å) while Th-$O_a$ (2.53(1) Å) is considerably longer than Th-$O_b$ (2.27(1) Å). The $C_b$-$C_b$' distance of 1.35(4) Å suggests appreciable double bond character.

At the present, the most straightforward mechanism for the formation of 5 from 1 is via insertion of CO into the Th-C(acyl) bond to form a ketene (H, I) (eq.(4)) which subsequently dimerizes. Presumably, initial CO interaction could involve coordination either to the metal ion as shown or to the electrophilic vacant "carbene" p atomic orbital. Considering the affinity of the Th(IV) ion for oxygenated ligands, interaction of the ketene oxygen atom with the metal ion seems reason-

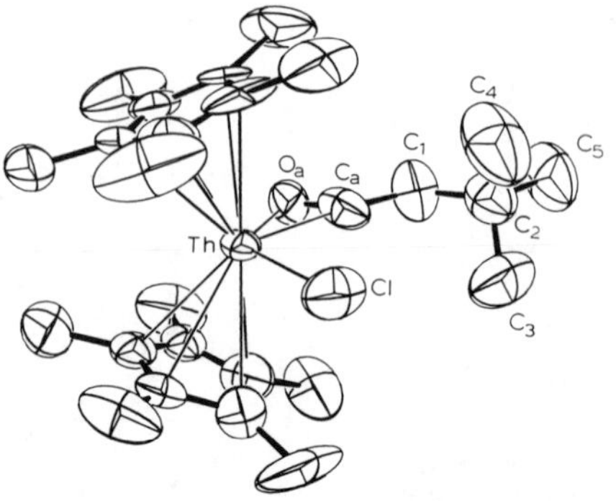


*Figure 1. ORTEP drawing of the nonhydrogen atoms of $Th[(CH_3)_5C_5]_2[\eta^2\text{-}COCH_2C(CH_3)_3]Cl$ molecule, **1**; all atoms are represented by thermal-vibration ellipsoids drawn to encompass 50% of the electron density (15)*

Journal of the American Chemical Society

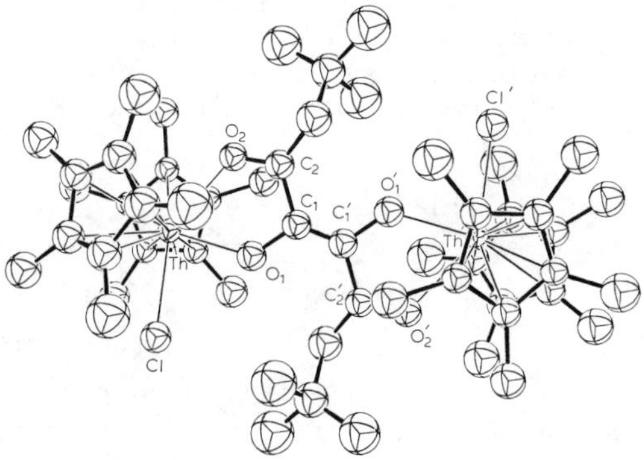


Journal of the American Chemical Society

*Figure 2. ORTEP drawing of the nonhydrogen atoms of one of the two crystallographically independent $\{Th[(CH_3)_5C_5]_2[\mu\text{-}CO(CH_2C(CH_3)_3)CO]Cl\}$ molecules in the unit cell of **5**. The stereochemistry of the second molecule differs from this one primarily in the orientation of the* t*-butyl groups. All atoms are represented by thermal-vibration ellipsoids drawn to encompass 50% of the electron density (15).*

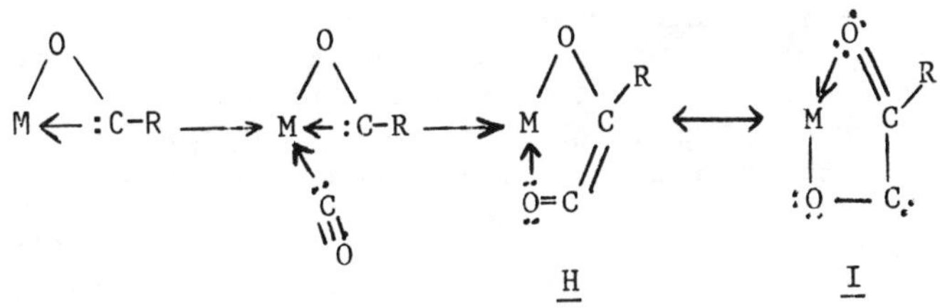

(4)

able. There is ample precedent for the reaction of carbenes with CO to form ketenes (35,36), the transfer of coordinated carbenes to carbon monoxide (37,38), and the formation of stable complexes between transition metal ions and ketenes (37-43). The precise manner in which the ketene units formally couple in the present case to form a dimer has not been observed for free ketenes (44). Important information on the mechanism of CO tetramerization is provided by $^{13}$C nmr experiments. Thus 1a reacts with $^{13}$CO to yield 5c with > 95% of the label incorporated at $^{\ddagger}$C ($\delta$ = 158.6 ppm in $C_6D_6$), while 1b reacts with $^{12}$CO to yield 5b with > 95% of the label retained at $^{*}$C ($\delta$ = 216.8 ppm in $C_6D_6$)). It is evident that the thorium ion has activated the inserted carbon monoxide, and that the resulting chemistry has a distinctly carbenoid character. Further studies of the mechanism of this unique carbon monoxide homologation process are under way.

## Rearrangement Reactions of Organoactinide Acyls

As already noted, the carbonylation of bis(pentamethylcyclopentadienyl) actinide hydrocarbyls is irreversible in the cases studied thus far. Thus, thermolysis does not result in CO loss, but rather in interesting chemical reactions. Thermolysis of 1 (15) in toluene solution results in hydrogen atom migration to yield an enolate (eq.(5)). NMR studies establish that eq.(5) is essentially quantitative, and that the stereochemical course

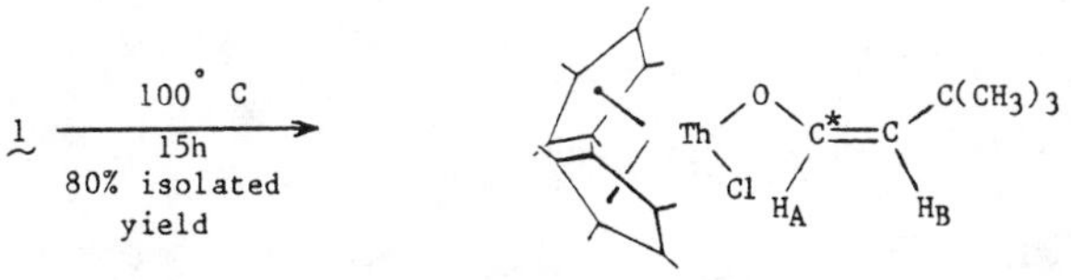


(5)

6a C* = $^{12}$C

6b C* = $^{13}$C

of the rearrangement is greater than 95% cis. The proton nmr spectrum exhibits singlets at $\delta$ 2.03 (30H) and at 1.34 (9H), and doublets at $\delta$ 6.30 ($H_A$, $J_{H_A-H_B}$ = 7.2 Hz) and 4.14 ($H_B$, $J_{H_A-H_B}$ =7.2 Hz). NMR studies on 6b demonstrate that the *C-O bond of 1b remains intact during the rearrangement process. Thus, the $^1$H nmr spectrum of 6b in the olefinic region exhibits a doublet of doublets at $\delta$ 6.30 ($H_A$, $J_{^{13}C-H}$ = 175 Hz, $J_{H_A-H_B}$ = 7.2 Hz) and a pseudotriplet at $\delta$ 4.14 ($H_B$, $J_{^{13}C-H} \approx J_{H_A-H_B}$=7.2 Hz).

The hydrogen atom migration observed on thermolysis of 1 is reminiscent of 1,2-hydrogen atom migrations in carbene chemistry (45,46,47). The stereochemistry of such processes is now relatively well-understood and involves initial hyperconjugative interaction between a gauche C-H bond and the carbene unoccupied p atomic orbital, followed by a low activation energy 1,2 shift (eq.(6)) (47,48,49,50).

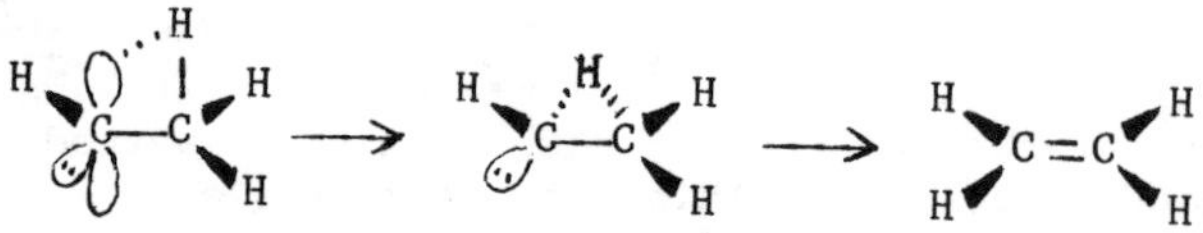


(6)

Application of this picture to the present thorium acyl system (1) suggests that a hydrogen atom migration to the electron deficient carbenoid p orbital must occur from a conformation (J) other than that found in the crystal structure (H) (Figure 1) to yield the cis product. These relationships are illustrated in eqs.(7) and (8).

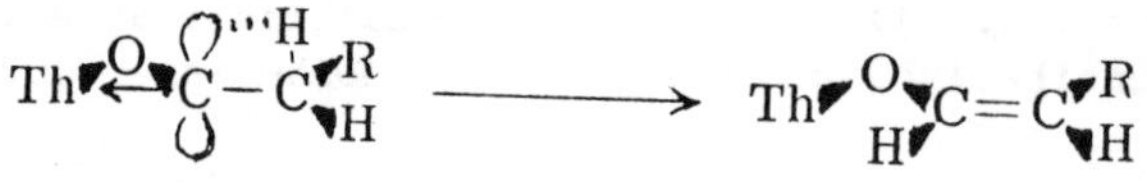


(7)

J

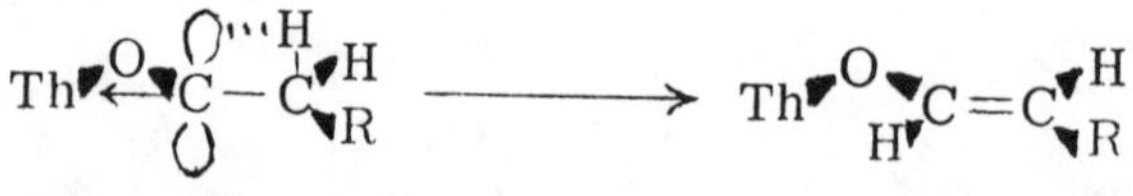


(8)

K

Additional evidence for the oxycarbene character of the inserted carbon monoxide is derived from studies of the chlorotrimethylsilylmethyl compounds 7 and 8 (eq.(9)).

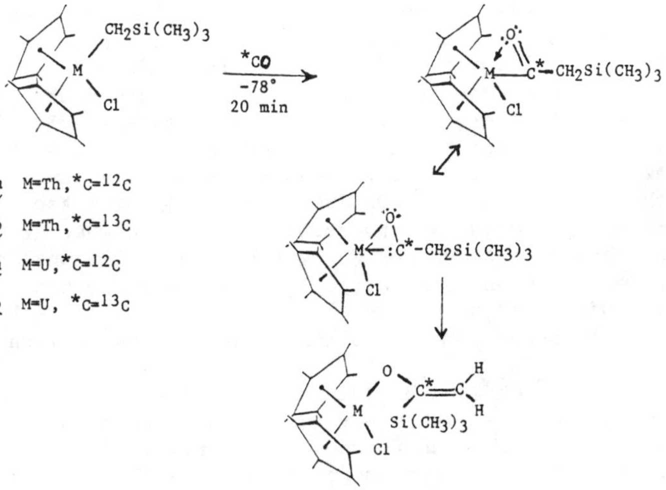


7a M=Th, *C=$^{12}$C

7b M=Th, *C=$^{13}$C

8a M=U, *C=$^{12}$C

8b M=U, *C=$^{13}$C

(9)

9a M=Th, *C=$^{12}$C

9b M=Th, *C=$^{13}$C

10a M=U, *C=$^{12}$C

10b M=U, *C=$^{13}$C

Carbonylation at low temperatures yields unstable intermediates which, on the basis of nmr spectra, are ascribed to dihaptoacyl complexes. On warming to room temperature, these intermediates rearrange, and compounds 9 and 10 are formed in essentially quantitative yield. The structures of these rearrangement products as well as the integrity of the *C-O bond during the trimethylsilyl migration were established by the usual analytical techniques as well as $^{1}$H and $^{13}$C nmr spectroscopy. In particular, 9a exhibits proton nmr signals ($C_6D_6$) at $\delta$ 2.01 (30H, s), 0.24 (9H, s), 4.54 (1H, s), and 4.88 (1H, s). In 9b, the latter three resonances are doublets with $J_{^{13}C-H}$ = 2.0, 9.6, and 6.6 Hz, respectively. The migration of $(CH_3)_3Si$ in preference to H is not unexpected in carbene chemistry. Third-row elements are known to exhibit substantially greater migratory aptitudes than hydrogen atoms (47,51,52,53).

## Reaction of Organoactinide Acyls with Hydrides. Catalytic Isomerization and Hydrogenation

Metal hydrides and acyl-like CO insertion products are two types of species likely to be present in any homogeneous or heterogeneous process for the catalytic reduction of carbon monoxide. The discovery and understanding of new types of reactivity patterns between such species are of fundamental interest. As discussed elsewhere (11,22,54-57), bis(pentamethylcyclopentadienyl) actinide hydrides (58) are highly active catalysts for olefin hydrogenation as well as H-H and C-H activation. Thus, the reaction of $\{Th[(CH_3)_5C_5]_2H_2\}_2$ (11) with the organoactinide dihaptoacyls was investigated to learn whether the inserted carbon monoxide was susceptible to any unusual modes of hydride reduction. In particular, analogues to the well-known insertion of carbenes into metal and metalloid hydride bonds (59,60) would offer a means to functionalize the acyl carbon atom.

The reaction of 1 with 11 in benzene solution at room temperature is complete within several hours. Compound 11 is unchanged while 1 is transformed quantitatively into an enolate 12 (eq. (10)). This rearrangement product was characterized by

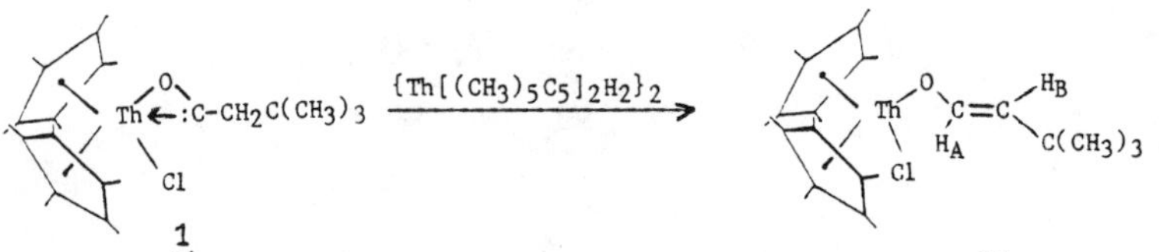


1 12 (10)

standard techniques, with the trans stereochemistry (> 95% isomeric purity) being established by nmr ($J_{H_A-H_B}$ = 12.0 Hz in 12 versus 7.2 Hz in the cis isomer 6) as illustrated in Figure 3. The reaction is catalytic in thorium hydride and, at 35°C in $C_6D_6$ with [1] = $6.3 \times 10^{-3}$ M and [11] = $3.8 \times 10^{-4}$ M, occurs with a turnover frequency per $ThH_2$ moiety of ca. 8 $h^{-1}$. The mechanism of this catalytic dihaptoacyl isomerization is proposed to involve initial insertion of the acyl carbon atom into the Th-H bond, followed by β-hydride elimination. This process is illustrated in eq. (11). There is precedent in recent transition metal chemistry for the formation of stable MOC(R)HM' species analogous to 13 from MH and M'($\eta^2$-COR) precursors (61). In the present case, the trans stereochemistry of the enolate

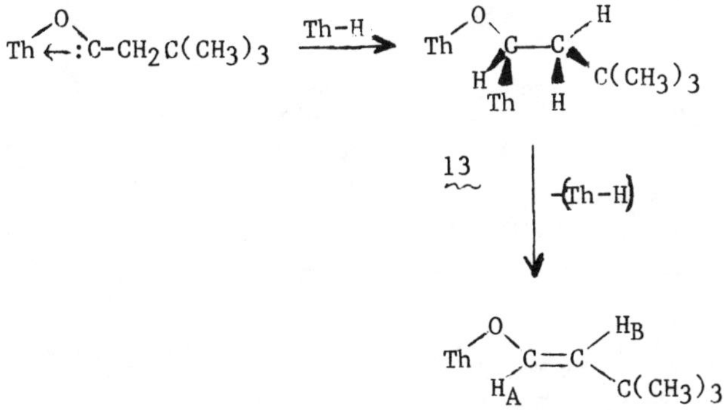


(11)

product can be readily understood in terms of the sterically most favorable conformation from which Th-H elimination in 13 can occur. Thus, conformation L maintains the greatest distance

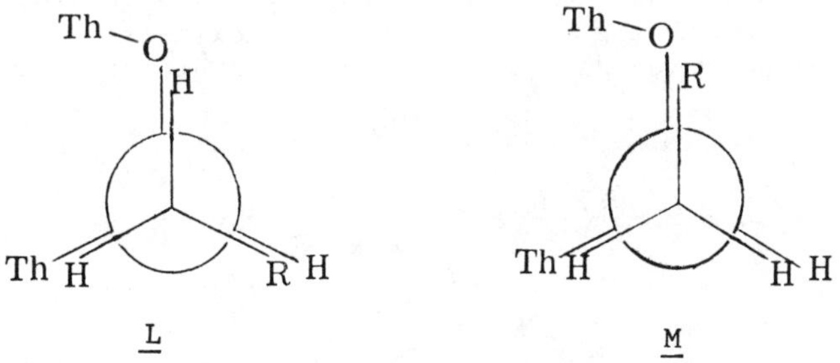
Th
O
H
Th
H
R
H
L
Th
O
R
Th
H
H
H
M

between the bulky $Th[(CH_3)_5C_5]_2(Cl)O-$ and $-C(CH_3)_3$ groups. Extrusion of the favorably eclipsed (62,63) Th-H moiety from the preferred conformation then yields the observed trans product.

Further support for the proposed mechanism of hydride-catalyzed dihaptoacyl isomerization is derived from deuterium labelling studies. When the reaction is conducted with an excess of $\{Th[(CH_3)_5C_5]_2D_2\}_2$, the enolate product is selectively deuterated at the $H_A$ position (Figure 3b) as expected from eq.(12). Furthermore, nmr studies confirm the production of

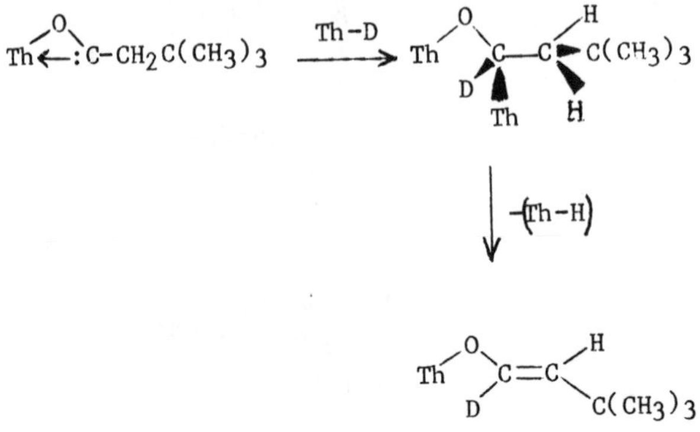


(12)

$\{[(CH_3)_5C_5]_2Th(H)D\}_2$ as required by eq.(12).

If hydrogen gas is added to the reaction mixture of 1 and 11 the hydrogenolysis reaction of thorium-to-carbon sigma bonds (11,22) allows interception of species 13 and thus, catalytic hydrogenation of the inserted carbon monoxide functionality. At 35°C under 0.75 atm initial $H_2$ pressure with [1] = $9.0 \times 10^{-3}$ M and [11] = $6.5 \times 10^{-4}$ M, hydrogenation and isomerization are competitive and both the enolate and the alkoxide reduction product 14 are produced (eq.(13)). Under these conditions, turnover frequencies per $ThH_2$ moiety for isomerization and hydrogenation (initial) are ca. 8 $h^{-1}$ and 4 $h^{-1}$, respectively. A typical nmr spectrum of such a reaction mixture is illustrated in Figure 4. Reduction product 14 was independently synthesized via the reaction shown in eq.(14). If the reaction in eq.(13) is conducted

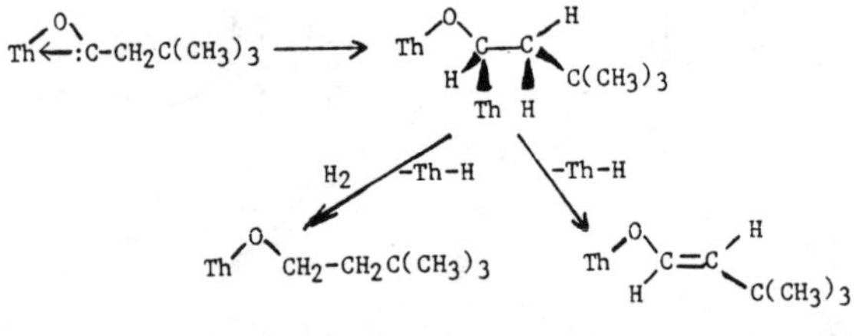


14 12

(13)

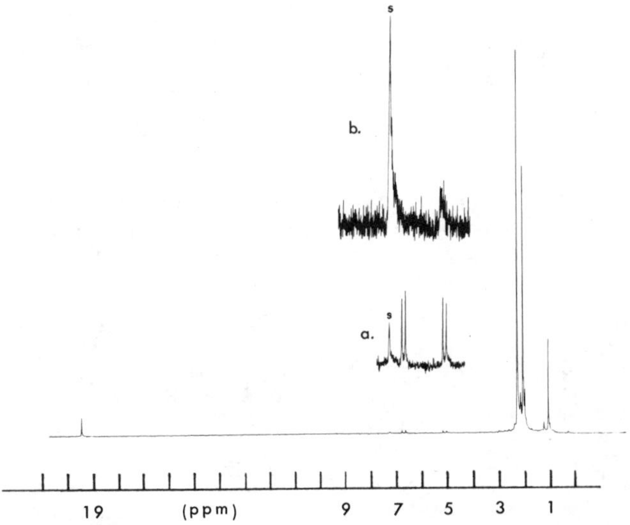


*Figure 3. A. ¹H NMR spectrum (90 MHz, FT, $C_6D_6$) of a mixture of $\{Th[(CH_3)_5C_5]_2H_2\}_2$ and $Th[(CH_3)_5C_5]_2$[trans-$OC(H)$=$C(H)C(CH_3)_3]Cl$, (**12**); the latter was produced by catalytic isomerization of $Th[(CH_3)_5C_5]_2[\eta^2$-$COCH_2C(CH_3)_3]Cl$, **1a.** The peak at δ 19.3 is the hydride resonance; the inset shows the olefinic AB pattern of **12.** B. ¹H NMR spectrum of the olefinic region of **12** prepared with an excess of $\{Th[(CH_3)_5C_5]_2D_2\}_2$ i.e., $Th[(CH_3)_5C_5]_2$[trans-$OC(D)$=$C(H)C(CH_3)_3]Cl$: S = $C_6D_5H$.*

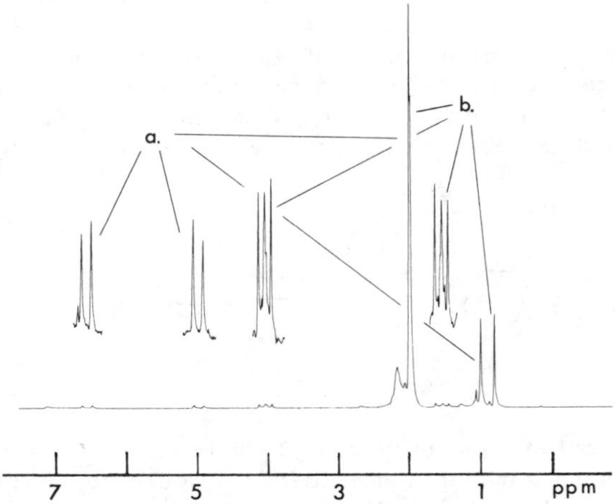


*Figure 4. ¹H NMR spectrum (90 MHz, FT, $C_6D_6$) of a mixture of $Th[(CH_3)_5C_5]_2$[trans-$OC(H)$=$C(H)C(CH_3)_3]Cl$, **12**(a), and $Th[(CH_3)_5C_5]_2(OCH_2CH_2C(CH_3)_3$-$Cl$, **14**(b) prepared by the $\{Th[(CH_3)_5C_5]_2H_2\}_2$-catalyzed competitive isomerization and hydrogenation of $Th[(CH_3)_5C_5]_2[\eta^2$-$COCH_2C(CH_3)_3]Cl$.*

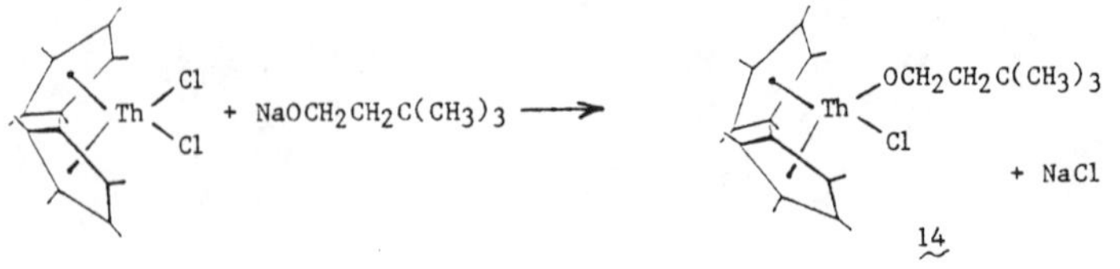

(14)

with excess $[Th[(CH_3)_5C_5]_2D_2]_2$ under $D_2$, the alkoxide product 15 is > 85% deuterated (by nmr) in the $\alpha$ position (eq.(15)). The

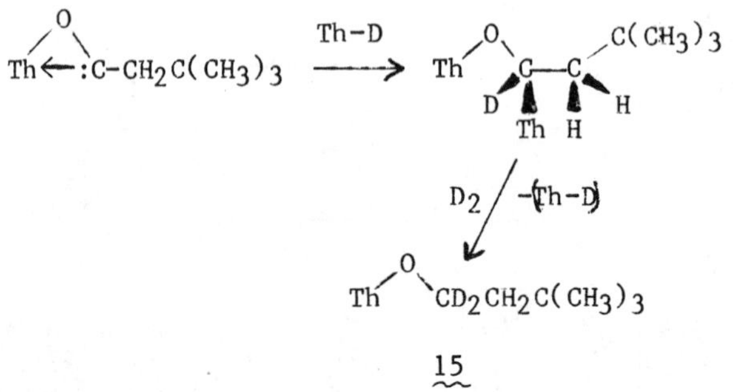

(15)

activity of $\{Th[(CH_3)_5C_5]_2H_2\}_2$ as a hydrogenation catalyst is also illustrated by the interesting observation that the hydrogenation of enolate 12 to produce 14 occurs as a secondary reaction in eq.(13). Under the conditions cited above, an approximate turn-over frequency of 0.01 $h^{-1}$ is calculated.

The catalytic hydrogenation of inserted carbon monoxide is by no means limited to nonconjugated thorium alkyl precursors. Thus, the uranium benzoyl compound 4 is readily hydrogenated (eq.(16)); in this case, the intermediate cannot undergo $\beta$-

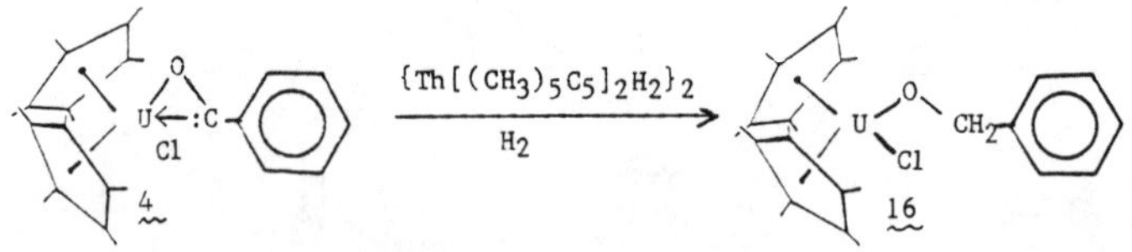

(16)

hydride elimination, and only benzyloxy product 16 is formed in the reaction. An authentic sample of 16 could be synthesized by the reaction of $U[(CH_3)_5C_5]_2Cl_2$ with one equivalent of $C_6H_5CH_2ONa$ in diethyl ether. Under conditions comparable to those in eq.(12), the turnover frequency per $ThH_2$ unit for hydrogenation of 16 is ca. 1 $h^{-1}$. If $Th[(CH_3)_5C_5]_2D_2$ $_2/D_2$ is used in eq.(16), the product is > 90% deuterated in the $\alpha$-position (eq.(17)).

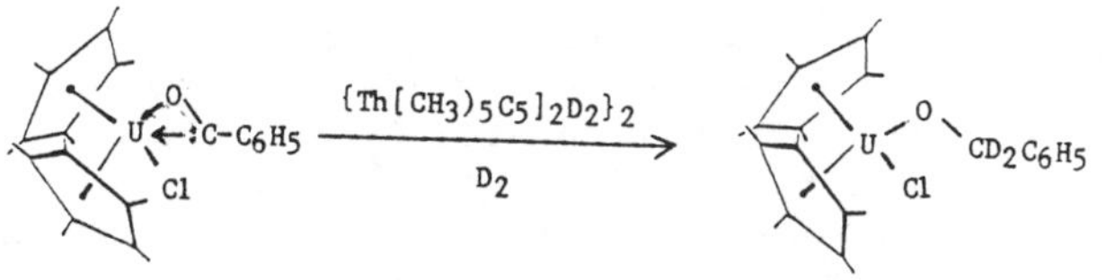


(17)

Although not the central subject of this review, several thorium dihaptocarbamoyl complexes (14) Th($\eta^2$-$CONR_2$), were also examined with respect to thorium hydride-catalyzed reduction. Under 0.75 atm $H_2$ and over the course of several days at temperatures as high as 100°C, no hydrogenation was observed. These results are in accord with other spectral, structural, and chemical data (14) indicating the importance of carbamoyl resonance hybrids O and P, and that the carbene-like reactivity is significantly reduced in comparison to the acyls (14).

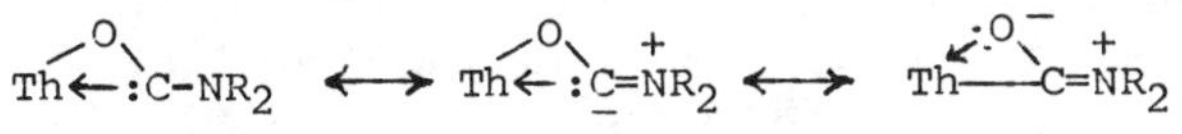


N O P

In related transition metal chemistry, it has been noted that $Zr(C_5H_5)_2(\eta^2$-$COCH_3)CH_3$ can be stoichiometrically reduced to $Zr(C_5H_5)_2(OCH_2CH_3)CH_3$ by $Mo(C_5H_5)_2H_2$ (61). The source of hydrogen atoms is largely but not exclusively the hydride ligands. It is also known that carbene complexes of the type M←:C(R)OM' can be hydrogenated to $H_2$(R)COM' alkoxides (65). All attempts to bring about the uncatalyzed hydrogenation of the organoactinide dihaptoacyls have so far been unsuccessful.

## Discussion

It is intriguing to speculate upon the degree to which the organoactinide carbonylation results may constitute homogeneous representations of key transformations in heterogeneous catalytic CO reduction. Both systems exhibit high unsaturation and oxygen affinity. In the heterogeneous systems, high oxygen affinity is demonstrated by evidence for dissociative CO adsorption (8-10,66-72), labelled alcohol and ketone deoxygenation (8,9,10,73), labelled ketene deoxygenation (74,75), as well as surface alkoxide, carboxylate, and possibly dihaptoacyl formation (76,77). Further connection between the two areas may exist in the role which actinide elements play in heterogeneous CO reduction catalysis. The "isosynthesis" reaction is a catalytic process for converting synthesis gas into branched paraffins, olefins, alcohols, and aromatics over thoria ($ThO_2$) alone or promoted with $K_2CO_3$ or $Al_2O_3$ (78,79,80,81,82). Thoria is also commonly used as a support in transition metal-catalyzed CO reduction (8,9,10). Actinides are components in a number of other heterogeneous CO reduction catalysts (83,84,85).

To the extent that mechanistic similarities exist, it is of interest to examine several crucial transformations in catalytic CO reduction and to see whether the organoactinide carbonylation results contribute to a better understanding of what may be occurring. The insertion of CO into a surface metal-hydrogen bond to produce a formyl (eq.(18)) has been discussed at length

$$\text{M-H} + \text{CO} \rightleftharpoons \text{M-}\overset{\overset{\text{O}}{\|}}{\text{C}}\text{-H} \qquad (18)$$

as an initial step in heterogeneous CO reduction (86). A conceptual problem in invoking this process has centered around the apparent endothermic character of the reaction for many transition metals (the metal-hydrogen bond strength is probably substantially greater than analogous M-C bond strengths). One means by which such a barrier may be surmounted is by involving a dihapto bonding mode for the inserted CO (eq.(19)). Such thermodynamic leverage was already discussed for thorium in this

$$\text{M-H} + \text{CO} \rightleftharpoons \overset{\text{O}}{\overset{/\ \backslash}{\text{M}\leftarrow\text{:C-H}}} \qquad (19)$$

review and is a possible reason why CO insertion into actinide-hydrogen (87) and early transition metal-hydrogen (65) bonds is facile. This type of bonding also appears to drive the insertion of CO into actinide-dialkylamide bonds (14). Of course, a cluster of later transition metal atoms, each with a weaker metal-oxygen interaction, might serve the same function (6).

The present results indicate that catalytic hydrogenation of a dihaptoformyl subsequent to CO insertion should be facile (eq.(20)). Furthermore, the organoactinide studies demonstrate

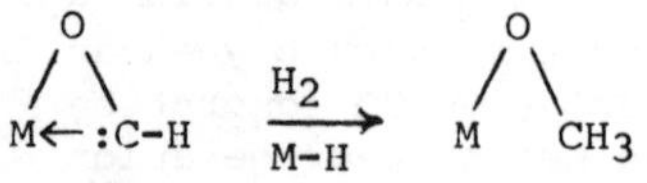

$$\overset{\text{O}}{\overset{/\ \backslash}{\text{M}\leftarrow\text{:C-H}}} \xrightarrow[\text{M-H}]{H_2} \overset{\text{O}}{\overset{/\ \ \ \backslash}{\text{M}\quad CH_3}} \qquad (20)$$

that such processes can be homometallic and do not require the agency of both a "hard" and "soft" metal ion. Although the sequence of eqs.(19) and (20) provides a plausible route to methoxy functionalities, it is still necessary to inquire as to how chain growth occurs in processes yielding higher molecular weight organics. Coupling of dihaptoformyl species (e.g., eq.(21)) is one possible vehicle for C-C bond formation.

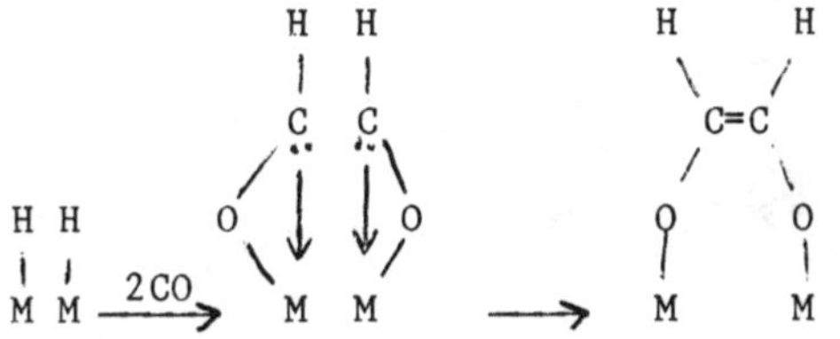


(21)

Enediolates are now known for organoactinides (14,16,87) and early transition metal organometallics (65). The results presented here and elsewhere (54,55) suggest that actinide hydrides should catalytically hydrogenate enediolates to glycolates (eq.(22)). Liberation of ethylene (*vide infra*) would then produce

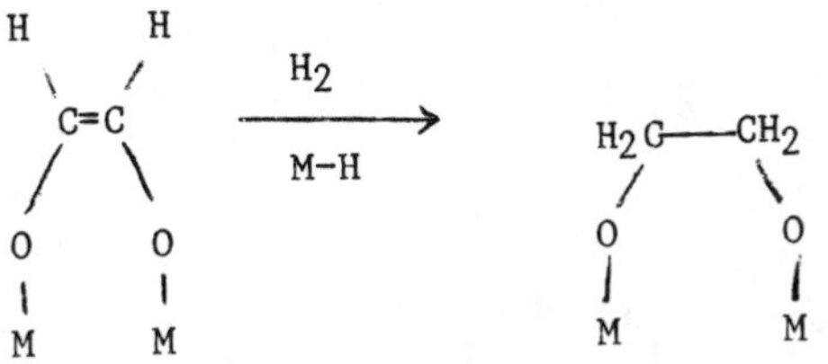

(22)

an attractive building block for chain growth. As shown elsewhere, supported organoactinide hydrides are active catalysts for ethylene polymerization (54,55). Ethylene has been implicated as a basic building unit of long-chain hydrocarbons in certain Fischer-Tropsch reactions (70). It is also conceivable that ketene-forming reactions as in eq.(4) could play a role in chain growth. Combined with the observation that organoactinide hydrides readily add to organic carbonyl groups (88), plausible schemes for saturated oxocarbon complexes can be generated (eqs.(23) and (24)). Another possible process for chain growth

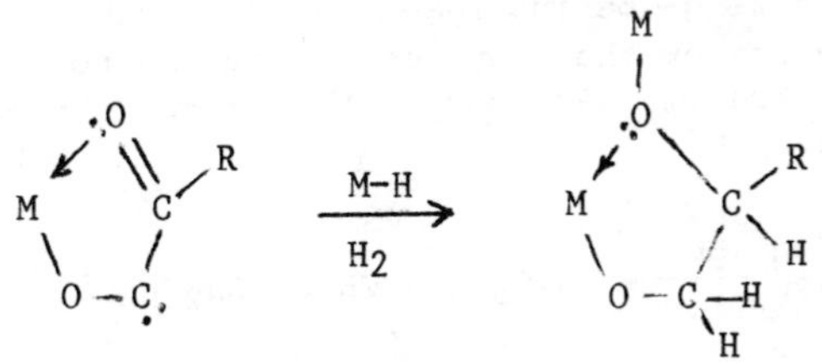

(23)

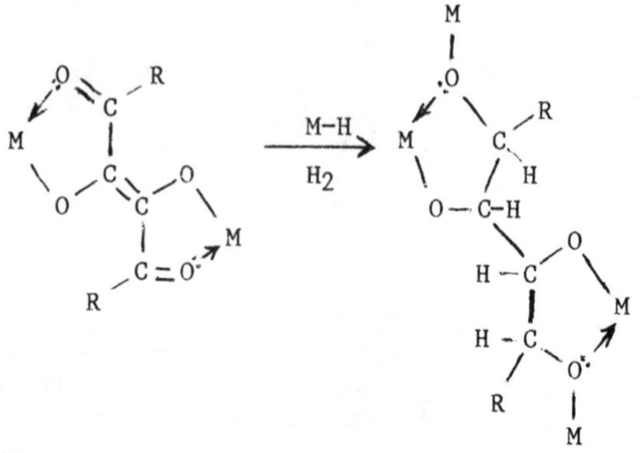


(24)

involves chemistry for transforming C-O into M-O bonds (eqs.(25) and (26)) and thus creating M-C bonds for further functionalization.

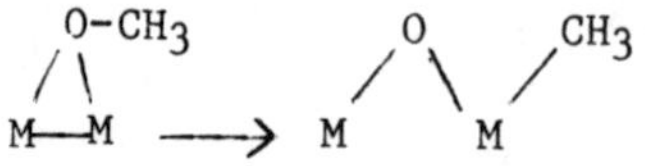

(25)

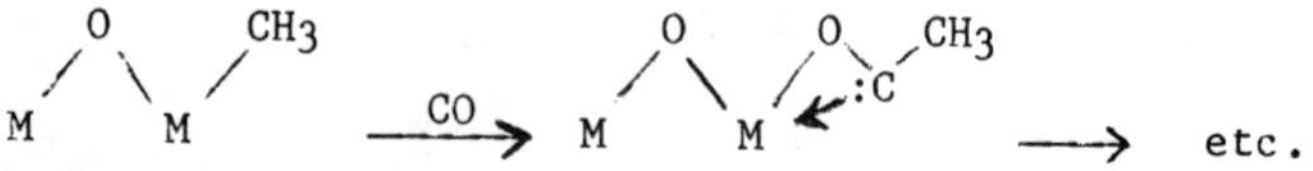


(26)

Equation (25) introduces another conceptual question in catalytic CO reduction--how C-O bond scission occurs. One possible process is, as shown, driving an M-O-C → M-O-M-C transposition with the high oxygen affinity of the metal involved. If such a process occurred prior to hydrogenation, carbide formation would result (6-10,68,69). It is thought that such species may play a major role in some CO reduction catalyses as methane, methylidene, methylene, and methyl precursors. The process shown in eq.(25) can also be viewed as an oxidative addition of a C-O bond to the metal (89,90,91). A relevant observation in organouranium chemistry is a ring-opening reaction of tetrahydrofuran (92) (eq.(27)). The source of the $\delta$ hydrogen atom has not yet been elucidated. An instructive analogy to the M-O-C → M-O-M-C

$$UCl_3 + 3\ NaC_5H_5 \xrightarrow{THF} U(C_5H_5)_3OCH_2CH_2CH_2CH_3 \qquad (27)$$

transformation also exists in borane-carbonyl chemistry (93) (eq. (28)) and in borane, alane reductions of metal carbonyls (86). The

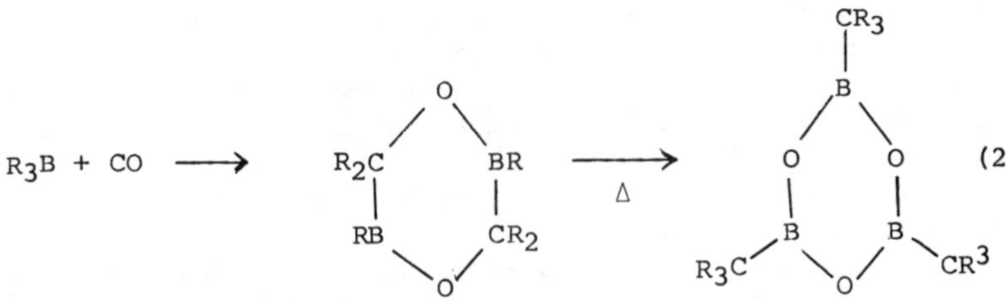

$$R_3B + CO \longrightarrow \underset{\underset{O}{\overset{O}{\bigcirc}}}{\overset{R_2C\quad\quad BR}{\underset{RB\quad\quad CR_2}{}}} \xrightarrow{\Delta} \quad (28)$$

high oxygen affinities of boron and aluminum are doubtless a major driving force for this reaction.  Once formed, the M-C bond could readily undergo hydrogenolysis or further carbonylation.  Redox processes, as demonstrated for low-valent, early transition metal complexes (94,95) offer yet another means to form C-C bonds and to simultaneously break C-O bonds (eqs.(29) and (30)).  There also appears to be precedent for eq.(30) in low-valent uranium chemistry (96).

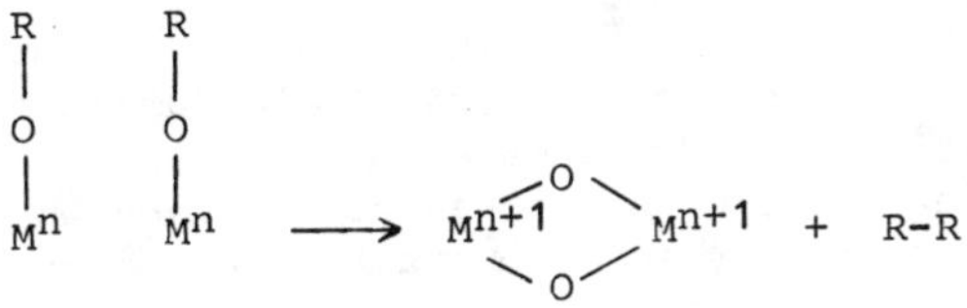

$$\underset{\underset{O}{\overset{R}{\mid}}\;\underset{O}{\overset{R}{\mid}}}{M^n\quad M^n} \longrightarrow M^{n+1}\underset{O}{\overset{O}{\diagup\diagdown}}M^{n+1} + R\text{-}R \qquad (29)$$

$$\underset{\underset{M^n}{\overset{O}{\mid}}\;\underset{M^n}{\overset{O}{\mid}}}{\overset{R_2\;R_2}{\underset{}{C\text{---}C}}} \longrightarrow M^{n+1}\underset{O}{\overset{O}{\diagup\diagdown}}M^{n+1} + R_2C{=}CR_2 \qquad (30)$$

Two additional problems in understanding heterogeneous catalytic CO reduction concern the means by which M-O bonds are cleaved to produce alcohols and the fate of the metal oxides once formed.  In regard to the former issue, most metal alkoxides (including those of actinides) readily undergo protolysis to form alcohols (97) (eq.(31)).  The source of the protons could be

$$\text{M-OR} \xrightarrow{H^+} M^+ + \text{HOR} \qquad (31)$$

competing methanation and/or alcohol dehydration chemistry.  In catalytic systems which only produce alcohols, "heterolytic" $H_2$ activation (98,99) (eq.(32)) may provide both metal hydride

$$\text{-M-O-} \xrightarrow{H_2} \underset{\text{-M}\quad\text{O-}}{\overset{H\quad H}{\mid\quad\mid}} \qquad (32)$$

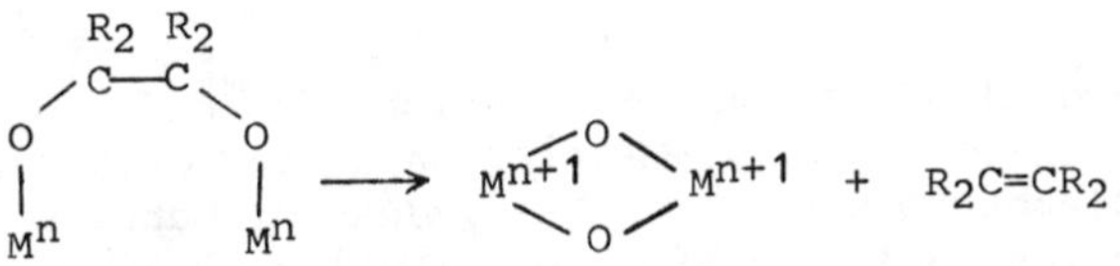

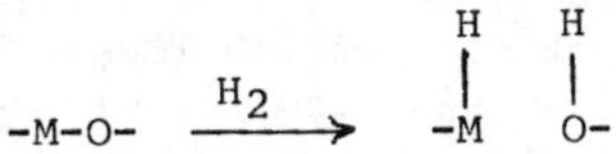

and protonic functionalities. This process currently has very little precedent in conventional homogeneous chemistry (100-102). For organoactinides, and for most early transition metal hydrides in solution, the equilibrium doubtless lies far to the left. Clearly this is an important reaction pattern to better understand through future "modelling" experiments. Likewise, the mechanism by which metal oxide functionalities produced in the various deoxygenation processes are returned, via hydrogenation, to the reduced state bears little analogy to existing solution organometallic chemistry. Again more information is needed.

The present results provide an informative picture of how carbon monoxide interacts with relatively weak two-center, two-electron metal-ligand sigma bonds in a coordinatively unsaturated metallic environment having high oxygen affinity and high kinetic lability. Migratory insertion leads to dihaptoacyl species in which coordination of the carbon monoxide oxygen atom is a major component of the metal-ligand bond. Subsequent reaction patterns of the dihaptoacyls evidence a pronounced carbene-like reactivity. Processes in which a bonding or nonbonding electron pair attacks the acyl carbon atom to break the metal-carbon bond (with retention of the metal-oxygen bond) appear to be facile and widespread. Thus, it has been demonstrated for the first time that such activated acyl carbon atoms can be involved in CO oligomerization as well as catalytic isomerization and/or reduction.

## Acknowledgments

We thank the National Science Foundation for generous support of this research through grants CHE76-84494A01 and CHE8009060. We thank our collaborators Victor W. Day, Juan M. Manriquez, and Kenneth G. Moloy for their valuable contributions to this program.

## Abstract

This article reviews recent results on the chemical, spectral and structural properties of bis(pentamethylcyclopentadienyl) thorium and uranium dihaptoacyl complexes produced by migratory insertion of carbon monoxide into actinide-carbon sigma bonds. The high coordinative unsaturation and oxygen affinity of the ligation environment produces a marked perturbation of the bonding and reactivity toward that of a coordinated oxycarbene: $M(\eta^2\text{-O}\ddot{C}R)$. Reactivity patterns observed include hydrogen atom and trimethylsilyl migration to the acyl carbon, as well as coupling with additional carbon monoxide to produce a dimeric complex of the enedionediolate ligand, $OC(R)(\bar{O})C{=}C(\bar{O})(R)CO$. The dihaptoacyls insert into the Th-H bond of $\{Th[(CH_3)_5C_5]_2H_2\}_2$. For $Th[(CH_3)_5C_5]_2[\eta^2\text{-}COCH_2C(CH_3)_3]Cl$, this results, via β-hydride elimination, in catalytic isomerization to $Th[(CH_3)_5C_5]_2$-

[trans-OC(H)=C(H)C($CH_3$)$_3$]. In the presence of hydrogen gas, the hydride catalytically hydrogenates the dihaptoacyls to alkoxides (M($\eta^2$-COR) → M-$OCH_2$R). Mechanistic studies include kinetic measurements as well as isotopic labelling and stereochemical analysis.

## Literature Cited


1. Parshall, G. W. "Homogeneous Catalysis," Wiley-Interscience, NY, 1980, Chap. 5
2. Eisenberg, R.; Hendricksen, D.E. Advan. Catal., 1979, 28, 79-172.
3. Heck, R.F.; "Organotransition Metal Chemistry," Academic Press, NY, 1974, Chap. IX.
4. Wojcicki, A.; Advan. Organometal. Chem., 1973, 11, 87-145.
5. Calderazzo, F. Angew. Chem. Int. Ed. Engl., 1977, 16, 299-311, and references therein.
6. Muetterties, E.L.; Stein, J. Chem. Rev., 1979, 79, 479-490.
7. Masters, C. Advan. Organometal. Chem., 1979, 17, 61-103.
8. Denny, P.J.; Whan, D.A. Chemical Society Specialist Periodical Report, Catalysis, 1978, 2, 46-86.
9. Ponec, V. Catal. Rev.-Sci. Eng., 1978, 18, 151-171.
10. Schulz, H.J. Erdöl, Kohle, Erdgas, Petrochem., 1977, 30, 123-131.
11. Fagan, P.J.; Manriquez, J.M.; Marks, T.J. in Marks, T.J.; Fischer, R.D. Eds., "Organometallics of the f-Elements," Reidel Publishing Co., Dordrecht, Holland, 1979, Chap. 4.
12. Marks, T.J.; Manriquez, J.M.; Fagan, P.J.; Day, V.W.; Day, C.S.; Vollmer, S.H. A.C.S. Sympos. Series, in press.
13. Marks, T.J. Prog. Inorg. Chem., 1979, 25, 224-333.
14. Fagan, P.J.; Manriquez, J.M.; Marks, T.J.; Day, V.W.; Vollmer, S.H.; Day, C.S., J. Am. Chem. Soc., in press.
15. Fagan, P.J.; Manriquez, J.M.; Marks, T.J.; Day, V.W.; Vollmer, S.H.; Day, C.S. J. Am. Chem. Soc., 1980, 102, 5393-5396.
16. Manriquez, J.M.; Fagan, P.J.; Marks, T.J.; Day, C.S.; Day, V.W. J. Am. Chem. Soc., 1978, 10, 7112-7114.
17. Keller, C. "The Chemistry of the Transuranium Elements," Verlag Chemie, Weinheim/Bergstr., 1971, pp. 151-152.
18. Navrotsky, A. in MTP International Review of Science, Inorganic Chemistry Series Two, Vol. 5, D.W.A. Sharp, Ed., University Park Press, Baltimore, 1975, Chap. 2.
19. Huheey, J.E. "Inorganic Chemistry," 2nd Ed., Harper and Row, NY, 1978, Appendix F.
20. Connor, J.A. Topics Curr. Chem., 1977, 71, 71-110.
21. Kochi, J.K. "Organometallic Mechanisms and Catalysis," Academic Press, NY, 1978, Chap. 11.
22. Manriquez, J.M.; Fagan, P.J.; Marks, T.J.; J. Am. Chem. Soc., 1978, 100, 3939-3941.
23. Fagan, P.J.; Manriquez, J.M.; Marks, T.J., manuscript in preparation.

24. Lappert, M.F.; Juong-Thi, N.T.; Milne, C.R.C. J. Organometal. Chem., 1979, 74, C35-C37.
25. Fachinetti, G.; Floriani, C.; Stoeckli-Evans, H. J. Chem. Soc., Dalton Trans., 1977, 2297-2302.
26. Fachinetti, G.; Fochi, G.; Floriani, C. J. Chem. Soc., Dalton Trans., 1977, 1946-1950.
27. Manriquez, J.M.; McAlister, D.R.; Sanner, R.D.; Bercaw, J.E. J. Am. Chem. Soc., 1978, 100 2716-2724.
28. Roper, W.R.; Taylor, G.D.; Waters, J.M.; Wright, L.J. J. Organometal. Chem., 1979, 182, C46-C48.
29. Carmona-Guzman, E.; Wilkinson, G.; Atwood, J.L.; Rogers, R.D.; Hunter, W.E.; Zaworotko, M.J. J. Chem. Soc., Chem. Comm., 1978, 465-466.
30. Chisholm, M.H.; Godleski, G. Prog. Inorg. Chem., 1976, 20, 299-436.
31. Schrock, R.R. Acct. Chem. Res., 1979, 12, 98-104.
32. Lauher, J.W.; Hoffmann, R., J. Am. Chem. Soc., 1976, 98, 1729-1742.
33. Brintzinger, H.H. J. Organometal. Chem., 1979, 171, 37-344.
34. Erker, G.; Rosenfeldt, F., Angew. Chem. Int. Ed. Engl., 1978, 17, 605-606.
35. Kirmse, W., "Carbene Chemistry," Second Edition, Academic Press, NY, 1971, Chap. 3, pp. 14-16.
36. Wilson, T.B.; Kistiakowaky, G.B. J. Am. Chem. Soc. 1978, 80, 2934-2939.
37. Herrmann, W.A.; Plank, J.; Ziegler, M.L.; Weidenhammer, K. J. Am. Chem. Soc., 1979, 101, 3133-3135.
38. Herrmann, W.A.; Plank, J. Angew. Chem. Int. Ed. Engl., 1978, 17, 525-526.
39. Dorrer, B.; Fischer, E.O. Chem. Ber., 1974, 107, 2683-2690.
40. Fachinetti, G.; Biran, C.; Floriani, C.; Chiesi-Villa, A.; Guastini, C. J. Am. Chem. Soc., 1978, 100, 1921-1922, and references therein.
41. Redhouse, A.D.; Herrmann, W.A. Angew. Chem. Int. Ed. Engl., 1976, 15, 615-616.
42. Herrmann, W.A. Angew. Chem. Int. Ed. Engl., 1974, 13, 335-336.
43. Hoberg, H.; Korff, J. J. Organometal. Chem., 1978, 152, 255-264.
44. March, J. "Advanced Organic Chemistry: Reactions, Mechanisms, and Structure," McGraw-Hill, NY, 1968, pp. 636, 723.
45. Baron, W.J.; DeCamp, M.R.; Hendrick, M.E.; Hones, M., Jr.; Levin, R.H.; Sohn, M.B.; in "Carbenes," Jones, M., Jr.; Moss, R.A.; Eds., Wiley-Interscience, NY, 1973, Vol. I, p. 128.
46. Moss, E.A; in reference 45, p. 280.
47. Wentrup, C.; Topics Curr. Chem., 1976, 62, 173-251.
48. Nickon, A.; Huang, F.-C.; Weglein, R.; Matsuo, K.; Yagi, H. J. Am. Chem. Soc., 1974, 96, 5264-5265.

49. Bodor, M.; Dewar, M.J.S. J. Am. Chem. Soc., 1972, 94, 9103-9106.
50. Menendez, V.; Figuera, J.M. Chem. Phys. Lett., 1973, 18, 426-430.
51. Reference 45, p. 72.
52. Reference 35, Chap. 12.
53. Robson, J.H.; Schechter, H., J. Am. Chem. Soc., 1967, 89, 7112-7113.
54. Bowman, R.G.; Nakamura, R.; Fagan, P.J.; Burwell, R.L., Jr.; Marks, T.J., Abstract, Spring Meeting of the American Chemical Society, Houston, TX, March 23-28, 1980, INOR 5.
55. Bowman, R.G.; Nakamura, R.; Fagan, P.J.; Burwell, R.L., Jr.; Marks, T.J., submitted for publication.
56. Maatta, E.A.; Marks, T.J. manuscript in preparation.
57. Fagan, P.J.; Jones, N.L.; Marks, T.J., unpublished results.
58. Broach, R.W.; Schultz, A.J.; Williams, J.M.; Brown, G.M.; Manriquez, J.M.; Fagan, P.J.; Marks, T.J. Science, 1979, 203, 173-174.
59. Reference 35, pp. 407-409.
60. Cooke, J.; Cullen, W.R.; Green, M.; Stone, F.G.A. Chem. Comm. 1968, 170-171.
61. Marsella, J.A.; Caulton, K.G. J. Am. Chem. Soc., 1980, 102, 1747-1748.
62. Thorn, D.L.; Hoffmann, R. J. Am. Chem. Soc., 1978, 100, 2079-2090.
63. Reference 21, pp. 247-261 and references therein.
64. Manriquez, J.M.; Fagan, P.J.; Marks, T.J.; Vollmer, S.H.; Day, C.S.; Day, V.W. J. Am. Chem. Soc., 1979, 101, 5075-5078.
65. Wolczanski, P.J.; Bercaw, J.E. Acc. Chem. Res., 1980, 13, 121-127, and references therein.
66. King, D.L. J. Catal., 1980, 61, 77-86.
67. Kroeker, R.M.; Kaska, W.C.; Hansma, K. J. Catal., 1980, 61, 87-95.
68. Bioloen, P.; Helle, J.N.; Sachtler, W.H.M. J. Catal., 1979, 58, 95-107.
69. Krebs, H.J.; Bonzel, H.P. Surf. Sci., 1979, 88, 269-283.
70. Dwyer, D.J.; Somorjai, G.A. J. Catal., 1979, 56, 249-257.
71. Dwyer, D.J.; Somorjai, G.A. J. Catal., 1978, 52, 291-301, and footnote 9 therein.
72. Sexton, B.A.; Somorjai, G.A. J. Catal., 1977, 46, 167-188.
73. Kummer, J.T.; Emmett, P.H. J. Am. Chem. Soc., 1953, 75, 5177-5183.
74. Blyholder, G.; Emmett, P.H. J. Phys. Chem., 1959, 63, 962-965.
75. Blyholder, G.; Emmett, P.H. J. Phys. Chem., 1960, 64, 470-472.
76. Blyholder, G.; Goodsel, A.J. J. Catal., 1971, 23, 374-378.
77. Blyholder, G.; Shihabi, D.; Wyatt, W.V.; Bartlett, R. J. Catal., 1976, 43, 122-130.

78. Natta, G.; Colombo, U.; Pasquon, I. in "Catalysis," Emmettt, P.H., Ed., Reinhold, NY, 1957, Vol. 5, Chap. 3.
79. Cohn, E.M. in "Catalysis," Emmett, P.H., Ed., Reinhold, NY, 1956, Vol. 4, Chap. 3
80. Pichler, H.; Ziesecke, H-H.; Traiger, B. Brennstoff-Chem., 1950, 31, 361-374.
81. Pichler, H.; Ziesecke, K.-H.; Fitzenthaler, E. Brennstoff-Chem., 1949, 30, 333-347.
82. Pichler, H.; Ziesecke, K.-H. Brennstoff-Chem., 1949, 30, 13-22.
83. Elatta, A.; Wallace, W.E.; Craig, R.S. A.C.S. Sympos. Series 1979, 178, 7-14.
84. Ellgen, P.C.; Bhasin, M. U.S. Pat., 1979, 4162262.
85. Baglin, E.G.; Atkinson, G.B.; Nicks, L.J. U.S. Pat. Appl., 1979, 964860.
86. Casey, C.P.; Andrews, M.A.; McAlister, D.R.; Rinz, J.E. J. Am. Chem. Soc., 1980, 102, 1927-1933, and references therein.
87. Fagan, P.J.; Maatta, E.A.; Manriquez, J.M.; Marks, T.J. manuscript in preparation.
88. Fagan, P.J.; Marks, T.J., unpublished results.
89. Schlodder, R.; Ibers, J.A.; Lenorda, M.; Graziani, M. J. Am. Chem. Soc., 1974, 96, 6893-6900.
90. Reference 3, pp. 255-260.
91. Noyori, R. in "Transition Metal Organometallics in Organic Synthesis," Alper, H., Ed., Academic Press, NY, 1976, Vol. 1, pp. 145-146.
92. Ter Haar, N.; Dubeck, M. Inorg. Chem., 1964, 3, 1648-1650.
93. Onak, T. "Organoborane Chemistry," Academic Press, NY, 1975, p. 122.
94. McMurry, J.E. Acc. Chem. Res., 1974, 7, 281-286, and references therein.
95. Walborsky, H.M.; Murari, M.P. J. Am. Chem. Soc., 1980, 102, 426-428, and references therein.
96. Rieke, R.D.; Rhyne, L.D. J. Org. Chem., 1979, 44, 3445-3446.
97. Bradley, D.C.; Mehrotra, R.C.; Gaur, D.P. "Metal Alkoxides," Academic Press, London, 1978, pp. 149-298.
98. Kung, H. Catal. Rev.-Sci. Eng., in press.
99. Herman, R.G.; Klier, K.; Simmons, G.W.; Finn, B.P.; Bulko, J.B.; Kobylinski, T.P. J. Catal., 1974, 56, 407-429, and references therein.
100. Reference 21, pp. 312-314.
101. Parshall, G.W. J. Am. Chem. Soc., 1972, 94, 8716-8719.
102. White, C.; Oliver, A.J.; Maitlis, P.M. J. Chem. Soc., Dalton Trans., 1973, 1901-1907.

RECEIVED December 8, 1980.

# 5

# Chemistry of the Water Gas Shift Reaction Catalyzed by Rhodium Complexes

T. YOSHIDA, T. OKANO, and SEI OTSUKA

Department of Chemistry, Faculty of Engineering Science, Osaka University, Toyonaka, Osaka, Japan 560

Transition metal compounds in various form such as metal carbonyls (1), carbonyl clusters (2), Pt(II) chloride/tin chloride (3), $PtL_n$ ($L=PR_3$) (4), etc. have been proposed as homogeneous catalysts for the water gas shift (wgs) reaction (eq. 1). Some of them are reportedly active at relatively low temperature (<150°)

$$H_2O + CO \rightleftharpoons H_2 + CO_2 \qquad (1)$$

where the thermodynamic equilibrium is favored (e.g., $K=1.45x10^3$ at 127° vs. 26.9 at 327°C) (5). Their catalytic activities, however, appear not to be as high as practical catalysts would have to be. Their mechanistic details also still remain to be elucidated. Even with the apparently simple catalyst, $Pt(PR_3)_3$, we recognized a number of component reaction steps (4).

The logical basis for employing metal carbonyls as catalysts would be the CO activation through coordination which facilitates nucleophilic attack by water or $OH^-$ (6). The key step then may be the formation of a hydroxy-carbonyl species followed by β-hydrogen elimination reaction (eq. 2,3). Another important elemental re-

$$M^+CO + H_2O \rightleftharpoons MCOOH + H^+ \qquad (2)$$

$$MCOOH \rightleftharpoons MH + CO_2 \qquad (3)$$

action associated with $H_2$ generation would be reduction of protons

0097-6156/81/0152-0079$05.00/0
© 1981 American Chemical Society

represented by eq. 4.

$$MH + H^{+} \rightleftharpoons M^{+} + H_2 \qquad (4)$$

We proposed a new approach based on a different strategy to induce two electron transfer from a low valent metal compound to a water molecule leading to a hydrido-hydroxo-metal species (eq. 5). The nucleophilic attack of $OH^-$ on a coordinated CO is expected to

$$M + H_2O \rightleftharpoons MH(OH) \qquad (5)$$

be more facile compared to the neutral water molecule.

In this paper, first we will briefly describe the $Pt[P(i\text{-}Pr)_3]_3$-catalyzed wgs reaction (4). Recently we have studied $RhHL_3$ ($L{=}PR_3$)-catalyzed wgs reaction in much more details. In comparison with the $PtL_3$ reaction, a perspective view of $RhHL_3$ (here we confine $L{=}P(i\text{-}Pr)_3$)-catalyzed wgs reaction will be given below.

## $Pt[P(i\text{-}Pr)_3]_3$-Catalyzed WGS Reaction

We have shown that the reaction of $PtL_3$ ($L{=}P(i\text{-}Pr)_3$) with water in acetone or pyridine produces a strong hydroxy base (7). The reaction is described in terms of equilibria (eq. 6-9). By adding $NaBF_4$ to the solution of $PtL_3$ in aqueous pyridine, trans-

$$PtL_3 \rightleftharpoons PtL_2 + L \qquad (6)$$

$$PtL_2 + H_2O \overset{K_o}{\rightleftharpoons} PtH(OH)L_2 \qquad (7)$$

$$PtH(OH)L_2 + S \overset{K_s}{\rightleftharpoons} [PtH(S)L_2]OH \qquad (8)$$

$$[PtH(S)L_2]OH \overset{K_d}{\rightleftharpoons} [PtH(S)L_2]^{+} + OH^{-} \qquad (9)$$

$[PtH(py)L_2]BF_4$ ($\nu$(Pt-H) 2230 $cm^{-1}$) could be isolated as crystals. After quenching the $PtL_3$-catalyzed wgs reaction in pyridine followed by addition of $NaBF_4$ , the same ionic compound was isolated in good yield (70 %). From the wgs reaction in acetone was isolated trans-$[PtH(CO)L_2]BPh_4$ ($\nu$(Pt-H) 2178, $\nu$(CO) 2058 $cm^{-1}$). These results imply that substitution of the coordinated pyridine with CO (eq. 10) requires a considerable activation energy, a fea-

$$[PtH(S)L_2]OH + CO \rightleftharpoons [PtH(CO)L_2]OH \qquad (10)$$

ture consistent with the observed solvent effect, a faster rate in acetone than in pyridine (Table I).

Table I. $PtL_3$- and $RhHL_3$-Catalyzed WGS Reaction at 100°[a]

| Catalyst[c] | Turnover[b] in pyridine | Turnover[b] in acetone |
|---|---|---|
| $PtL_3$ | 0.6 | 5.2 |
| $RhHL_3$ | 33 | 28 |
| $Rh(OH)(CO)L_2$ | 24 | |
| $Rh_2(CO)_3L_3$ | 15 | |
| $Rh_2(CO)_3L_3$-3L | 30 | |
| $Rh_2(\mu\text{-}O_2CO)(CO)_2L_4$ | 28 | |

[a] Catalyst, 0.1 mmol; $H_2O$, 2 ml; CO, 20 Kg/$cm^2$. [b] Moles/g atom of catalyst/h. The molar ratio of $H_2$ and $CO_2$ is unity within experimental errors (±5 %). [c] L=P(i-Pr)$_3$.

The key process following reaction (eq. 10) would be the nucleophilic attack of $OH^-$ on the coordinated CO (eq. 11). Analogous reactions have been observed (eq. 12,13). The fact that

$$[PtH(CO)L_2]OH \rightleftharpoons PtH(COOH)L_2 \qquad (11)$$

$$[PtH(CO)L_2]^+ + KOH \rightleftharpoons [PtH(COOK)L_2]^+ \qquad (12)$$

$$[PtH(CO)L_2]^+ + NaOMe \longrightarrow [PtH(COOMe)L_2]^+ \qquad (13)$$

PtMe(COOH)(diphos) (diphos=$Ph_2PCH_2CH_2PPh_2$) (8) and trans-PtCl-($CO_2H$)$(PEt_3)_2$ (9) are thermally quite stable whereas trans-[PtH-(COOH)$L_2$] in solution is unstable manifests a dramatic trans-effect for the thermal stability of $Pt^{II}$-COOH moiety.

The thermal decomposition of PtH(COOH)$L_2$ should produce trans-$PtH_2L_2$ (eq. 14). Reductive elimination of $H_2$ from $PtH_2L_2$ leads to $PtL_2$. The $H_2$ evolution should be accelerated by the CO attack on

$$PtH(COOH)L_2 \rightleftharpoons PtH_2L_2 + CO_2 \qquad (14)$$

$PtH_2L_2$ as was observed for the reaction with $PtH_2$(diphos) (diphos= $(t\text{-}Bu)_2PCH_2CH_2CH_2P(t\text{-}Bu)_2$) (10). The role of CO for the $H_2$ evolution from metal dihydride species will be discussed for the $RhHL_3$-catalyzed reaction (see the next section).

Involvement of trans-$PtH_2L_2$ in the catalytic cycle was confirmed by the wgs reaction, employing trans-$PtH_2L_2$ as the catalyst precursor, from which was also isolated trans-[PtH(CO)L]OH as its $BPh_4$ salt. The following two processes (eq. 15,16) would complete the catalytic cycle. The formation of $Pt_3(CO)_3L_4$ (ν(CO) 1840,1770 $cm^{-1}$) in the reaction of $PtH_2L_2$ with CO is considered indirect evidence for the intermediacy of the coordinatively unsaturated Pt-(CO)$L_2$. A simplified scheme of the cycle may then be depicted

$$PtH_2L_2 + CO \rightleftharpoons Pt(CO)L_2 + H_2 \qquad (15)$$

$$Pt(CO)L_2 + H_2O \xrightleftharpoons{\text{in acetone}} [PtH(CO)L_2]OH \qquad (16)$$

(Scheme I). In the catalysis in acetone, the addition of $H_2O$ to Pt(CO)$L_2$ would give [PtH(CO)$L_2$]OH (eq. 6) rather than the solvated species [PtH(S)$L_2$]OH which is the case in pyridine.

## $RhH[P(i\text{-}Pr)_3]_3$-Catalyzed WGS Reaction

$RhHL_3$ is more efficient as the catalyst than the corresponding $PtL_3$ (L=P(i-Pr)$_3$). In contrast to $PtL_3$, the wgs reaction ef-

Scheme I

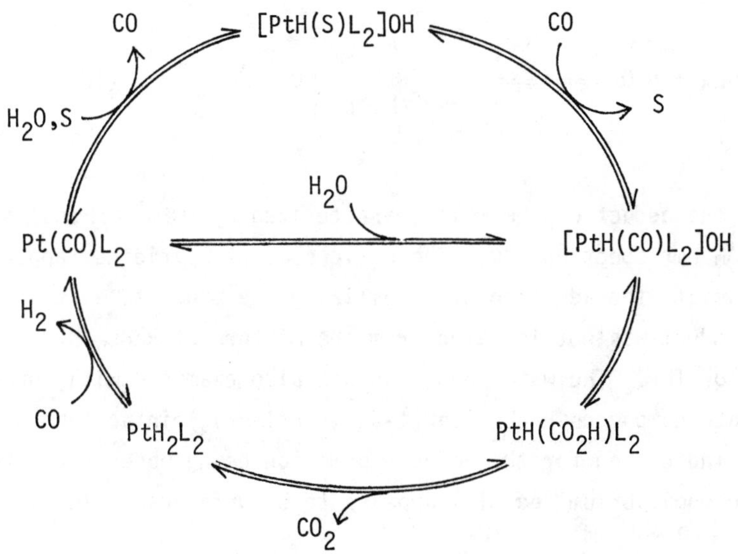
CO
[PtH(S)L2]OH
CO
S
H2O,S
H2O
Pt(CO)L2
[PtH(CO)L2]OH
H2
CO
PtH2L2
PtH(CO2H)L2
CO2

fected with $RhHL_3$ shows a faster rate in pyridine than in acetone (Table I). Under the same condition (100°), $RhCl(PPh_3)_3$ and $RhCl(CO)(PPh_3)_2$ were practically inactive. This is likely due to their inability of producing the hydroxo species, an oxidative adduct of water.

$RhHL_3$ (L=P(i-Pr)$_3$) is capable of forming the water adduct (11). The water addition to $RhHL_3$ takes place in pyridine readily at room temperature to give an ionic product $[RhH_2(py)_2L_2]OH$ (eq. 17), which can be isolated as its $BPh_4$ salt ($\nu$(Rh-H) 2076, 2112

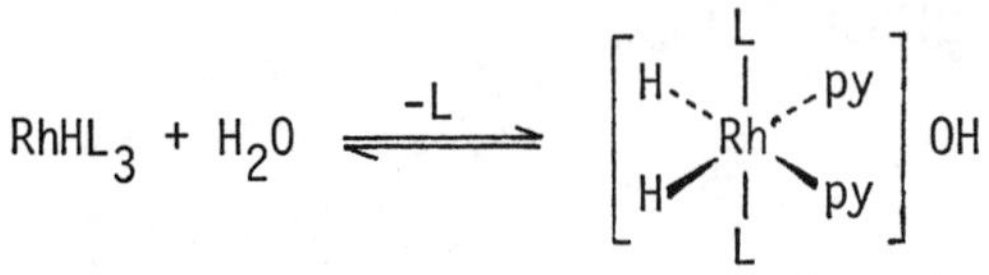

$$RhHL_3 + H_2O \; \underset{}{\overset{-L}{\rightleftharpoons}} \; \begin{bmatrix} H_{\cdots} & \overset{L}{\underset{L}{|}} & _{\cdots}py \\ & Rh & \\ H^{\diagup} & | & {}^{\diagdown}py \end{bmatrix} OH \qquad (17)$$

cm$^{-1}$). The adduct can be well characterized by elemental analysis, IR and $^1$H nmr spectroscopy. The basic solvent, pyridine, apparently assists the addition by stabilizing the product, as the oxidative adduct was not isolated from the mixture of $RhHL_3/H_2O$ in acetone or THF. The water addition was also examined with three coordinate compounds $RhHL_2$ (L=P(t-Bu)$_3$, P(i-Pr)$_3$) in acetone or THF, no indication for the adduct formation being observed. The addition equilibrium (eq. 18) appears to be unfavorable in such a

$$RhHL_2 + H_2O \; \rightleftharpoons \; RhH_2(OH)L_2 \qquad (18)$$

solvent, a result in parallel with the observed solvent effect for the wgs reaction rate (Table I).

From the wgs reaction in pyridine was isolated trans-[Rh(CO)-(py)L$_2$]$^+$ ($\nu$(CO) 1985 cm$^{-1}$), while a neutral compound trans-[Rh(OH)-(CO)L$_2$] ($\nu$(CO) 1925; $\nu$(OH) 3644 cm$^{-1}$) was detected by IR in the wgs reaction carried out in acetone. trans-[RhH(CO)L$_2$] ($\nu$(Rh-H) 1980; $\nu$(CO) (in benzene-d$_6$) 1944 cm$^{-1}$) was isolated for L=P(c-C$_6$H$_{11}$)$_3$ from the wgs reaction in acetone but not for L=P(i-Pr)$_3$.

Hence, the hydrido-carbonyl compound is apparently stabilized with bulky phosphines. trans-$[RhH(CO)L_2]$ ($L=P(i\text{-}Pr)_3$) ($\nu$(Rh-H) 1980; $\nu$(CO) 1920, 1942 $cm^{-1}$), which can be prepared separately by treating $RhHL_3$ with methanol, was found to be very reactive toward CO. We confirmed that the reaction of trans-$[RhH(CO)L_2]$ produces a Rh(0) carbonyl ($\nu$(CO) 1732, 1769, 1957 $cm^{-1}$) (eq. 19). An analogous compound $Rh_2(CO)_4L_2$ ($L=P(t\text{-}Bu)_3$) ($\nu$(CO) 1785,1940,1985 $cm^{-1}$)

$$RhH(CO)L_2 + CO \xrightarrow[-L]{-H_2}$$ 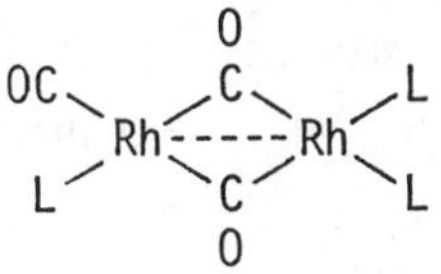 (19)

was actually isolated from the wgs reaction effected by $RhHL_2$.

$Rh_2(CO)_3L_3$ appears to participate in the catalytic cycle since we observed that it can react with $H_2O$ resulting in trans-[RhH-$(CO)L_2]$ and trans-$Rh(OH)(CO)L_2$ (eq. 20). As the former with CO

$$Rh_2(CO)_3L_3 + H_2O \xrightarrow[-CO]{L} \text{trans-}[RhH(CO)L_2] + \text{trans-}[Rh(OH)(CO)L_2] \quad (20)$$

readily transforms into $Rh_2(CO)_3L_3$ (eq. 19), the isolation of trans-$[Rh(OH)(CO)L_2]$ from the wgs reaction is reasonable. In addition, the turnover rate with a mixture of $Rh_2(CO)_3L_3$ and 3 moles of free $L(P(i\text{-}Pr)_3)$ was comparable with that obtained with $RhHL_3$ (Table I).

We have confirmed the transformation of trans-$[Rh(OH)(CO)L_2]$ into trans-$[RhH(CO)L_2]$ which occurs upon treatment with CO. Thus, treating trans-$[Rh(OH)(CO)L_2]$ ($L=P(c\text{-}C_6H_{11})_3$) with CO, we observed $CO_2$ and $Rh_2(CO)_4L_2$, the latter being presumably derived from RhH-$(CO)L_2$. The following steps (eq. 21,22,23) are most likely to be involved. Note that the nucleophilic attack of $OH^-$ on the CO

$$Rh(OH)(CO)L_2 + CO \longrightarrow Rh(COOH)(CO)L_2 \quad (21)$$

$$Rh(COOH)(CO)L_2 \longrightarrow RhH(CO)L_2 + CO_2 \quad (22)$$

$$2RhH(CO)L_2 + 2CO \longrightarrow Rh_2(CO)_4L_2 + H_2 + 2L \quad (23)$$

ligand is facilitated by an extra CO molecule, a result contransting with the $PtL_3$ catalyst system (cf. eq. 11).

For the $P(i\text{-}Pr)_3$ complex, trans-[$Rh(OH)(CO)L_2$], similar processes could be involved. The product from the last step, however, is $Rh_2(CO)_3L_3$ but not $Rh_2(CO)_4L_2$. In fact, the formation of $Rh_2(CO)_3L_3$ ($L=P(i\text{-}Pr)_3$) was confirmed in the reaction of trans-[$Rh(OH)(CO)L_2$] with CO in THF. Further, the assumption of reaction (eq. 21) received support from the following experiment. trans-[$Rh(OMe)(CO)L_2$] was treated with CO to give trans-[$Rh(COOMe)(CO)L_2$] ($\nu$(CO) 1949; $\nu$(C=O) 1613 $cm^{-1}$) as yellow crystals (75 %).

Based on the component reactions described above, a catalytic cycle responsible primarily for the $CO_2$ production may be depicted (Scheme II).

## Elemental Reactions Associated with $H_2$ Production

In addition to the component reaction (eq. 23), there are several reactions responsible for $H_2$ generation. The hexa-coordinate water adduct [$RhH_2(py)_2L_2$]OH ($L=P(i\text{-}Pr)_3$) and its bipyridyl analog $[RhH_2(bipy)L_2]^+$ ($\nu$(Rh-H) 2080, 2135 $cm^{-1}$) are thermally stable. Upon contact with CO, the dihydride [$RhH_2(py)_2L_2$]OH immediately releases $H_2$ which probably occurs through a transient species (eq. 24) in which the CO ligand is coplanar with the hydride ligands (12). By contrast, a reaction of CO with $[RhH_2(PEt_3)_3]^+$, a water adduct of $RhH(PEt_3)_3$, gave $[RhH_2(CO)(PEt_3)_3]^+$ ($\nu$(Rh-H) 2005, 2030, $\nu$(CO) 1960 $cm^{-1}$) where the CO ligand is cis to the two hydrides (eq. 25). This dihydrido carbonyl compound is stable toward reductive elimination of the dihydrido ligands at room temperature. The dramatic effect of electron-withdrawing CO ligand reducing the M-H bond strength, therefore, can best be accounted for with this coplanar geometry of the CO and two hydrido

Scheme II

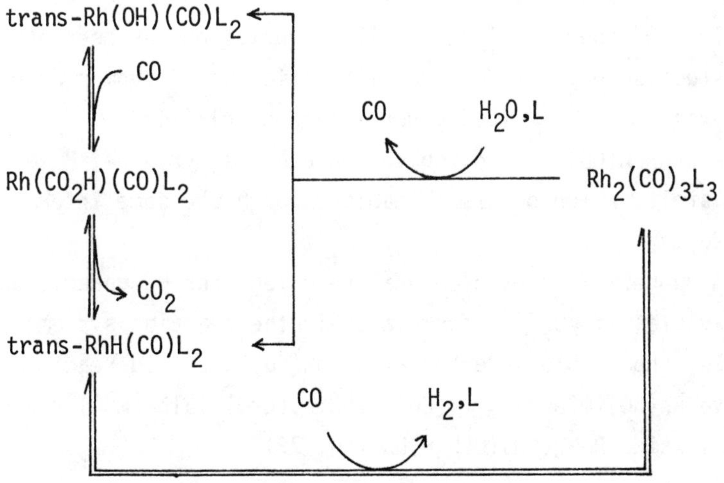
trans-Rh(OH)(CO)L2
CO
Rh(CO2H)(CO)L2
CO2
trans-RhH(CO)L2
CO
H2O,L
Rh2(CO)3L3
CO
H2,L

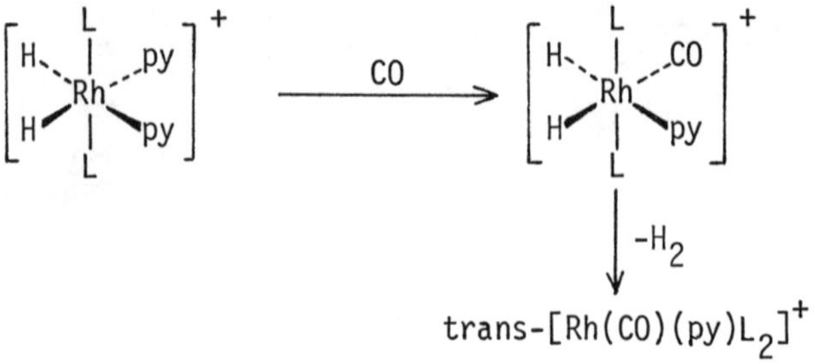

$$\left[\begin{array}{c} L \\ H\cdots \overset{|}{\underset{|}{Rh}}\cdots py \\ H\quad\overset{|}{L}\quad py \end{array}\right]^{+} \xrightarrow{\;CO\;} \left[\begin{array}{c} L \\ H\cdots \overset{|}{\underset{|}{Rh}}\cdots CO \\ H\quad\overset{|}{L}\quad py \end{array}\right]^{+}$$

$$\Big\downarrow -H_2$$

$$\text{trans-}[Rh(CO)(py)L_2]^{+} \qquad (24)$$

$$\left[\begin{array}{c} PEt_3 \\ H\cdots\overset{|}{Rh}\!-\!PEt_3 \\ H\quad\overset{|}{PEt_3} \end{array}\right]^{+} + \; CO \longrightarrow \left[\begin{array}{c} CO \\ H\cdots\overset{|}{Rh}\cdots PEt_3 \\ H\quad\overset{|}{PEt_3}\;PEt_3 \end{array}\right]^{+} \qquad (25)$$

ligands. This effect was studied by *ab initio* MO-SCF-CI calculation on $NiR_2$ species (13). Similar conclusion has been drawn by EHMO studies on the reductive elimination of $D_2$ from planar $d^8$ complexes, $cis\text{-}MD_2A_2$ (D=$\sigma$-donor, A=acceptor) (14).

When $RhH(CO)L_2$ is dissolved in aqueous pyridine, $H_2$ evolved immediately, which occurs probably through the same intermediate (eq. 26).

A number of other elemental reactions for $H_2$ generation are conceivable, if $CO_2$ is accumulated in the reaction system. For example, the viable intermediate, $RhH(CO)L_2$ should react with $H_2CO_3$ to give $Rh(CO)(OCO_2H)L_2$ ($\nu$(CO) 1952; $\nu$(C=O) 1615 cm$^{-1}$) via postulated species $RhH_2(CO)(OCO_2H)L_2$ (eq. 27).

$$RhH(CO)L_2 + H_2O \xrightarrow[\;py\;]{} \left[\begin{array}{c} L \\ H\cdots\overset{|}{\underset{|}{Rh}}\cdots CO \\ H\quad\overset{|}{L}\quad py \end{array}\right] OH$$

$$\Big\downarrow -H_2$$

$$[Rh(CO)(py)L_2]OH \qquad (26)$$

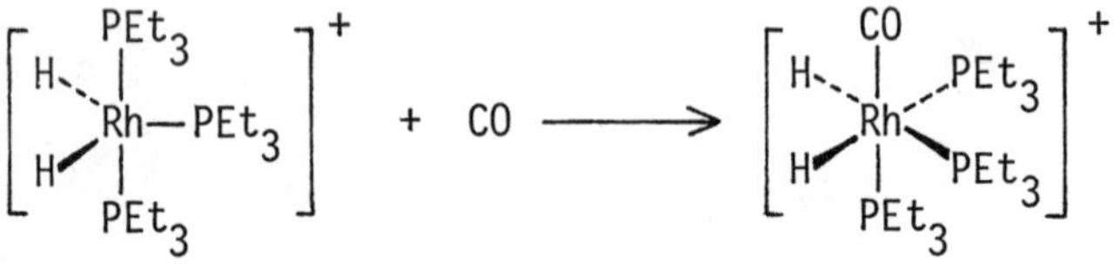

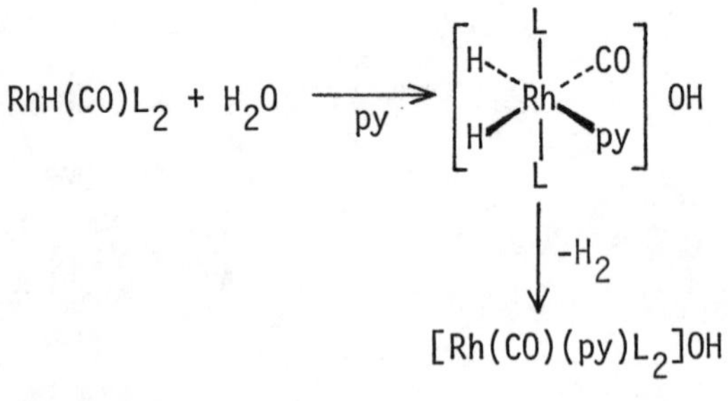

$$RhH(CO)L_2 + H_2CO_3 \longrightarrow RhH_2(CO)(OCO_2H)L_2 \xrightarrow{-H_2} Rh(CO)(OCO_2H)L_2 \quad (27)$$

The reaction of $H_2CO_3$ with the other viable intermediate Rh(OH)(CO)$L_2$ is expected to give the same product releasing merely $H_2O$. The product actually isolated was a μ-carbonato compound (ν(CO) 1934; ν(C=O) 1533 $cm^{-1}$) which is presumably derived from Rh(CO)($OCO_2H$)$L_2$ (eq. 28) (15). Remarkably the μ-carbonato compound can be hydrolyzed readily in pyridine affording [Rh(CO)(py)$L_2$]OH (eq. 29) which was isolated as its $BPh_4$ salt. Since we

$$2Rh(CO)(OCO_2H)L_2 \xrightarrow{-H_2CO_3} [Rh(CO)L_2]_2(\mu\text{-}CO_3) \quad (28)$$

$$[Rh(CO)L_2]_2(\mu\text{-}CO_3) + H_2O \xrightarrow{py} Rh(CO)(OCO_2H)L_2 + [Rh(CO)(py)L_2]OH \quad (29)$$

isolate solely $[Rh(CO)(py)L_2]^+$ but not the bicarbonato compound Rh(CO)($OCO_2H$)$L_2$ from the hydrolysis, the equilibrium (eq. 30) ap-

$$Rh(CO)(OCO_2H)L_2 \underset{CO_2}{\overset{py,\ -CO_2}{\rightleftharpoons}} [Rh(CO)(py)L_2]OH \quad (30)$$

pears to be favored toward the formation of [Rh(CO)(py)$L_2$]OH. With this information we summarize steps associated with $H_2$ generation (Scheme III).

## The Catalytic Cycle

With the information on the component reactions described above the construction of the whole catalytic cycle is in order. It is highly unlikely that the catalyst precursor $RhHL_3$ carries the

Scheme III

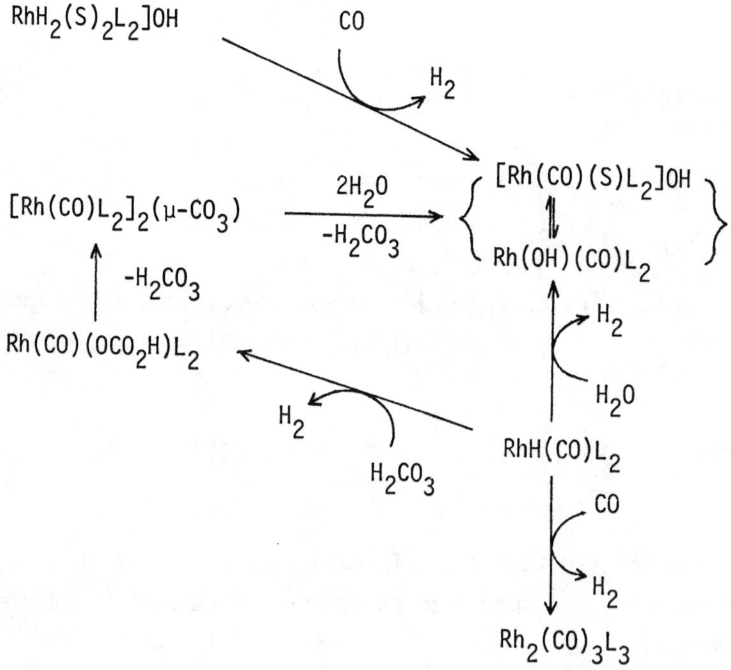
RhH2(S)2L2]OH
CO
H2
[Rh(CO)(S)L2]OH
Rh(OH)(CO)L2
[Rh(CO)L2]2(μ-CO3)
2H2O
-H2CO3
-H2CO3
Rh(CO)(OCO2H)L2
H2
H2O
H2
H2CO3
RhH(CO)L2
CO
H2
Rh2(CO)3L3

Scheme IV

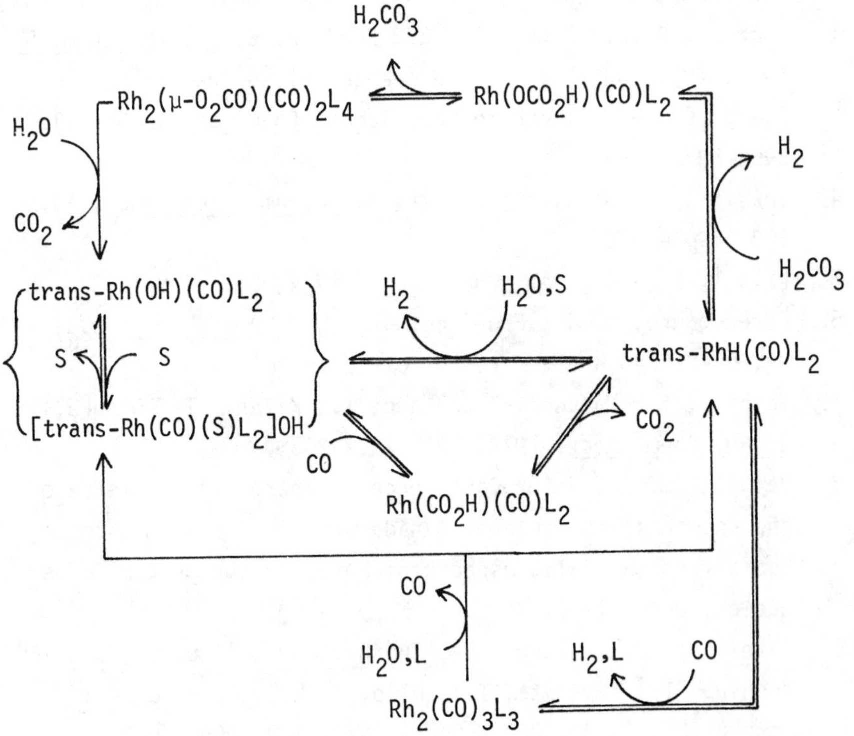
$H_2CO_3$
$Rh_2(\mu\text{-}O_2CO)(CO)_2L_4$
$Rh(OCO_2H)(CO)L_2$
$H_2O$
$CO_2$
$H_2$
$H_2CO_3$
trans-Rh(OH)(CO)L$_2$
S
S
[trans-Rh(CO)(S)L$_2$]OH
$H_2$
$H_2O$,S
trans-RhH(CO)L$_2$
$CO_2$
CO
$Rh(CO_2H)(CO)L_2$
CO
$H_2O$,L
$H_2$,L
CO
$Rh_2(CO)_3L_3$

catalytic cycle, as it readily reacts with CO or $H_2O$, both processes being found low-energy processes. Therefore, $RhHL_3$ is not included in the cycle shown in Scheme IV. All the isolated species $Rh(OH)(CO)L_2$, $Rh_2(CO)_3L_3$ and $Rh_2(\mu\text{-}CO_3)(CO)_2L_4$ were tested for the catalysis to confirm their participation in the catalysis (Table I).

Literature Cited

1. King, A. D.; King, R. B.; Yang, D. B., J. Am. Chem. Soc., 1980, 102, 1028-1032 and references cited therein.
2. Ungermann, C.; Landis, V.; Moya, S. A.; Cohen, H.; Walker, M.; Pearson, R. G.; Rinker, R. G.; Ford, P. C., J. Am. Chem. Soc., 1979, 101, 5922-5929 and references cited therein.
3. Cheng, C. -H.; Eisenberg, R., J. Am. Chem. Soc., 1978, 100, 5968-5970.
4. Yoshida, T.; Ueda, Y.; Otsuka, S., J. Am. Chem. Soc., 1978, 100, 3941-3942.
5. Kassel, L. S., J. Am. Chem. Soc., 1934, 56, 1838-1842.
6. Darensbourg, D. J.; Froelich, J. A., J. Am. Chem. Soc., 1977, 99, 5940-5946 and references cited therein.
7. Yoshida, T.; Matsuda, T.; Okano, T.; Kitani, T.; Otsuka, S., J. Am. Chem. Soc., 1979, 101, 2027-2038.
8. Bennett, M. A., Joint Conference of Inorganic Chemistry of the Chemical Institute of Canada and the American Chemical Society on Catalytic Aspects of Metal Phosphine Complexes, Guelph, June, 1980.
9. Catellani, M.; Halpern, J., Inorg. Chem., 1980, 19, 566-568.
10. Yoshida, T.; Yamagata, T.; Tulip, T. H.; Ibers, J. A.; Otsuka, S., J. Am. Chem. Soc., 1978, 100, 2063-2073.
11. Yoshida, T.; Okano, T.; Saito, K.; Otsuka, S., Inorg. Chim. Acta., 1980, 44, L135-L136.
12. Yoshida, T.; Okano, T.; Otsuka, S., J. Am. Chem. Soc., 1980, 102, 5966-5967.
13. Åkermark, B.; Johansen, H.; Roos, B.; Wahlgren, U., J. Am. Chem. Soc., 1979, 101, 5876-5883.

14. Tatsumi, K.; Hoffmann, R., J. Am. Chem. Soc., submitted for publication.
15. Yoshida, T.; Thorn, D. L.; Okano, T.; Ibers, J. A.; Otsuka, S., J. Am. Chem. Soc., 1979, 101, 4212-4221.

RECEIVED December 8, 1980.

6

# The Water Gas Shift Reaction as Catalyzed by Ruthenium Carbonyl in Acidic Solutions

PETER C. FORD, PAUL YARROW, and HAIM COHEN

Department of Chemistry, University of California, Santa Barbara, CA 93106

The past several years have seen renewed interest in the catalyst chemistry of the water gas shift reaction (WGSR, Eq. 1).

$$CO + H_2O \rightleftharpoons CO + H_2 \qquad (1)$$

This has been largely stimulated by the recognition that the shift reaction is a key step in the production of the copious hydrogen and/or synthesis gas ($H_2/CO$) required for the gasification or liquifaction of coal. In 1977, we reported that ruthenium carbonyl in alkaline aqueous ethoxyethanol solution formed a homogeneous WGSR catalyst (1,2). Subsequently, a number of other reports of homogeneous shift reaction catalysts have appeared (3-12). Our rationale for choosing an alkaline solution reaction medium for our initial studies derived from the historical precedent by Heiber (14) that metal carbonyls undergo reactions with aqueous bases to give metal carbonyl hydride anions (e.g., Eq. 2). Acidification of these solutions released both

$$Fe(CO)_5 + 3OH^- \rightleftharpoons HFe(CO)_4^- + CO_3^= + H_2O \qquad (2)$$

$CO_2$ and $H_2$, presumably from the respective neutralizations of carbonate and of the metal carbonyl anions to give metal hydrides (e.g., $H_2Fe(CO)_4$) which undergo reductive elimination of $H_2$.

In this context we postulated that the shift reaction might proceed catalytically according to a hypothetical cycle such as Scheme I. There are four key steps in Scheme I: a) nucleophilic attack of hydroxide or water on coordinated CO to give a hydroxycarbonyl complex, b) decarboxylation to give the metal hydride, c) reductive elimination of $H_2$ from the hydride and d) coordination of new CO. In addition, there are several potentially crucial protonation/deprotonation equilibria involving metal hydrides or the hydroxycarbonyl. The mechanistic details have been worked out (but only incompletely) for a couple of the alkaline solution WGSR homogeneous catalysts. In these cases,

0097-6156/81/0152-0095$05.00/0
© 1981 American Chemical Society

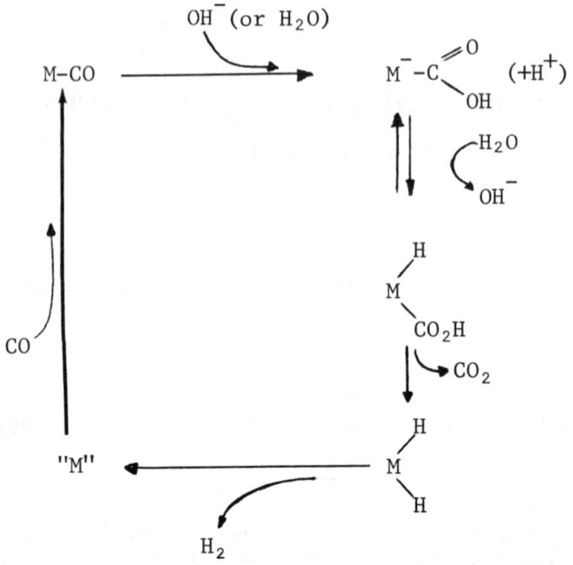


Scheme I

the general features of this cycle are apparently followed, although there is some question as to whether steps (c) and (d) occur sequentially as proposed for the catalyst based on $Fe(CO)_5$ (12) or in a concerted fashion as suggested for the alkaline ruthenium carbonyl catalyst (2). Regardless, it is notable that in basic solutions WGSR catalysts are formed from a number of the simple transition metal carbonyls $M_x(CO)_y$ including those of ruthenium, iridium, iron, osmium, rhodium, rhenium, platinum, molybdenum, tungsten, and chromium (1,4,5,13,15,16) as well as from mixed metal carbonyls and more complicated systems with additional ligands such as pyridines, cyclopentadienyls and phosphines (7,15,16). Thus, one can conclude that the ability to form such shift reaction catalysts is a quite general reactivity property of metal carbonyls in basic solutions containing water.

A report in 1977 (3) of an active system prepared from $[Rh(CO)_2Cl]_2$, $CH_3CO_2H$, conc. HCl and NaI in water demonstrated that a basic medium is not a necessary condition for WGSR catalysis. This result stimulated us to examine the potential activity of several simple metal carbonyls in acidic solution as well. Attempts with $Fe(CO)_5$ and $Ir_4(CO)_{12}$ (17), both active in alkaline and amine solutions, proved unfruitful. However $Ru_3(CO)_{12}$ in acidic (0.5 N $H_2SO_4$) aqueous ethoxyethanol gave WGSR activity substantilly larger than found in basic solutions under otherwise analogous conditions ($P_{CO}$=0.9 atm, T=100°C, $[Ru]_{Total}$=0.036 mol/L) (15). This solution proved unstable and an insoluble red solid (a ruthenium carbonyl polymer) formed over a period of several days at the cooler neck of the glass batch reactor. Solutions prepared using diglyme as the principal solvent proved both more stable and more active. Summarized here are experimental studies aimed at characterizing this catalytic system.

## Activities and Kinetics:

The catalytic activity of the ruthenium carbonyl system in acidic aqueous diglyme has been examined by batch reactor methods in our laboratory and is the subject of flow reactor studies now in progress in collaboration with colleagues in the UCSB Chemical Engineering Department. A key feature of these studies is the observation that $Ru_3(CO)_{12}$ itself is not an effective catalyst. The typical run at 100°C under CO ($P_{CO}$=0.9 atm) has an induction period of several hours during which some $CO_2$ production is seen but little $H_2$ formation occurs. A period of about 6 hours is required before full activitiy of $H_2$ production (~50 turnovers/day) and good stoichiometry (according to Eq. 1) is attained. Over the same time frame, the $Ru_3(CO)_{12}$ initially added undergoes complete conversion to other species as reflected by the decrease in the characteristic electronic absorption band of this cluster at $\lambda_{max}$=394 nm (Fig. 1). Although these spectral changes are also accompanied by absorbance increases at wavelengths below 300 nm, band maxima were not discerned. Heating the $Ru_3(CO)_{12}$ in acidic aqueous diglyme for several hours under either an argon or a hydrogen atmosphere leads to similar spectral changes. Furthermore, the resulting solutions showed no induction period in forming effective WGSR catalysts when the solutions were then charged with an atmosphere of CO. Similar spectral changes to those described above are seen when $Ru_3(CO)_{12}$ in octane under CO is irradiated with visible light giving $Ru(CO)_5$ as the photochemical product (18,19). Given that the assignment of the 394 nm band of $Ru_3(CO)_{12}$ as a $\sigma \rightarrow \sigma^*$ transition of the cluster metal-metal bond framework (20) and that other ruthenium clusters show similar near UV or visible absorption bands, it seems likely that the ruthenium carbonyl species in the acidic diglyme catalyst are mononuclear or dinuclear.

The kinetics results of the batch reactor runs lead to the following qualitative observations: At low CO pressures (less than about 1 atm) the catalysis appears to be first order in ruthenium over the range 0.018 M to 0.072 M and also in $P_{CO}$ as illustrated by the log $P_{CO}$ vs time plots of Fig. 2 and also shown by the method of initial rates. Changes in the sulfuric acid and water concentrations over the respective ranges 0.25 M to 2.0 M and 4 M to 12 M have relatively small effects on the catalysis rates, although the functionalities are complicated and show concave rate vs concentration curves with maximum rates

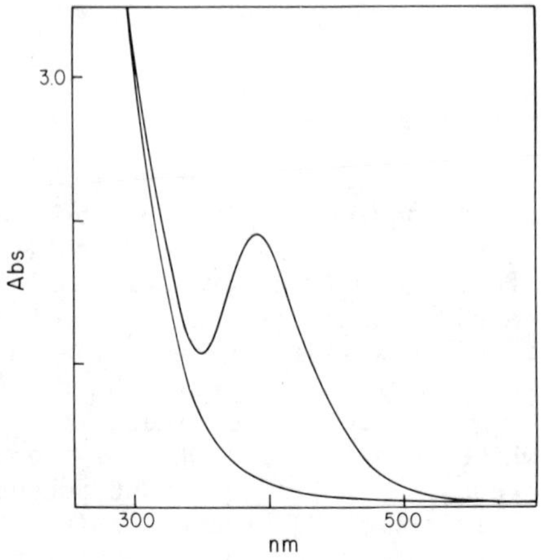


*Figure 1. UV–visible spectrum (0.10-cm cell) of a catalyst solution prepared from $Ru_3(CO)_{12}$ ($3 \times 10^{-3}$M), $H_2SO_4$ (0.5M), $H_2O$ (8.0M) under a CO atmosphere ($P_{CO} = 1$ atm) in diglyme at 100°C. The upper curve represents the spectrum 5 min after the solution was prepared at this temperature, the lower curve is the spectrum after 6 h.*

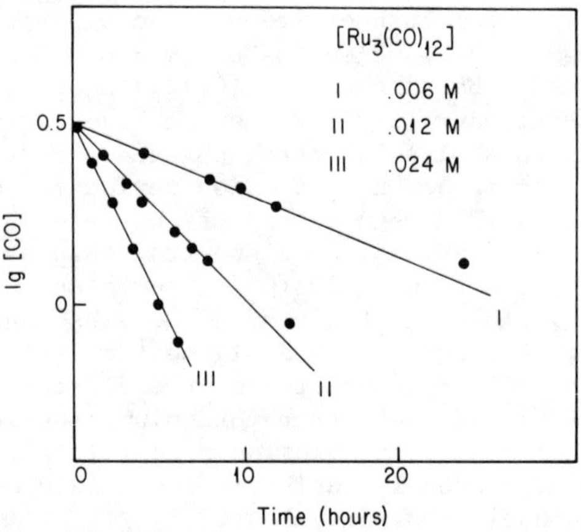


*Figure 2. First-order rate plots for the consumption of CO in a 100-mL batch reactor (catalyst solution is 5 mL of aqueous diglyme with 8.5M $H_2O$, 1.0M $H_2SO_4$, T = 100°C and $P_{CO}$ (initial) = 0.9 atm). Slopes of the three linear plots are $2 \times 10^{-2}$, $4.4 \times 10^{-2}$, and $9.3 \times 10^{-2}$ $h^{-1}$ for the respective $Ru_3(CO)_{12}$ initial concentrations of (I) 0.006M, (II) 0.012M, and (III) 0.024M.*

found at ~4 M $H_2O$ and at ~0.5 M $H_2SO_4$ with $P_{CO}$=1 atm. However, using $H_3PO_4$, $CH_3CO_2H$ or $CF_3CO_2H$ instead as the added acid decreased the activity markedly. The system is temperature sensitive with an activation energy of about 14 kcal/mole derived from a linear Arrhenius plot for the catalysis rates over the temperature range 90-140°C in the low $P_{CO}$ region. A dramatic turnaround in activity occurs at CO pressures much larger than 1 atm with the production of $H_2$ and $CO_2$ being inhibited by increasing $P_{CO}$ under the conditions. Notably, a batch reactor run initiated at low pressures and demonstrated to be active displays a much lower rate when the bulb is charged with a high $P_{CO}$. The initial catalytic activity is regenerated when the system is recharged at the lower $P_{CO}$, thus showing the inhibition at higher $P_{CO}$ to be reversible.

Another characteristic of the batch reactor runs is that after a number of flushing/recharging cycles (see Experimental) over a period of days there is a marked degradation of the system's catalytic activity. Whether this is the result of irreversible transformations of the catalyst to inactive species (for example introduction of air to a hot catalyst solution causes irreversible destruction of the activity) or of the loss of volatile ruthenium species during the freeze/thaw, degassing/recharging cycles is not clear. The latter is certainly a major contributor to the slow degradation of the activity in the flow reactor runs where, despite the presence of a condensor designed to return solvent and catalyst to the reaction vessel, volatile ruthenium carbonyl species are trapped downstream from the reactor (see below). If a fresh, active catalyst in acidic diglyme is cooled to room temperature after operating under a low $P_{CO}$, the solution is light yellow and undergoes a slow transformation to give $Ru_3(CO)_{12}$ which precipitates from solution over a period of several days. As much as 95% of the original $Ru_3(CO)_{12}$ can be recovered under these conditions. In contrast a solution operating under a higher $P_{CO}$ (2.7 atm) precipitates $Ru_3(CO)_{12}$ quickly upon cooling indicating that the principal ruthenium species present under such conditions is $Ru_3(CO)_{12}$ or one easily converted to this cluster.

## In Situ Spectroscopic Studies:

Besides the electronic spectral studies noted above, we have also carried out in situ studies of the acidic ruthenium catalyst using nmr and infrared spectral techniques. A key set of observations derive from the $^1H$ and $^{13}C$ nmr spectra of an operating catalyst at 90° and $P_{CO}$ 1 atm which indicate the presence of only one major ruthenium species. The proton spectrum shows a sharp singlet at 24.0 τ which remains such when the solution is cooled to room temperature, although the slow formation of other species was observed over a period of hours at the latter conditions. The $^1H$-decoupled $^{13}C$ spectrum of the operating catalyst also shows a singlet at 198.2 ppm downfield from TMS) which becomes a doublet ($J_{C-H}$=10 Hz) when proton coupled. The same spectrum is seen when the solution is cooled to room temperature.

Notably these nmr spectra are inconsistent with those of $H_2Ru(CO)_4$ or $HRu(CO)_5^+$ (Table I) which should be key species in a catalysis cycle based solely on mononuclear complexes. For example, the proton resonance at 24.0 τ is considerably higher field than those seen for the mononuclear species with terminal hydrides (17.6 and 17.2 τ, respectively) and falls in the region where bridging hydrides are normally seen. Further comparison of the spectra in Table I shows that the catalyst solution $^{13}C$ resonance occurs at a position downfield from those found for cationic ruthenium carbonyl hydrides such as $HRu(CO)_5^+$ and $HRu_3(CO)_{12}^+$ and in a region more consistent with a neutral or anionic complex. Thus we conclude that the principal species present in the acidic catalyst solution has a single hydride, is neutral or anionic and is fluxional at room temperature and above. Given the conclusion from the UV-visible spectra that the nuclearity of the complex is less than three (see above) and the conclusion from the nmr data that the hydride is bridging, the circumstantial evidence is that the principal ruthenium species under the catalysis conditions is a dinuclear complex. A logical proposition is that this is the dinuclear anion $HRu_2(CO)_8^-$ which is unknown for ruthenium, although the iron analog is known and has been shown to be fluxional even to low temperatures (21).

Attempts to obtain *in situ* infrared spectra of this catalyst system utilizing a high temperature infrared cell similar to that described by King (25) have met with mixed success owing to the strong absorption of the solvent medium in the carbonyl region. Broad peaks at 2084, 2040, 2013 and 1960(br) $cm^{-1}$ all of medium to strong intensity were observed for the reaction solution at 100°C under an atmosphere of CO. A survey of ruthenium carbonyl infrared spectra indicate that these bands are not consistent with those expected for $Ru(CO)_5$, $Ru_3(CO)_{12}$, $H_2Ru(CO)_4$ or $Ru_2(CO)_9$ among the simpler known species of this type. Lowering the temperature of the reaction solution to 25°C does not lead to major differences in the spectrum although there are some changes in the relative peak heights. Whether this is the result of shifts in the concentrations of several species present in solution or of medium effects on the band shapes is not clear; however, the former is an unlikely prospect given the nmr results noted above that the proton and carbon-13 spectra do not undergo immediate changes upon lowering the catalyst solution temperature from 90° to 25°.

## A Proposed Mechanism for Catalysis:

The information currently available for the acidic ruthenium catalyst system, is consistent with a cyclic mechanism such as

Table I: $^1H$ and $^{13}C$ N.M.R. Data for Ruthenium Carbonyl Complexes.

| Complex | $^{13}CO$(ppm) | $^1H(\tau)$ | $J^{13}C-^1H$(Hz) | References |
|---|---|---|---|---|
| $Ru_4(CO)_{13}^{2-}$ | 223.7 | - | - | 22 |
| $H_2Ru_4(CO)_{12}^{2-}$ | 220.0 | 29.3 | 10.3 (trans)<br>5.9 (cis) | 22 |
| $HRu_4(CO)_{13}^{-}$ | 203.7 | 25.8 | - | 22 |
| $HRu_3(CO)_{11}^{-}$ | 202.2 | 22.6 | 6 (average) | 2 |
| $H_3Ru_4(CO)_{12}^{-}$ | 198.2 | 27.0 | 7.3 | 2,22 |
| $HRu_3(CO)_{10}NO$ | 202.9,195.5<br>194.5,185.8 | 21.9 | | 23 |
| $Ru_3(CO)_{12}$ | 198.0 | - | - | This work |
| $H_2Ru(CO)_4$ | 192.5<br>190.1 | 17.6 | 7 (cis)<br>(trans) | 24 |
| $HRu(CO)_5^+$ | 180.4<br>178.4 | 17.2 | 4 (cis)<br>24 (trans) | This work |
| $HRu_3(CO)_{12}^+$ | 191.0,188.0<br>184.5,178.9 | 29.8 | - | This work |

illustrated by Scheme II. The key features of this scheme are that at low $P_{CO}$ the ruthenium is present largely as the $HRu_2(CO)_8^-$

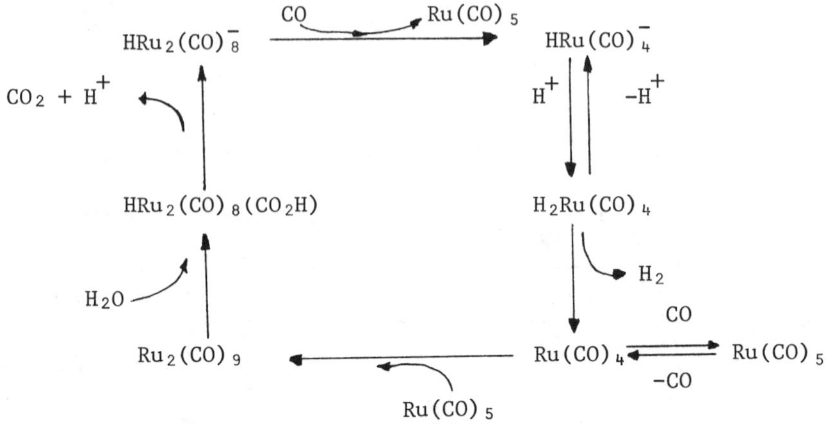


Scheme II

anion and the rate limiting step is the reaction of this ion with CO to cleave the metal-metal linkage giving $Ru(CO)_5$ plus $HRu(CO)_4^-$. Under such conditions the rate should be first order in both [Ru] and $P_{CO}$ since the concentration of $HRu_2(CO)_8^-$ would be proportional to the total concentration of ruthenium present. The next step would be protonation of $HRu(CO)_4^-$ to give the dihydride which undergoes reductive elimination of dihydrogen. Although the first $pK_a$ of $H_2Ru(CO)_4$ is as yet unknown, values for the iron and osmium analogs (26) clearly indicate that $HRu(CO)_4^-$ should be sufficiently strong a base to be fully protonated under the solution conditions. The dihydride is reported to be unstable in toluene solution above -20°C; however in that case excess CO was not present and the products presumably were ruthenium clusters (24) (see below).

Inhibition of catalysis at high $P_{CO}$ may be explained as reflecting conditions where the equilibrium

$$CO + Ru_2(CO)_9 \rightleftharpoons 2\ Ru(CO)_5 \qquad (3)$$

becomes a dominant factor in the catalysis. Support for this proposition comes from the flow reactor kinetics currently in progress. At very low ruthenium concentrations, first order behavior in [Ru] apparently no longer holds and the reaction kinetics indicate orders closer to two than one, thus supporting the possible importance of Eq. (3) to the overall catalysis rate under these conditions. Further supporting evidence comes from a complication in the flow reactor kinetics. These systems show slow decreases in activity over a period of time owing to the loss of ruthenium from the solution, a problem especially apparent at high $P_{CO}$. Examination of a low temperature trap downstream from the catalysis vessel showed the presence of a clear solution, mostly aqueous diglyme, which when warmed to room temperature turned yellow and slowly precipitated $Ru_3(CO)_{12}$. Thus the gas stream of the flow system served to sweep a volatile ruthenium species out of the reaction solution, probably $Ru(CO)_5$ but possibly $H_2Ru(CO)_4$. There is another potential source of the CO inhibition in Scheme II. Studies in progress in this laboratory (27) have shown that the initial step in the decomposition and clusterification of $H_2Ru(CO)_4$ in solution is not $H_2$ elimination but is CO dissociation. Thus it is possible that the elimination of $H_2$ from $H_2Ru(CO)_4$ requires prior CO dissociation *via* a mechanism similar to that proposed for $H_2$ elimination from $H_2Os(CO)_4$ (28) and thus would be inhibited at the higher $P_{CO}$. This question is currently being investigated.

## Experimental Procedures

Infrared spectra were recorded on a Perkin-Elmer model 283 spectrophotometer. Proton and carbon-13 nuclear magnetic resonance spectra were recorded on Varian XL-100 and CFT-20 spectrometers, respectively, operating in the pulsed mode. UV-visible spectra were recorded on a Cary 118C recording spectrometer equipped with a thermostated cell compartment. Gas sample analyses were performed on a Hewlett-Packard 5830A programmable gas chromatograph, calibrated for the appropriate substrates. The columns used were Carbosieve B (Mesh 80-100) columns obtained from Hewlett Packard and the carrier gas used was a Linde prepared 8.5% $H_2$/91.5% He mixture. Gas samples were taken with Analytical Pressure Lok gas syringes obtained from Precision Sampling Corporation. Calibration curves for the chromatographs and sampling procedures were prepared periodically for CO, $CH_4$, $CO_2$, and $H_2$ for gas sample sizes ranging from 0.05 to 1.5 mL STP of the gas. These calibration curves were linear for CO, $CH_4$, and $CO_2$ but not for $H_2$. Catalytic activity and kinetics runs were largely done in all-glass batch reactors (100 mL) consisting of round bottom flasks with sidearm stopcocks designed for attachment to a vacuum line and for periodic gas phase sampling. Typically, the $Ru_3(CO)_{12}$ and solvent medium were added to the reactor vessel (at room temperature) which was then attached to the vacuum line, and the solution was degassed by freeze-pump-thaw cycles then charged with a $CO/CH_4$ (94/6) gas mixture (Linde) at the desired pressure. The reactors were suspended in thermostated oil baths and the solutions stirred magnetically. The systems were periodically flushed and recharged with the $CO/CH_4$ mixture in a manner similar to that described above. Gas samples were removed by gas syringe and the compositions were analyzed with methane serving as an internal calibrant, thus allowing for the calculation of the absolute quantities of $H_2$ and $CO_2$ produced and CO consumed. These values were corrected for the small background signals noted when gas samples from control reactions in the absence of added catalyst were analyzed.

Acknowledgements:

This research was supported by the Department of Energy, Office of Basic Energy Sciences. Initial studies on the acidic ruthenium carbonyl catalyst system were carried out by Dr. Charles Ungermann in this group, Professor R.G. Rinker and his research group of the UCSB Chemical Engineering Department contributed significantly to the discussion and interpretation of these results.

Abstract:

Solutions prepared from $Ru_3(CO)_{12}$ in acidic aqueous diglyme solutions are shown to be catalysts for the water gas shift reaction under reasonably mild conditions (100°C, $P_{CO}$=1 atm). This system shows an induction period of about six hours before constant activity is attained during which the $Ru_3(CO)_{12}$ undergoes complete conversion to another ruthenium carbonyl complex. *In situ* nmr studies suggest this species to be the $HRu_2(CO)_8^-$ ion. Kinetic studies show complex rate profiles; however, a key observation is that the catalysis rate is first order in $P_{CO}$ at low pressures ($P_{CO}$<1 atm) but is sharply inhibited by increasing $P_{CO}$ at higher pressures. A catalysis scheme consistent with these observations is proposed.

Literature Cited:

1. Laine, R.M.; Rinker, R.G.; Ford, P.C.; *J. Amer. Chem. Soc.*, 1977, 99, 252.
2. Ungermann, C.; Landis, V.; Moya, S.A.; Cohen, H.; Walker, H.; Pearson, R.G.; Rinker, R.G.; Ford, P.C.; *J. Amer. Chem. Soc.* 1979, 101, 5922.
3. Cheng, C.H.; Hendriksen, D.E.; Eisenberg, R.; *J. Amer. Chem. Soc.*, 1979, 99, 2791.
4. Kang, H.; Mauldin, C.H.; Cole, T.; Slegeir, W.; Cann, K; Pettit, R.; *J. Amer. Chem. Soc* 1977, 99, 8323.
5. King, R.B.; Frazier, C.C.; Hanes, R.M.; King, A.D.; *J. Amer. Chem. Soc.*, 1978, 100, 2925.
6. Cheng, C.H.; Eisenberg, R.; *J. Amer. Chem. Soc*. 1978, 100, 5968.
7. Yoshida, T.; Ueda, Y.; Otsuka, S.; *J. Amer. Chem. Soc.* 1978, 100, 3941.
8. Darensbourg, D.J.; Darensbourg, M.Y.; Burch, R.R.; Froelich, J.A.; Incorvia, M.J.; *ACS Adv. in Chem.*, 1979, 173, 106.
9. Nuzzo, R.G., Feitler, D.; Whitesides, G.M.; *J. Amer. Chem. Soc.*

1979, 101, 3683.
10. Singleton, T.C.; Park, L.J.; Price, J.C.; Forster, D.; Preprints of the Division of Petroleum Chemistry, Amer. Chem. Soc.; 1979, 24, 329.
11. Baker, E.C.; Hendricksen, D.E.; Eisenberg, R., J. Amer. Chem. Soc.; 1980, 102, 1020.
12. King, A.D.; King, R.B.; Yang, D.B.; J. Amer. Chem. Soc.; 1980, 102, 1028.
13. Moya, S.A.; M.A. Dissertation, University of California, Santa Barbara, 1979.
14. Herter, W.; Leutert, F.; Z. Anorg. Allg. Chem., 1932, 204, 145.
15. Ford, P.C.; Rinker, R.G.; Ungermann, C.; Laine, R.M.; Landis, V.; Moya, S.A.; J. Amer. Chem. Soc. 1978, 100, 4595.
16. Ford, P.C.; Rinker, R.G.; Laine, R.M.; Ungermann, C.; Landis, V.; Moya, S.A.; ACS Adv. Chem. Ser. 1979, 173, 81.
17. Suzuki, T.M.; Ford, P.C.; work in progress.
18. Johnson, B.F.G.; Lewis, J.; Twigg, M.V.; J. Organometal. Chem.; 1974, 67, C75.
19. Desrosiers, M; Ford, P.C.; work in progress.
20. Tyler, D.R.; Levenson, R.A.; Gray, H.B.; J. Amer. Chem. Soc., 1978, 100, 7888.
21. Collman, J.P.; Fineke, R.G.; Matlock, P.L.; Wahren, R.; Kanote, R.G.; Brauman, J.I.; J. Amer. Chem. Soc., 1978, 100, 1119.
22. Nagel, C.C.; Shore, S.G.; JCS Chem. Commun., 1980, 530.
23. Johnson, B.F.J.; Raithley, P.R.; Zuccano, C.; J. Chem. Soc., Dalton, 1980, 99.
24. Vancea, L; Graham, W.A.G.; J. Organometal. Chem. 1977, 134, 219.
25. King, R.G.; King, A.D.; Iqbal, M.Z.; Frazier, C.C.; J. Amer. Chem. Soc. 1978, 100, 1687.
26. Walker, H.W.; Kresge, C.; Ford, P.C.; Pearson, R.G.; J. Amer. Chem. Soc. 1979, 101, 7428.
27. Walker, H.W.; Ford, P.C.; unpublished data.
28. Evans, J; Norton, J.R.; J. Amer. Chem. Soc. 1974, 96, 7577.

RECEIVED December 8, 1980.

# The Importance of Reactions of Oxygen Bases with Metal Carbonyl Derivatives in Catalysis

## Homogeneous Catalysis of the Water Gas Shift Reaction

DONALD J. DARENSBOURG and ANDRZEJ ROKICKI[1]

Department of Chemistry, Tulane University, New Orleans, LA 70118

Base catalysis of ligand substitutional processes of metal carbonyl complexes in the presence of oxygen donor bases may be apportioned into two distinct classifications. The first category of reactions involves nucleophilic addition of oxygen bases at the carbon center in metal carbonyls with subsequent oxidation of CO to $CO_2$, eqns. 1 and 2 (1, 2). Secondly, there are

$$Ir_4(CO)_{12} + Me_3NO \rightleftharpoons \left[Ir-\overset{\overset{O}{|}^{NMe_3}}{C}O\right] \longrightarrow Ir_4(CO)_{11}NMe_3 + CO_2 \quad (1)$$

$$Cr(CO)_6 + OH^- \rightleftharpoons \left[Cr-\overset{\overset{O}{|}^H}{C}O\right]^- \longrightarrow Cr(CO)_5H^- + CO_2 \quad (2)$$

reactions involving coordination of the oxygen base, with the thus formed metal-oxygen bond greatly lowering the energetics for dissociative carbon monoxide displacement (eq. 3) (3, 4).

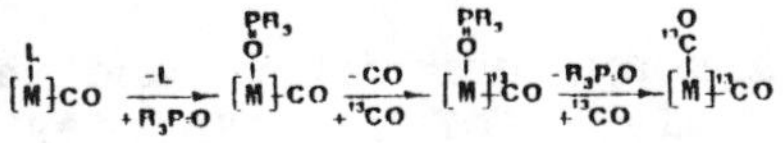

$$[M]CO \xrightarrow[+R_3PO]{-L} [M]CO \xrightarrow[+CO]{-CO} [M]CO \xrightarrow[+CO]{-R_3PO} [M]CO \quad (3)$$

An essential step in processes utilizing soluble transition metal catalysts is the coordination of the substrate to the transition metal (5). A corequisite is the availability of a vacant site in the coordination sphere of the metal for substrate binding, a provision often met by dissociation of a bonded

[1] Current address: Institute of Organic Chemistry and Technology, Technical University of Warsaw (Politechnika), 00-662 Warszawa, ul. Koszykowa 75, Poland.

0097-6156/81/0152-0107$05.00/0
© 1981 American Chemical Society

ligand. Hence, processes such as those described in eqns. 1 and 3 are not only useful from a synthetic viewpoint but also can serve to activate the metal carbonyl in catalytic reactions.

More pertinent to the tenor of this Symposium, the carbon monoxide oxidation reaction depicted in eq. 2 has received renewed attention largely because of its pivotal role during the homogeneous catalysis of the water gas shift reaction (WGSR) by a variety of metal carbonyls (2,6). In this correspondence we wish to discuss reaction processes of importance to the homogeneous catalysis of the WGSR utilizing metal carbonyls. Particular emphasis will be placed on the relative rates of oxygen-exchange vs. metal-hydride bond formation for several metal carbonyls; including group 6b metal carbonyls and derivatives thereof, $Fe(CO)_5$, and $Ru_3(CO)_{12}$. Summarized are our recent investigations on the species present in solution and their reactivity patterns during the homogeneous catalysis of the WGSR by group 6b metal carbonyls under mild reaction conditions (1 atmosphere CO pressure and temperature $\lesssim$ 100°C).

## Results and Discussion

The lability of oxygen atoms in the activated carbon monoxide ligands of $Re(CO)_6^+$ was demonstrated by Muetterties in 1965 (7). Rhenium hexacarbonyl cation underwent a facile exchange process with the oxygen atoms in oxygen-18 enriched water, and the intermediacy of a metallocarboxylic acid was proposed (eq. 4). Deeming and Shaw isolated a metallocarboxylic acid a few years later from the reaction of a

$$Re(CO)_6^+ + H_2O \rightleftharpoons \{Re(CO)_5COOH\} \quad (4)$$

cationic iridium carbonyl derivative and water, and showed that upon pyrolysis of this species the corresponding metal hydride and $CO_2$ were produced (8). In addition, it has long been known that some metal carbonyls readily react with hydroxide ions to yield metal carbonyl hydride anions and $CO_2$, e.g., $Fe(CO)_5 + OH^- \rightarrow HFe(CO)_4^- + CO_2$ (9).

These observations illustrate that there are two transformations open to metallocarboxylic acid intermediates; reversible loss of $OH^-$ accompanied by oxygen exchange, and metal-hydride formation with expulsion of $CO_2$. Our entry into this area of chemistry was in 1975 when extensive studies of oxygen lability in metal carbonyl cations were initiated (10). These investigations were centered around defining the factors that determined whether the oxygen exchange process predominated over the production of metal hydrides and carbon dioxide or vice versa (11, 12, 13, 14). For example, the reaction of $Mn(CO)_6^+$ with water led to metal hydride formation concomitantly with oxygen exchange, with metal hydride production being much slower (eq. 5). Some of the other germane observations noted during these studies were the following:

$$Mn(CO)_6^+ + H_2O \rightleftharpoons \{Mn(CO)_5COOH\} \rightarrow Mn(CO)_5H + CO_2 \qquad (5)$$

(1) The rate of nucleophilic addition of $^-OH$ to the metal bound carbon monoxide ligand, as viewed by oxygen exchange, decreased with increasing substitution at the metal center with electron donating ligands, i.e., $M(CO)_6^+ > M(CO)_5L^+ \gg M(CO)_4L_2^+$. (2) The more electron-rich $L_n(CO)_{5-n}M(COOH)$ intermediates were less disposed to $CO_2$ elimination with M-H bond formation. (3) Metal-hydride formation was enhanced over oxygen exchange as the basicity of the solution increased.

In the case of the bis phosphine derivatives of the group 7b metals, the lability of the oxygen atoms was so markedly retarded by substitution that it was necessary to enhance their reactivity by means of base catalysis (13, 14), a process having mechanistic features common with general base catalysis of the hydration of ketones (eq. 6) (15).

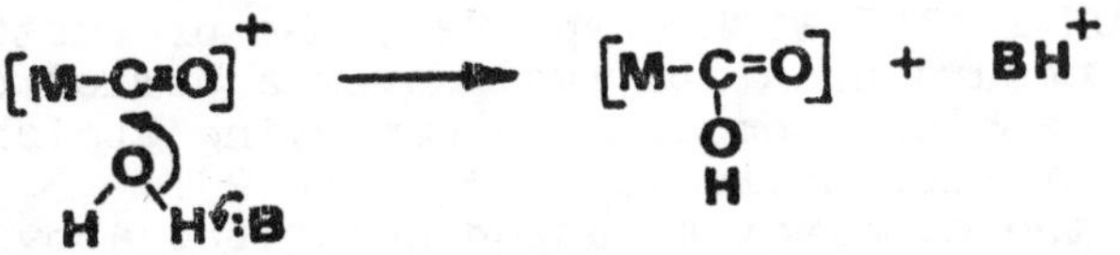

$$[\text{M-C≡O}]^+ \longrightarrow [\text{M-C=O}] + BH^+ \qquad (6)$$

We have exploited this base catalysis of the oxygen exchange process to effect oxygen lability in the less electrophilic carbonyl sites of neutral metal carbonyl species. Because [MCOOH] intermediates are readily decarboxylated in the presence of excess hydroxide ion, in order to observe oxygen exchange processes in neutral metal carbonyl complexes it was convenient to carry out these reactions in a biphasic system employing phase transfer catalysis (<PTC>) (16, 17, 18). Under <PTC> conditions (eq. 7), the

(organic phase) $[MCO] + \underline{n}\text{-Bu}_4N^+OH^- \rightleftharpoons [M\text{-}\overset{O}{\overset{\|}{C}}\text{-OH}]^- \underline{n}\text{-Bu}_4N^+$

$$------\||------ \qquad (7)$$

(aqueous phase) $Na^+I^- + \underline{n}\text{-Bu}_4N^+OH^- \rightleftharpoons Na^+OH^- + \underline{n}\text{-Bu}_4N^+I^-$

hydroxide ion concentration is small in the organic phase which contains the metal carbonyl, since $^{-}OH$ is more highly hydrated than the halide ion. In general, the <PTC> oxygen exchange reactions of neutral mononuclear metal carbonyl species with hydroxide ions paralleled the observations summarized for the cationic species (*vide supra*) with, however, the significant additional developments discussed hereafter.

When the <PTC> reactions of $M(CO)_6$ (M = Cr, Mo, W) with hydroxide ions were carried out under an atmosphere of CO, these systems were observed to generate hydrogen catalytically with a low turnover rate (mol. $H_2$/mol. catalyst per day), on the order of a week at 75° (18). However, in more strongly alkaline solutions (e.g., aqueous KOH in 2-ethoxyethanol) at 100°C the $M(CO)_6$ species were observed to be quite active catalysts for the WGSR with a turnover rate of ~ 30 for $Cr(CO)_6$. The principal metal carbonyl components in an alkaline 2-ethoxyethanol solution of a catalyst prepared from $Cr(CO)_6$ were determined to be $Cr(CO)_6$ and $Cr(CO)_5H^-$ by infrared spectroscopy in the $\nu(CO)$ region (see Figure 1). Identification of the chromium pentacarbonyl monohydride species was based on spectral comparisons with an authentic sample prepared by protonation of $Cr(CO)_5^{=}$ (19). The reaction of hydroxide ions with $Cr(CO)_6$ in aqueous 2-ethoxyethanol to afford $Cr(CO)_5H^-$ occurs over several hours at 40-50°, even in the presence of a large excess of hydroxide. However, under conditions where the $^{-}OH$ is not hydrated, KOH in THF with Crypt 222, the production of $Cr(CO)_5H^-$ occurs quantitatively over a period of a few minutes at ambient temperature employing stoichiometric quantities of reagents (Figure 2).

When the catalyst solution described above (Figure 1A) was prepared using $H_2{}^{18}O$, $Cr(CO)_6$ was shown to undergo the oxygen exchange process at a much faster rate than formation of metal hydride, an observation consistent with the <PTC> results. See Figure 1B where the various $Cr(CO)_{6-n}(C^{18}O)_n$ species are seen in advance of the highly oxygen-18 enriched $Cr(CO)_5H^-$ species. Hence even in the highly alkaline catalyst solution the metallocarboxylic acid derivative has a long enough lifetime to effect oxygen exchange prior to anionic metal hydride production. Indeed in the reaction of $W(CO)_6$ with KOH/Crypt 222 in THF we have preliminary spectral evidence for the existence of $W(CO)_5COOH^-$ in solution at room temperature. The fact that the $W(CO)_5COOH^-$ species is less prone to proceeding to metal-hydride and $CO_2$ than $Cr(CO)_5COOH^-$ is anticipated based on the earlier results obtained on the isoelectronic $Mn(CO)_6{}^+$ and $Re(CO)_6{}^+$ species.

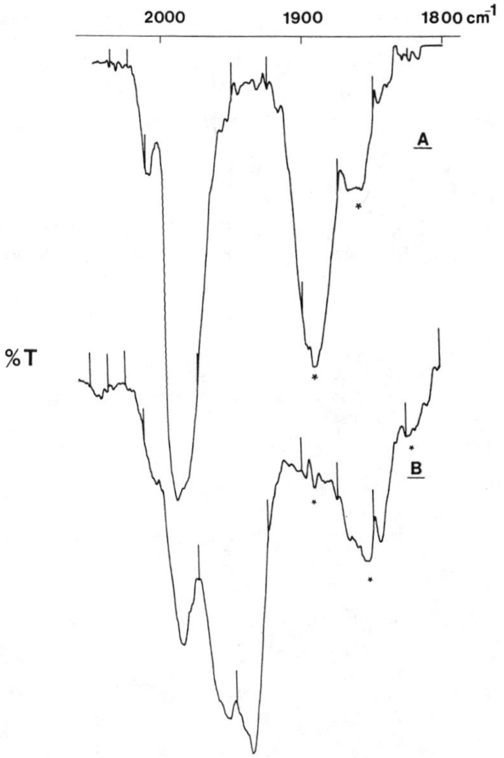


*Figure 1. IR spectra in ν(CO) region of catalyst components; $Cr(CO)_6$ (0.40 mmol), KOH (5.0 mmol), $H_2O$ (11.1 mmol), and 2-ethoxyethanol (15 ml solution). Peaks with asterisks are those assigned to $Cr(CO)_5H^-$, others are for $Cr(CO)_6$: A, $H_2{}^{16}O$, heated at 45°C for 300 min; B, $H_2{}^{18}O$, heated at 45°C for ~ 200 min.*

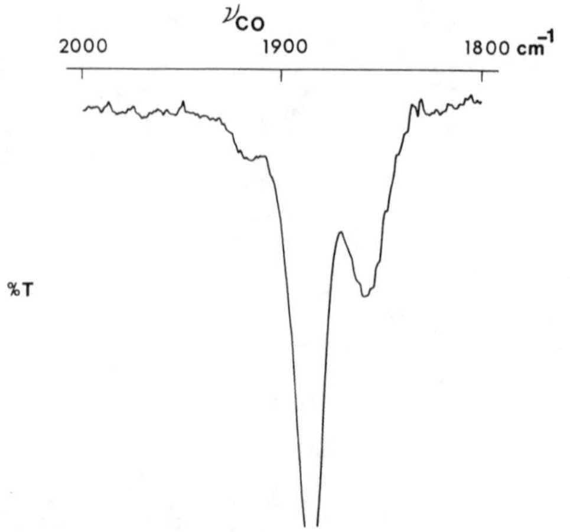


*Figure 2. IR spectrum in ν(CO) region of $Cr(CO)_6$ plus two equivalents of [K-Crypt-222]OH in THF after 5 min at ambient temperature*

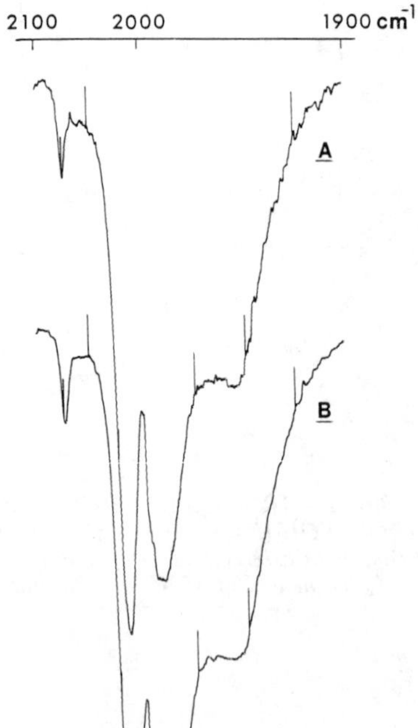


*Figure 3. IR spectra in ν(CO) region of $[Et_4N][HRu_3(CO)_{11}]$ in $CH_2Cl_2$: A, sample prepared using $H_2{}^{16}O$; B, sample prepared using $H_2{}^{18}O$.*

In contrast to the group 6b metal carbonyls the group 8 metal carbonyls which have been successfully employed as catalysts for the WGSR, $Fe(CO)_5$ and $Ru_3(CO)_{12}$, exhibited little tendency to undergo oxygen exchange prior to affording metal hydrides upon reaction with hydroxide ion (17, 20). This is illustrated by the synthesis of $[Et_4N][HRu_3(CO)_{11}]$ from $Ru_3(CO)_{12}$ and aqueous KOH in 2-ethoxyethanol under a CO atmosphere in both $H_2{}^{16}O$ and $H_2{}^{18}O$ where no significant difference in oxygen isotope composition of the purified products was observed (see Figure 3).

The Scheme depicts our most comprehensive understanding of the processes operative during the WGSR catalyzed by group 6b metal carbonyls at temperature $\lesssim$ 100°C. As noted in the Scheme when the reaction is carried out in the presence of $^{13}CO$ both $Cr(CO)_6$ and $Cr(CO)_5H^-$ are enriched in $^{13}C$-carbon monoxide, the latter species to a greater extent (see Figure 4). This is a useful occurrence for it commodiously allows

SCHEME

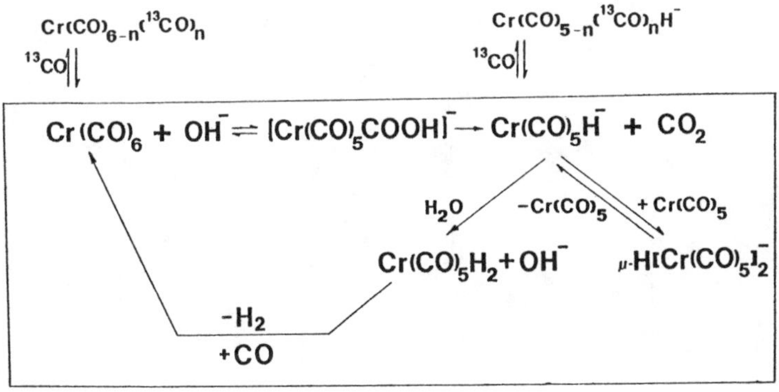


for the examination of the catalytic solution by $^{13}C$ NMR.

The catalyst system is not poisoned by thiophene, for $Cr(CO)_5SR_2$ species readily revert to $Cr(CO)_6$ in the presence of CO. Similarly, the group 6b metal hexacarbonyls have also been shown to be active in the presence of NaSH (21). However, this catalyst system

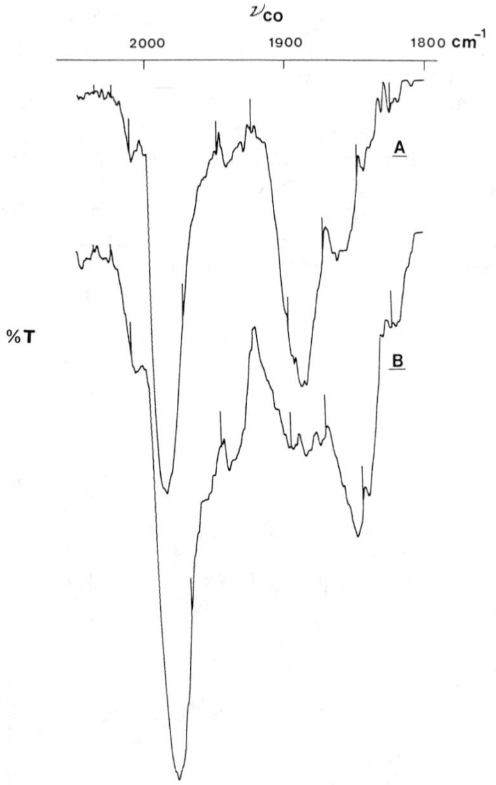


*Figure 4. IR spectra in ν(CO) region of catalyst components as in Figure 1A: A, under $^{12}CO$ atmosphere; B, under $^{13}CO$ (93%) atmosphere. Peaks above 1900 $cm^{-1}$ are attributable to $Cr(CO)_6$ and those below 1900 $cm^{-1}$ are assigned to $Cr(CO)_5H^-$.*

requires completely anaerobic conditions or the catalyst is irreversibly converted to $Cr_2O_3$. Hence the CO gas stream is scrubbed for removal of trace oxygen by passing it through a column containing MnO supported on silica gel. This is of particular importance in our system which employs a continuous, constant carbon monoxide gas flow coupled to an online gas sampling valve for efficient chromatographic monitoring of $H_2$ and CO.

Figure 5 displays a typical time dependent trace of the hydrogen production during catalysis of the WGSR by $Cr(CO)_6$. The decrease in activity of mature catalyst solutions is due to the consumption of KOH by $CO_2$, i.e., the formation of bicarbonate ($CO_2 + OH^- \rightleftharpoons HCO_3^-$). Reaction solutions prepared from $Cr(CO)_6$ with $KHCO_3$ as the added alkaline were much less active than their KOH counterparts. Experiments are planned at higher reaction temperatures in an effort to minimize this behavior. However, at 100° the $Cr(CO)_6$ catalyst is quite active for the decomposition of formate ion to $H_2$ plus $CO_2$ (*vide infra*).

As seen in Figure 5 addition of excess triphenylphosphine to an active catalyst solution results in a rapid quenching of $H_2$ production. The chromium carbonyl components in solution were rapidly converted to $Cr(CO)_5PPh_3$ and *trans*-$Cr(CO)_4[PPh_3]_2$ (22). Consistent with these observations aqueous alkaline solutions of $Cr(CO)_5PPh_3$ or *trans*-$Cr(CO)_4[PPh_3]_2$ in ethoxyethanol exhibited no activity for WGSR catalysis. There is a trace of $H_2$ produced due to the production of $Cr(CO)_6$ from loss of $PPh_3$ in $Cr(CO)_5PPh_3$ (23). Nevertheless, phase transfer catalyzed reactions of oxygen-18 labelled $^-OH$ with a large variety of group 6b metal carbonyl phosphine and phosphite complexes have demonstrated that nucleophilic addition of base to the carbonyl centers in these derivatives does indeed occur, albeit slower than in the parent hexacarbonyls (16,18). For example, Figures 6 and 7 illustrate $\nu(CO)$ and $^{13}C$ NMR spectra for a representative derivative, $Mo(CO)_5P(OMe)_3$, which has undergone extensive oxygen exchange with hydroxide ions for a reaction carried out under <PTC> conditions as defined in eq. 7. Hence, the failure of these phosphine substituted metal carbonyls to function as catalysts for the WGSR is due to their reluctance to afford metal-hydrides from reactions with alkali, i.e., $P\text{-}M\text{-}CO + OH^- \rightleftharpoons [P\text{-}M\text{-}COOH]^- \not\rightarrow P\text{-}M\text{-}H^-$.

Figures 6 and 7 also point out the two principal techniques we have employed in monitoring these oxygen exchange reactions, namely, frequency shifts in the

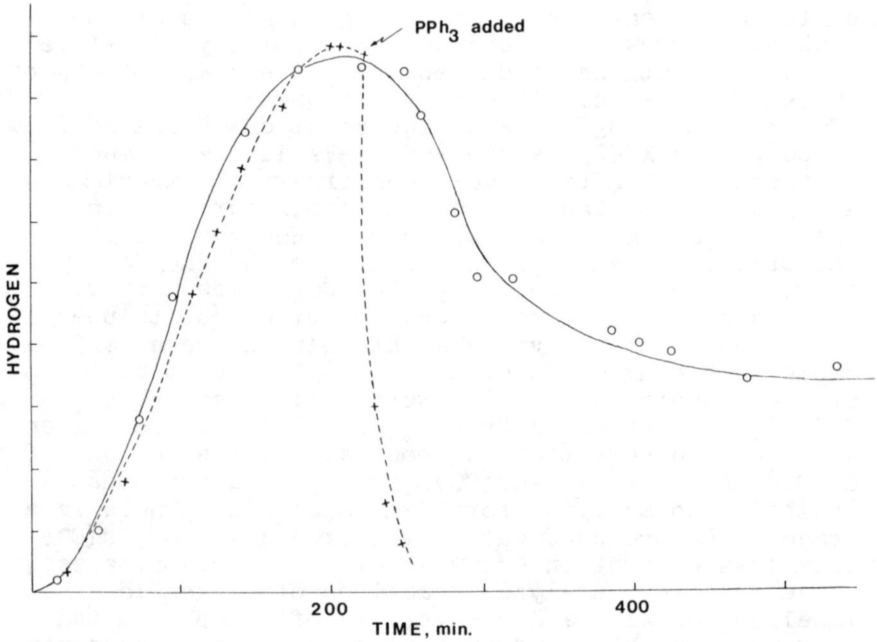


*Figure 5. Time-dependence of hydrogen in CO gas stream for a catalyst solution prepared from $Cr(CO)_6$ (7 mmol), KOH (87.5 mmol), $H_2O$ (195 mmol), and 2-ethoxyethanol (87.5 mL of solution) operated at 100°C. The units for the quantity of $H_2$ are arbitrary, representing the relative areas of the hydrogen peak in the chromatograms as a function of time.*

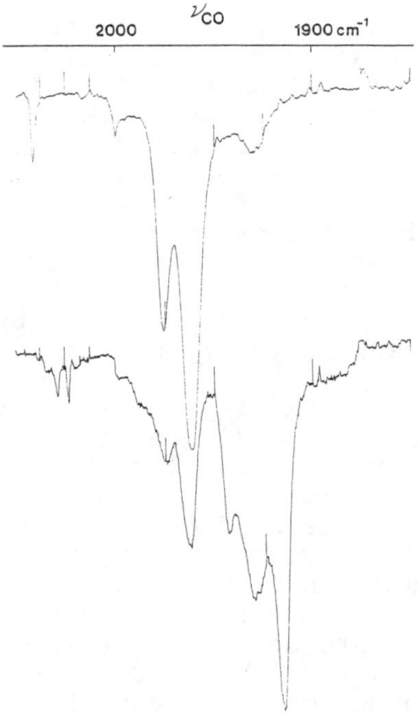


*Figure 6. IR spectral traces in the ν(CO) region during the monitoring of oxygen-18 incorporation into $Mo(CO)_5P(OMe)_3$ under <PTC> conditions (spectra observed in hexane solution): A, initial spectrum; B, after extensive oxygen exchange*

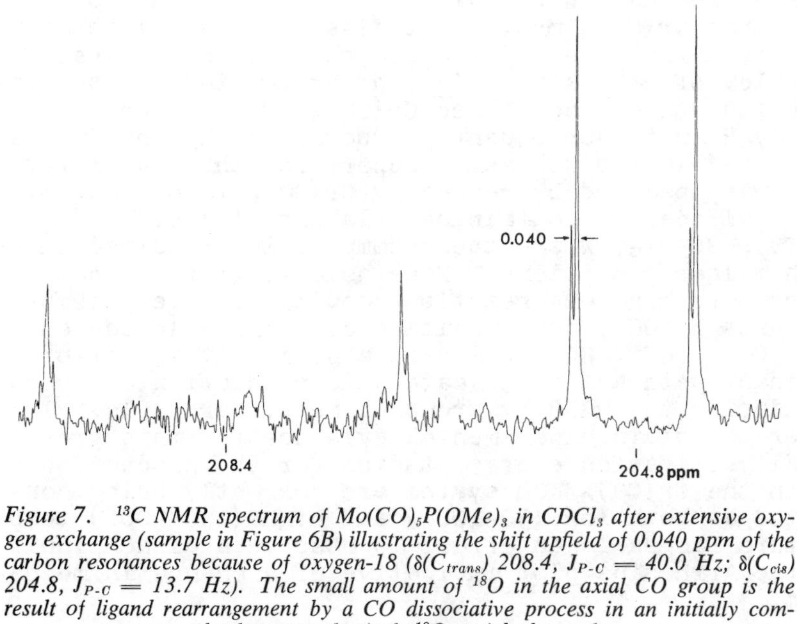


*Figure 7. $^{13}C$ NMR spectrum of $Mo(CO)_5P(OMe)_3$ in $CDCl_3$ after extensive oxygen exchange (sample in Figure 6B) illustrating the shift upfield of 0.040 ppm of the carbon resonances because of oxygen-18 ($\delta(C_{trans})$ 208.4, $J_{P\text{-}C}$ = 40.0 Hz; $\delta(C_{cis})$ 204.8, $J_{P\text{-}C}$ = 13.7 Hz). The small amount of $^{18}O$ in the axial CO group is the result of ligand rearrangement by a CO dissociative process in an initially completely stereoselectively $^{18}O$-enriched sample.*

$\nu$(CO) infrared spectra and the small upfield shifts in the natural abundance $^{13}C$ NMR resonances caused by oxygen-18. It is worthwhile reiterating here that in substituted metal-carbonyl derivatives, where electronically different carbonyl ligands are present, the oxygen-exchange reaction has been shown to occur preferentially at the more electrophilic carbon site, e.g., at the cis carbon monoxide ligands in $M(CO)_5L$ derivatives (18). Several of these thus formed stereoselectively oxygen-18 labelled species have been found to subsequently undergo ligand rearrangements by either non-dissociative or dissociative processes (18, 24).

We would like at this time to amplify the earlier brief comments on the catalytic decomposition of formate ion to $H_2$ and $CO_2$ by the group 6b metal carbonyls. This process requires the presence of a vacant coordination site on the metal for formate binding, i.e., formation of $M(CO)_5O_2CH^-$ species. Consequently, the reaction of $Cr(CO)_6$ with formate ion in 2-ethoxyethanol was found to take place under more rigorous conditions than those needed for the production of $H_2$ in the $Cr(CO)_6$/KOH system. That is, whereas $Cr(CO)_6$ in aqueous KOH/2-ethoxyethanol commences to produce $H_2$ at ~ 50°C (where the rds appears to succeed the formation of $Cr(CO)_5H^-$), the analogous $Cr(CO)_6$/formate ion reaction necessitates temperatures approaching 100°C. This difference in behavior is felt to be due to the fact that $Cr(CO)_6$ must first dissociatively lose a CO ligand, a fairly energetic process with an activation enthalpy of ~ 40 kcal. (25), prior to binding the formate ion. The once formed $Cr(CO)_5O_2CH^-$ proceeds to $Cr(CO)_5H^-$ with subsequent production of $H_2$ and $CO_2$ as described in the Scheme. Support for this interpretation was obtained by replacing $Cr(CO)_6$ with a chromium carbonyl species containing a labile ligand, $Cr(CO)_5NHC_5H_{10}$, where the decomposition occurred under much milder conditions. Nonetheless, we feel that under more rigorous reaction conditions (i.e., temperatures > 100°) decomposition of formate (produced from $CO + OH^- \rightleftharpoons HCO_2^-$) may be a prominent if not predominant pathway in the catalytic production of hydrogen during the WGSR by group 6b metal carbonyls. In order to obtain experimental evidence for this proposal, activation energy studies for the production of $H_2$ in the $Cr(CO)_6$/KOH system are currently being conducted at both low (50-100°) and high (150-200°) temperature ranges in an effort to observe a discontinuity in the Arrhenius plot indicative of a change in mechanism.

It is likely that during the decomposition of $M(CO)_5O_2CH^-$ (as indeed may also be the case for $M(CO)_5COOH^-$ (26)) loss of a CO ligand is requisite prior to hydride transfer to the metal with concomitant expulsion of carbon dioxide. Pertinent to this question we have obtained some initial results on the decomposition of $\eta^5$-$C_5H_5Fe(CO)_2O_2CH$ to $[\eta^5$-$C_5H_5Fe(CO)_2]_2$ plus $H_2$ and $CO_2$, a process presumably proceeding *via* the unstable $\eta^5$-$C_5H_5Fe(CO)_2H$ intermediate. The $\eta^5$-$C_5H_5Fe(CO)_2O_2CH$ derivative was observed to exchange CO ligands with free $^{13}CO$ in hydrocarbon solution at 50° at a rate faster than the formation of $[\eta^5$-$C_5H_5Fe(CO)_2]_2$. On the other hand, Pettit and co-workers (27) have reported that the metallocarboxylic acid analog, $\eta^5$-$C_5H_5Fe(CO)_2COOH$, spontaneously decomposes to $[\eta^5$-$C_5H_5Fe(CO)_2]_2$ plus $H_2$ and $CO_2$ at ambient temperature. Hence the pattern that is beginning to emerge is that of the two intermediates capable of affording $H_2$ and $CO_2$ described in eq. 8, the metallocarboxylic acid pathway is energetically more favorable.

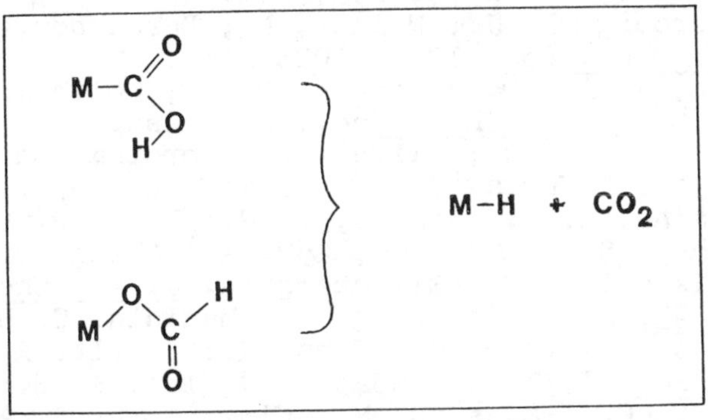


(8)

The observation of CO lability in the $\eta^5$-$C_5H_5Fe(CO)_2O_2CH$ derivative is an instance of the role of oxygen bases in catalysis as described in eq. 3. Other examples of metal-oxygen base bond formation resulting in labilizing CO dissociation are seen in our tri-*n*-butylphosphine oxide work (3, 28) and, in complexes quite similar to the $M(CO)_5O_2CH^-$ species discussed above, $M(CO)_5O_2CR^-$ (M = Cr, Mo, W; R = $CH_3$, $CF_3$). The latter derivatives have been shown to undergo rapid preferential exchange of equatorial CO ligands with free $^{13}CO$ in solution at ambient temperature (4, 29). These anionic metal carbonyl acetate derivatives may prove to be very active catalysts for a variety of processes, including disproportionation of olefins; i.e., providing a homogeneous analog to catalysis of olefin disproportionation by $W(CO)_6$ on alumina.

Acknowledgements

The financial support of this research by the National Science Foundation (Grants CHE 78-01758 and CHE 80-09233) is greatly appreciated. The authors are also grateful to Professor Marcetta Y. Darensbourg and Mr. Joseph Deaton for helpful discussions and for making available their very important results on the isolation and characterization of the $[Cr(CO)_5H][Et_4N]$ derivative. The expert secretarial assistance of Mrs. Helen George is gratefully acknowledged.

Literature Cited

1. Johnson, B. F. G.; Lewis, J.; Raithby, P. P.; Zuccaro, C. J. C. S. Chem. Commun. 1979, 916.
2. Darensbourg, D. J.; Darensbourg, M. Y.; Burch, R. R., Jr.; Froelich, J. A.; Incorvia, M. J. Adv. Chem. Ser. 1979, 173, 106.
3. Darensbourg, D. J.; Walker, N.; Darensbourg, M. Y. J. Am. Chem. Soc. 1980, 102, 1213.
4. Cotton, F. A.; Darensbourg, D. J.; Kolthammer, B. W. S. J. Am. Chem. Soc., in press.
5. Henrici-Olivé, G.; Olivé, S. "Coordination and Catalysis," Verlag Chemie, New York, 1977.
6. (a) Laine, R. M.; Rinker, R. G.; Ford, P. C. J. Am. Chem. Soc. 1977, 99, 252. (b) Cheng, C. H.; Hendriksen, D. E.; Eisenberg, R. J. Am. Chem. Soc. 1977, 99, 2791. (c) Kang, H.; Mauldin, C. H.; Cole, T.; Slegeir, W.; Cann, K.; Pettit, R. J. Am. Chem. Soc. 1977, 99, 8323. (d) King, R. B.; Frazier, C. C.; Hanes, R. M.; King, A. D. J. Am. Chem. Soc. 1978, 100, 2925. (e) Frazier, C. C.; Hanes, R.; King, A. D.; King, R. B. Adv. Chem. Ser. 1979, 173, 94. (f) Pettit, R.; Cann, K.; Cole, T.; Maudlin, C. H.; Slegeir, W. Adv. Chem. Ser. 1979, 173, 121. (g) Singleton, T. C.; Park, L. J.; Price, J. C.; Forster, D. Preprints of the Div. of Petroleum Chem., Am. Chem. Soc. 1979, 24, 329. (h) Laine, R. M. J. Am. Chem. Soc. 1978, 100, 6451. (i) Baker, E. C.; Hendriksen, D. E.; Eisenberg, R. J. Am. Chem. Soc. 1980, 102, 1020. (j) King, A. D.; King, R. B.; Yang, D. B. J. Am. Chem. Soc. 1980, 102, 1028. (k) Ungermann, C.; Landis, V.; Moya, S. A.; Cohen, H.; Walker, H.; Pearson, R. G.; Rinker, R. G.; Ford, P. C. J. Am. Chem. Soc. 1979, 101, 5922.

7. Muetterties, E. L. Inorg. Chem. 1965, 4, 1841.
8. Deeming, A. J.; Shaw, B. L. J. Chem. Soc. A 1969, 443.
9. See, e.g., the review by F. Calderazzo in Wender, I.; Pino, P. "Metal Carbonyls in Organic Synthesis", Interscience, New York, 1968.
10. Darensbourg, D. J.; Drew, D. J. Am. Chem. Soc. 1976, 98, 275.
11. Darensbourg, D. J.; Froelich, J. A. J. Am. Chem. Soc. 1977, 99, 4726.
12. Darensbourg, D. J. Isr. J. Chem. 1977, 15, 247.
13. Darensbourg, D. J.; Froelich, J. A. J. Am. Chem. Soc. 1977, 99, 5940.
14. Darensbourg, D. J.; Froelich, J. A. Inorg. Chem. 1978, 17, 3300.
15. Bell, R. P. Adv. Phys. Org. Chem. 1966, 4, 1.
16. Darensbourg, D. J.; Froelich, J. A. J. Am. Chem. Soc. 1978, 100, 338.
17. Darensbourg, D. J.; Darensbourg, M. Y.; Walker, N.; Froelich, J. A.; Barros, H. L. C. Inorg. Chem. 1979, 18, 1401.
18. Darensbourg, D. J.; Baldwin, B. J.; Froelich, J. A. J. Am. Chem. Soc. 1980, 102, 4688.
19. Darensbourg, M. Y.; Deaton, J. Inorg. Chem., submitted for publication.
20. Darensbourg, D. J.; Fischer, M., unpublished results.
21. King, A. D.; King, R. B.; Yang, D. B. J. Chem. Soc., Chem. Commun. 1980, 529.
22. (a) Darensbourg, M. Y.; Walker, N. J. Organomet. Chem. 1976, 117, C68. (b) Darensbourg, M. Y.; Walker, N.; Burch, R. R., Jr. Inorg. Chem. 1978, 17, 52.
23. Wovkulich, M. J.; Atwood, J. D. J. Organomet. Chem. 1979, 184, 77.
24. Darensbourg, D. J.; Baldwin, B. J. J. Am. Chem. Soc. 1979, 101, 6447.
25. Angelici, R. J. Organomet. Chem. Rev. 1968, 3, 173.
26. Brown, T. L.; Bellus, P. A. Inorg. Chem. 1978, 17, 3727.
27. Crice, N.; Kao, S. C.; Pettit, R. J. Am. Chem. Soc. 1979, 101, 1627.
28. Darensbourg, D. J.; Darensbourg, M. Y.; Walker, N. Inorg. Chem., submitted for publication.
29. Cotton, F. A.; Darensbourg, D. J.; Kolthammer, B. W. S., to be submitted for publication.


RECEIVED December 8, 1980.

# 8

# Homogeneous Catalysis of the Water Gas Shift Reaction Using Simple Mononuclear Carbonyls

R. B. KING, A. D. KING, JR., and D. B. YANG

Department of Chemistry, University of Georgia, Athens, GA 30602

The water gas shift reaction is used extensively in industry to increase the hydrogen content of water gas (synthesis gas) through the reaction of carbon monoxide (CO) with water according to the following equation:

$$CO + H_2O \rightleftharpoons H_2 + CO_2 \quad (1)$$

Current industrial practice for carrying out this reaction involves heterogeneous catalysts at relatively high temperatures, e.g. $Fe_2O_3/Cr_2O_3$ above 300°C (1). However, relatively recent work has shown that the water gas shift reaction can also be carried out at considerably lower temperatures (below 200°C) using various metal carbonyl complexes as homogeneous catalysts. Thus a variety of platinum metal derivatives are active water gas shift reaction catalysts including ruthenium carbonyls (2, 3, 4), rhodium carbonyls (3,5,6,7), platinum-tin complexes (8), and phosphine-platinum(0) complexes (9). In 1978 we reported (10) that several carbonyl derivatives of more abundant metals (iron, chromium, molybdenum, and tungsten) reacted with base to give active water gas shift reaction catalysts. Subsequent work led to a detailed study on the kinetics of the water gas shift reaction catalyzed by $Fe(CO)_5$ in the presence of base (11). More recently we have extended such detailed kinetic studies to similar catalysts derived from the Group VI metal carbonyls $M(CO)_6$ (M = Cr, Mo, and W) (12). This paper summarizes the results obtained with the mononuclear carbonyls of iron and the Group VI transition metals and compares the kinetics of these two water gas shift catalyst systems.

## Experimental Techniques

The water gas shift reactions were carried out in vertically mounted type 304 stainless steel autoclaves having an internal volume of 700 ml. The autoclaves were heated electrically and

0097-6156/81/0152-0123$05.00/0
© 1981 American Chemical Society

stirred magnetically using a Teflon-coated magnetic stirring bar. The temperature was controlled to within ±1°C by a proportional controller with a thermocouple sensor mounted in a thermocouple well extending into the interior of the autoclave. A digital thermocouple read-out meter provided a continuous temperature reading. The pressure was monitored using a 0-3000 psig test gauge accurate to 0.25% of full scale which was attached to each autoclave through the closure at the top.

Analyses of the gases in the bomb (CO, $CO_2$, Ar, $H_2$) were performed using a Fisher Model 1200 gas partitioner with a 6.5 ft. 80-100 mesh Columnpak PQ column and an 11 ft 13 X molecular sieve column in series. Helium was used as a carrier gas in all determinations. Care was taken to insure that all hydrogen analyses were performed at concentrations within the linear response region of the sensitivity curve for this gas. Argon was used as an internal standard. A Varian CDS-III digital integrator was used to integrate the output from the gas partitioner. Gas samples were taken by releasing a portion of the interior gas mixture into a sample chamber which uses a small balloon to maintain a low positive pressure against a septum. This chamber was purged three times with the gas mixture before removing a sample for injection into the gas partitioner by means of a Pressure-Lok syringe.

The following two methods were used to compute gas compositions expressed as partial pressures after determining the external sensitivity factors for each gas:

(a) The pressure in the autoclave was recorded at the time the gas sample was taken. Subtraction of the previously determined solvent vapor pressure gave the total pressure of non-condensible gases. Dalton's law was then used to determine the individual partial pressures of the three gases of interest (CO, $CO_2$, $H_2$) using the gas mole fractions obtained in the gas analysis.

(b) The CO initially loaded in the autoclave was mixed with argon for use as an internal standard. The composition of this gas mixture was then checked by a gas analysis. The resulting computed partial pressure of argon was corrected to the elevated temperatures at which the kinetic data were obtained, thereby allowing partial pressures of the gases of interest to be computed directly using argon as an internal standard.

The gas phase compositions obtained using methods (a) and (b) agreed with each other in every instance. Further details on the experimental techniques used in this work are given elsewhere (11,12).

## Results

In a typical experiment the autoclave was loaded with a methanol-water solution containing dissolved base (potassium hydroxide or sodium formate) and metal carbonyl ($Fe(CO)_5$ or $M(CO)_6$ where M = Cr, Mo, or W) and charged with a CO/argon gas mixture. The autoclave was then heated rapidly with stirring to the desired temperature where periodic pressure readings and gas analyses were made. Since one mole of $H_2$ was produced for each mole of CO consumed in equation 1, the partial pressure of $CO + H_2$ remained essentially constant throughout the reaction except for minor sampling losses. This indicates the absence of significant side reactions consuming CO without producing $H_2$. However, in the experiments using KOH as the base the initial pressure of CO or $CO + H_2$ at the reaction temperature was less than the loading pressure of CO by an amount corresponding to the quantitative formation of formate according to the following equation:

$$CO + OH^- \longrightarrow HCO_2^- \qquad (2)$$

This explanation for this pressure discrepancy was verified by experiments using different amounts of KOH and by the lack of such pressure discrepancies in experiments using formate as a base. These observations clearly indicate that when a base stronger than formate is used in the water gas shift reaction, the actual base present in the reaction is formate produced according to equation 2. During the course of a typical water gas shift reaction catalyzed by $Fe(CO)_5$ in the presence of base the pH starts at 8.6 and falls gradually to 7.4 as determined from fresh liquid samples withdrawn periodically from the bomb. Thus the formate acts as a buffer to keep the pH of the water gas shift reaction system in a relatively narrow range almost independent of the amount of base (e.g. KOH) originally loaded into the autoclave.

The water gas shift reactions were also run in methanol-water mixtures of varying compositions using KOH as the base. In the case of the system derived from $Fe(CO)_5$, a 25% water-75% methanol mixture gave the fastest rate (11) whereas in the cases of the systems derived from $M(CO)_6$ (M = Cr, Mo, and W), a 10% water-90% methanol mixture gave the fastest rates (12). These optimum methanol-water mixtures as solvents for the water gas shift reactions represent compromises between a high concentration of the reactant water and a high concentration of methanol to solubilize the CO and metal carbonyls. Furthermore, all of the solvent mixtures used in this work contain amounts of water which are large relative to that consumed in the water gas shift reaction. Therefore, the concentration of water may be regarded as a constant during the water gas shift reactions conducted in this research project.

The rates of the water gas shift reactions were compared using different amounts of the mononuclear metal carbonyl precursor for all four cases ($Fe(CO)_5$ and $M(CO)_6$ where M = Cr, Mo, and W). In all cases the rates of hydrogen production were found to double as the concentration of the metal carbonyl was doubled. Thus all of the water gas shift reactions investigated in this work are first order with respect to the mononuclear metal carbonyl precursor.

Major differences were noted between the systems derived from $Fe(CO)_5$ and $M(CO)_6$ (M = Cr, Mo, and W) with respect to the effect of the base concentration on the reaction rate. Thus in the case of the catalyst system derived from $Fe(CO)_5$ tripling the amount of KOH while keeping constant the amounts of the other reactants had no significant effect on the rate of $H_2$ production (11). However, in the case of the catalyst system derived from $W(CO)_6$ the rate of $H_2$ production increased as the amount of base was increased regardless of whether the base was KOH, sodium formate, or triethylamine (12). This increase may be interpreted as a first order dependence on base concentration provided some solution non-ideality is assumed at high base concentrations. Similar observations were made for the base dependence of $H_2$ production in catalyst systems derived from the other metal hexacarbonyls $Cr(CO)_6$ and $Mo(CO)_6$ (12). Thus the water gas shift catalyst system derived from $Fe(CO)_5$ has an apparent zero order base dependence whereas the water gas shift catalyst systems derived from $M(CO)_6$ (M = Cr, Mo, and W) have an approximate first order base dependence. Any serious mechanistic proposals must accommodate these observations.

Major differences were also noted between the catalyst systems derived from $Fe(CO)_5$ and those derived from $M(CO)_6$ (M = Cr, Mo, and W) with respect to the effect of CO pressure on the reaction rate. In the system derived from $Fe(CO)_5$ the rate of $H_2$ production in the early stages of the reaction was independent of the CO loading pressure in the range 10 to 40 atmospheres (11). However, $H_2$ production using this catalyst system was found to cease abruptly when enough CO was consumed so that the CO partial pressure fell to a threshold value between 3 and 7 atmospheres. Two independent experiments conducted with CO loading pressures around 1 atmosphere indicated excess $H_2$ productions relative to the CO consumed of $5.7 \pm 0.1$ mole $H_2$/mole $Fe(CO)_5$. These observations indicate that a minimum threshold pressure of CO is needed in order to prevent the iron carbonyl catalyst system from decomposing to carbonyl-free catalytically inactive iron(II) derivatives (11). The observation that 5.7 moles of excess $H_2$ are produced for each mole of $Fe(CO)_5$ may be interpreted on the basis of $Fe(CO)_5$ acting as an average 11.4 electron reducing agent for water under the reaction conditions in accord with reported (13) observations that $Fe(CO)_5$ in alkaline solution is an average 10.8 electron reducing agent for the reduction of nitrobenzene to aniline.

The rates of the water gas shift reactions catalyzed by the systems derived from $M(CO)_6$ (M = Cr, Mo, and W) were found to be inversely proportional to the CO pressure as indicated by straight line plots of rates of $H_2$ production versus $(1/P_{CO})_{init}$ (12). Furthermore, the catalysts derived from $M(CO)_6$ (M = Cr, Mo, and W) retain their catalytic activities at lower CO pressures than the catalyst derived from $Fe(CO)_5$. Thus, the water gas shift catalysts derived from $M(CO)_6$ (M = Cr, Mo, and W) appear to be more robust than those derived from $Fe(CO)_5$.

The increased chemical stability of the catalyst systems derived from $M(CO)_6$ relative to those derived from $Fe(CO)_5$ also result in an increased tolerance for sulfur, an important characteristic for a practical water gas shift catalyst system because of the possibility of using synthesis gas feedstocks derived from high sulfur coals. In order to evaluate the sulfur tolerance of water gas shift catalyst systems, the catalytic reactions were carried out as above but using sodium sulfide rather than potassium hydroxide or sodium formate as the base to generate the catalytically active species (14). Aqueous sodium sulfide is a strong enough base to generate formate through the following reactions:

$$S^{2-} + H_2O + CO \rightleftharpoons HS^- + HCO_2^- \qquad (3a)$$

$$HS^- + H_2O + CO \rightleftharpoons H_2S + HCO_2^- \qquad (3b)$$

The $H_2S$ by-product, representing a relatively reduced form of sulfur, is a reasonable model for the sulfur impurities in the synthesis gas obtained from sulfur-rich coal. This sodium sulfide test of sulfur resistance of water gas shift catalyst systems generated in basic solutions is a very severe test since the quantities of sulfur involved are much larger than those likely to be found in synthesis gas made from any sulfur-rich coals.

The water gas shift catalyst system derived from $Fe(CO)_5$ was found to be relatively sensitive towards sulfur poisoning since an aqueous methanol solution generated from sodium sulfide and $Fe(CO)_5$ using an S/Fe ratio of 26 was completely inactive as a catalyst for the water gas shift reaction. This sulfur poisoning of the $Fe(CO)_5$ catalyst system may arise from the formation of the iron carbonyl sulfide $Fe_3(CO)_9S_2$ which was detected in these reaction mixtures (14). However, aqueous methanol solutions generated from sodium sulfide and the metal hexacarbonyls $M(CO)_6$ (M = Cr, Mo, and W) retained 21% (M = Cr) to 67% (M = W) of the catalytic activity of the corresponding catalyst systems generated from KOH and the same metal hexacarbonyls even when S/M ratios as high as 400 were used. Thus the water gas shift catalyst systems derived from $M(CO)_6$, particularly M = W, have a high sulfur tolerance. Therefore they are potentially very useful for processing synthesis gas derived from high sulfur coals.

The activation energies of these water gas shift catalyst systems were determined by rate measurements as a function of temperature. Thus on the basis of rate measurements at the five temperatures 180, 160, 150, 140, and 130°C the activation energy of the catalyst system derived from $Fe(CO)_5$ was estimated at

22 kcal/mole. Similar measurements for the catalyst systems derived from $M(CO)_6$ gave estimated activation energies of 35, 35, and 32 kcal/mole for M = Cr, Mo, and W, respectively. These latter numbers are roughly similar to the activation energies of 39, 31, and 40 kcal/mole reported (15) for the replacement of CO by triphenylphosphine for $Cr(CO)_6$, $Mo(CO)_6$, and $W(CO)_6$, respectively.

The combined results of these studies are summarized in Table I.

## Discussion

The following catalytic cycle (3,13) is capable of explaining our experimental observations (11) on the water gas shift reaction catalyzed by $Fe(CO)_5$ in the presence of a base:

$$Fe(CO)_5 + OH^- \xrightarrow{k_1} Fe(CO)_4CO_2H^- \quad (4a)$$

$$Fe(CO)_4CO_2H^- \xrightarrow{k_2} HFe(CO)_4^- + CO_2 \quad (4b)$$

$$HFe(CO)_4^- + H_2O \xrightarrow{k_3} H_2Fe(CO)_4 + OH^- \quad (4c)$$

$$H_2Fe(CO)_4 \xrightarrow{k_4} Fe(CO)_4 + H_2 \quad (4d)$$

$$Fe(CO)_4 + CO \xrightarrow{k_5} Fe(CO)_5 \quad (4e)$$

The nucleophilic attack of a metal-bonded carbonyl group with hydroxide to give a metal-bonded carboxyl group (equation 4a) is well established in metal carbonyl cation chemistry (16,17,18) and has reasonable experimental support from studies on the rhodium carbonyl iodide catalyzed water gas shift reaction (3). Furthermore, carboxyl groups directly bonded to transition metals similar to that in the proposed intermediate $Fe(CO)_4CO_2H^-$ are well known (17,18,19,20) to undergo facile decarboxylation to give the corresponding metal hydride exactly as in equation 4b. The $H_2Fe(CO)_4$ intermediate formed by protonation of $HFe(CO)_4^-$ (equation 4c) is necessary to account for the observation made by Pettit and coworkers (3) that both Reppe hydroformylations and water gas shift reactions catalyzed by iron carbonyls do not proceed to any measurable extent at a pH greater than 10.7. The remaining steps of the catalytic cycle involve reductive elimination of $H_2$ (equation 4d) and coordinative saturation (equation 4e) completely analogous to steps found in many types of catalytic cycles (21).

A standard kinetic analysis of the mechanism 4a-4e using the steady state approximation yields a rate equation consistent with the experimental observations. Thus since equations 4a to 4e form a catalytic cycle their reaction rates must be equal for the catalytic system to be balanced. The rate of $H_2$ production

Table I. Comparison of Catalyst Systems Derived from $Fe(CO)_5$ and $M(CO)_6$ (M = Cr, Mo, W).

| | $Fe(CO)_5$ | $Cr(CO)_6$ | $Mo(CO)_6$ | $W(CO)_6$ |
|---|---|---|---|---|
| Optimum Solvent Composition (% $H_2O$, V/V in methanol) | 25 | 10 | 10 | 10 |
| Rate dependence on metal carbonyl concentration. | first order | first order | first order | first order |
| Rate dependence on carbon monoxide pressure. | zero order | inverse first order | inverse first order | inverse first order |
| Rate dependence on added base (formate) concentration. | zero order | approximately first order | approximately first order | approximately first order |
| Temperature dependence of rate expressed as activation energy (kcal/mole) | 22 | 35 | 35 | 32 |
| Minimum CO pressure required to maintain catalyst activity (atm) | 3-7 | none | none | none |
| Catalytic activity in presence of sulfide ion (expressed as percent of activity under sulfur free conditions) | none | 21 | 59 | 67 |

by equation 4d (namely $k_4[H_2Fe(CO)_4]$) must be equal to the rate of $CO_2$ production by equation 4b (namely $k_2[Fe(CO)_4CO_2H^-]$) which after applying the steady state approximation to $Fe(CO)_4CO_2H^-$ (namely $\frac{d[Fe(CO)_4CO_2H^-]}{dt} = 0$) leads to the following expression for $H_2$ production:

$$\frac{d[H_2]}{dt} = k_1 [Fe(CO)_5][OH^-] \qquad (5)$$

The rate of $H_2$ production using the catalyst system derived from $Fe(CO)_5$ is thus seen to have a first order dependence on $Fe(CO)_5$ concentration and to be independent of CO pressure in accord with the experimental observations outlined above. Furthermore, the formate buffer system generated by reaction of CO with the base by equation 2 keeps the $OH^-$ concentration essentially independent of the amount of base introduced into the system. Therefore the rate of $H_2$ production using the catalyst system derived from $Fe(CO)_5$, although having a first order dependence on $OH^-$ concentration, is essentially independent of the base concentration.

The following rather different type of catalytic cycle involving formate decomposition explains our observations on the water gas shift reactions catalyzed by $M(CO)_6$ (M = Cr, Mo, and W) in the presence of a base:

$$M(CO)_5 + HCO_2^- \xrightarrow{k_1} M(CO)_5OCOH^- \qquad (6a)$$

$$M(CO)_5OCOH^- \xrightarrow{k_2} HM(CO)_5^- + CO_2 \qquad (6b)$$

$$HM(CO)_5^- + H_2O \xrightarrow{k_3} H_2M(CO)_5 + OH^- \qquad (6c)$$

$$H_2M(CO)_5 \xrightarrow{k_4} H_2 + M(CO)_5 \qquad (6d)$$

In addition the following two external steps to this catalytic cycle are needed:

(a) Dissociation of the metal hexacarbonyl:

$$M(CO)_6 \underset{k_a}{\overset{k_d}{\rightleftharpoons}} M(CO)_5 + CO \qquad (6e)$$

(b) Generation of formate from CO and hydroxide

$$CO + OH^- \xrightarrow{k_o} HCO_2^- \qquad (2)$$

Equation 6e followed by equation 6a is analogous to a reported (22) preparation of the trifluoroacetate $W(CO)_5OCOCF_3^-$ by treatment of $W(CO)_6$ with tetraethylammonium trifluoroacetate at elevated temperatures. Equation 6b represents an unknown reaction but was shown to be reasonable by observing $CO_2$ as a product from the pyrolysis at 110°C of $[(C_6H_5)_3PNP(C_6H_5)_3]$-$[W(CO)_5OCOH]$, which was prepared by a standard method (22) using the reaction of $W_2(CO)_{10}^{2-}$ with silver formate. Equations 6c and 6d for the catalyst systems derived from $M(CO)_6$ are completely analogous to equations 4c and 4d for the catalyst system derived from $Fe(CO)_5$.

The kinetic analysis of the mechanism 6a-6e,2 is more complicated than that of the mechanism 4a-4e because of the external reaction 6e but nevertheless is feasible using the steady state approximation. By a procedure similar to the derivation of equation 5 the following equation can be derived:

$$\frac{d[H_2]}{dt} = k_4[H_2M(CO)_5] = k_1[M(CO)_5](HCO_2^-) \qquad (7)$$

However, the steady state concentration of $M(CO)_5$ depends upon the concentrations of the stable species $M(CO)_6$, $HCO_2^-$, and CO as well as $H_2M(CO)_5$ in the cycle. Thus applying the steady state approximation to $[M(CO)_5]$ one obtains the following equation where $k_{eq} = k_d/k_a$:

$$\frac{d[H_2]}{dt} = (k_d/k_a)k_1\frac{[M(CO)_6][HCO_2^-]}{[CO]} = K_{eq}k_1\frac{[M(CO)_6][HCO_2^-]}{[CO]} \qquad (8)$$

This mechanistic scheme agrees with the experimental observations of the first order dependences on $M(CO)_6$ and formate concentrations and the inverse first order dependence on CO pressure for the rate of $H_2$ production in the water gas shift reaction catalyzed by $M(CO)_6$ (M = Cr, Mo and W) in the presence of a base sufficiently strong to generate formate from CO by equation 2.

## Acknowledgment

The authors would like to express appreciation for the partial support for this research provided by the Division of Basic Sciences of the U. S. Department of Energy under Contract EY-76-S-09-0933.

## Literature Cited


1. Thomas, C. L. "Catalytic Processes and Proven Catalysts," Academic Press, New York, 1970.
2. Laine, R.M.; Rinker, R. G.; Ford, P. C. J. Am. Chem. Soc., 1977, 99, 252.
3. Kang, H.; Mauldin, C. H.; Cole, T.; Slegeir, W.; Cann, K.; Pettit, R. J. Am. Chem. Soc., 1977, 99, 8323.
4. Ungerman, C.; Landis, V.; Moya, S. A.; Cohen, H.; Walker, H. Pearson, R. G.; Rinker, R. G.; Ford, P. C. J. Am. Chem. Soc., 1979, 101, 5922.
5. Cheng, C.-H.; Hendriksen, D. E.; Eisenberg, R. J. Am. Chem. Soc., 1977, 99, 2791.
6. Laine, R. M. J. Am. Chem. Soc., 1978, 100, 6451.
7. Baker, E. C.; Hendriksen, D. E.; Eisenberg, R. J. Am. Chem. Soc., 1980, 102, 1020.
8. Cheng, C.-H.; Eisenberg, R. J. Am. Chem. Soc., 1978, 100, 5968.
9. Yoshida, T.; Ueda, Y.; Otsuka, S. J. Am. Chem. Soc., 1978, 100, 3941.
10. King, R. B.; Frazier, C. C.; Hanes, R. M.; King, A. D. J. Am. Chem. Soc., 1978, 100, 2925.
11. King, A. D.; King, R. B.; Yang, D. B. J. Am. Chem. Soc., 1980, 102, 1028.
12. King, A. D.; King, R. B.; Yang, D. B. submitted for publication.
13. Pettit, R.; Cann, K.; Cole, T.; Mauldin, C. H.; Slegeir, W. Advan. Chem. Ser., 1979, 173, 121.
14. King, A. D.; King, R. B.; Yang, D. B. Chem. Comm., 1980, 529.
15. Werner, H.; Prinz, R. Chem. Ber., 1966, 99, 3582.
16. Kruck, T.; Noack, M. Chem. Ber., 1964, 97, 1693.
17. Darensbourg, D. J.; Froelich, J. A. J. Am. Chem. Soc., 1977, 99, 4726.
18. Darensbourg, D. J.; Baldwin, B. J.; Froelich, J. A. J. Am. Chem. Soc., 1980, 102, 4688.
19. Hieber, W.; Kruck, T. Z. Naturforsch. B, 1961, 16, 709.
20. Clark, H. C.; Dixon, K. R.; Jacobs, W. J. Chem. Comm. 1968, 548.
21. Tolman, C. A. Chem. Soc. Revs., 1972, 1, 337.
22. Schlientz, W. J.; Lavender, Y.; Welcman, N.; King, R. B.; Ruff, J. K. J. Organometal. Chem., 1971, 33, 357.


RECEIVED December 8, 1980.

# Homogeneous Catalytic Reduction of Benzaldehyde with Carbon Monoxide and Water

## Applications of the Water Gas Shift Reaction

WILLIAM J. THOMSON[1] and RICHARD M. LAINE

Physical Organic Chemistry Department, SRI International, Menlo Park, CA 94025

With the advent of the energy crisis, the study of homogeneous catalytic CO hydrogenation has become very popular because CO represents one of the cheapest, readily available $C_1$ building blocks for production of synfuels and because homogeneous catalysts can be very efficient hydrogenation catalysts. To date, the majority of the research has been devoted to catalytically hydrogenating CO directly to hydrocarbons (Fischer-Tropsch synthesis) with little attention paid to other CO reduction reactions that could also be useful for synfuel production (1,2,3,4). For example, two areas that could be more thoroughly explored are CO homologation [reaction (1)] and methanol synthesis [reaction (2)], which can be catalyzed homogeneously (5,6,7,8).

$$CH_3\text{-}O\text{-}CH_3 + 2CO + 2H_2 \xrightarrow{\text{"Ru"}} CH_3\overset{O}{\overset{\|}{C}}\text{-}OCH_2CH_3 + H_2O \qquad (1)$$

$$CO + 2H_2 \xrightarrow{\text{catalyst}} CH_3OH \qquad (2)$$

$$\text{Catalyst} = Ru(CO)_5,\ HCo(CO)_4 \text{ or } Rh_5(CO)_{15}^-$$

With the recent development of zeolite catalysts that can efficiently transform methanol into synfuels, homogeneous catalysis of reaction (2) has suddenly grown in importance. Unfortunately, aside from the reports of Bradley (6), Bathke and Feder (7), and the work of Pruett (8) at Union Carbide (largely unpublished), very little is known about the homogeneous catalytic hydrogenation of CO to methanol. Two possible mechanisms for methanol formation are suggested by literature discussions of Fischer-Tropsch catalysis (9-10). These are shown in Schemes 1 and 2.

[1] On sabbatical leave from the University of Idaho.

0097-6156/81/0152-0133$05.00/0
© 1981 American Chemical Society

$$MH + CO \longrightarrow M-C\overset{\displaystyle O}{\underset{\displaystyle H}{\big\langle}} \xrightarrow{\ H_2\ } MH + CH_2O$$

$$MH + CH_3OH \xleftarrow{\ H_2\ } M-CH_2OH$$

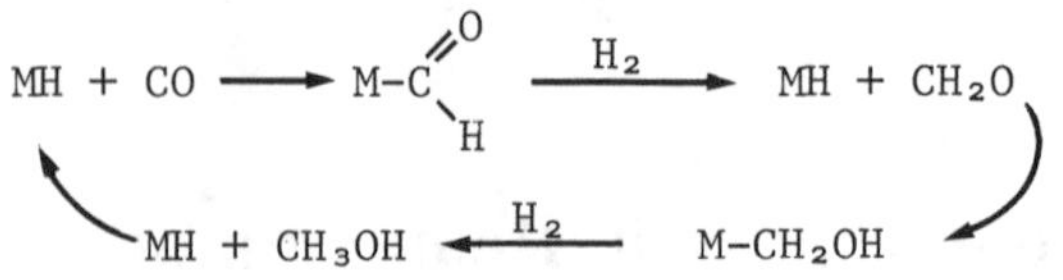

MH ≡ homogeneous metal hydride complex.

Scheme 1

$$MH + CO \longrightarrow M-C\overset{\displaystyle O}{\underset{\displaystyle H}{\big\langle}} \xrightarrow{\ H_2\ } HM=C\overset{\displaystyle OH}{\underset{\displaystyle H}{\big\langle}}$$

$$MH + CH_3OH \xleftarrow{\ H_2\ } M-CH_2OH$$

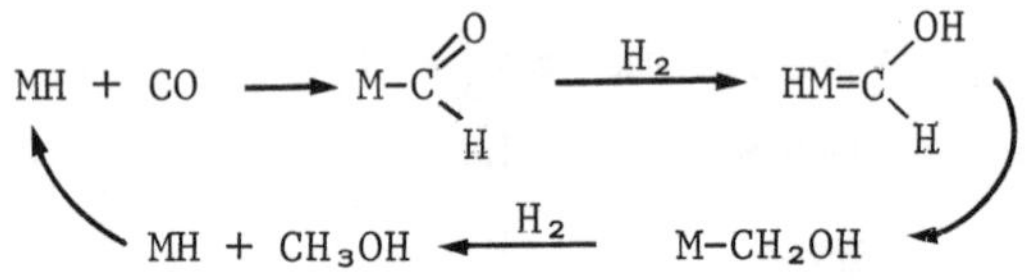

Scheme 2

We are interested in homogeneous catalytic methanol synthesis because of our previous work on the Reppe reactions ($\underline{11},\underline{12},\underline{13}$):

$$R-CH=CH_2 + 3CO + 2H_2O \xrightarrow[\text{MeOH}]{Rh_6(CO)_{16}/KOH} RCH_2CH_2CH_2OH + 2CO_2 \quad (3)$$

$$R-CH=CH_2 + 3CO + H_2O + R_2'NH \xrightarrow[\text{MeOH}]{Ru_3(CO)_{12}/KOH} R(CH_2)_3-NR_2' + 2CO_2 \quad (4)$$

A common theme in reactions (1)-(4) and Schemes 1 and 2 is the hydrogenation of carbon-oxygen multiple bonds to species with carbon-oxygen or carbon-nitrogen single bonds. In undertaking the work presented here it was our intention to determine whether or not information obtained in the study of the hydrogenation of aldehydic carbon-oxygen double bonds is applicable to understanding (a) reduction of carbon monoxide to methanol, (b) CO reduction to Fischer-Tropsch products, (c) homologation as in reaction (1), and (d) the validity of either Scheme 1 or 2.

We chose to investigate the $CO/H_2O/KOH$ and $Rh_6(CO)_{16}$ catalyzed reduction of benzaldehyde and substituted benzaldehydes because:

- Benzaldehydes are not subject to base catalyzed aldol condensations, and under the reaction conditions Cannizarro reactions are not important.
- From thermodynamic considerations $CO/H_2O$ must be a better reductant than $H_2$, thus milder reaction conditions might be possible.
- The catalyst derived from $Rh_6(CO)_{16}$ is the most active catalyst for aldehyde reduction of all the group 8 metals we have studied ($\underline{12},\underline{13}$).
- p-Substitution of the benzaldehydes should provide information about the electron density at the carbonyl carbon during the addition of metal hydride to the carbon-oxygen double bond.

## Experimental Procedures

General Methods. Methanol used in kinetic runs was distilled from sodium methoxide or calcium hydride in a nitrogen atmosphere before use. Freshly distilled cyclohexanol was added to the methanol in the ratio 6.0 ml cyclohexanol/200 ml MeOH and was used as an internal standard for gas chromatographic (GC) analysis. Benzaldehyde was distilled under vacuum and stored under nitrogen at 5°. Other aldehydes (purchased from Aldrich) were also distilled before use. The corresponding alcohols (purchased from Aldrich) were distilled and used to prepare GC standards. All metal carbonyl cluster complexes were purchased from Strem Chemical Company and used as received. Tetrahydrofuran (THF) was distilled from sodium benzophenone under nitrogen before use.

Analytical Methods. All the analyses were done by gas chromatography. Liquid products were analyzed on a Hewlett-Packard Model 5711 gas chromatograph equipped with FID using a 4.0 m by 0.318 cm column packed with 5% Carbowax acid-washed Chromosorb G. Gas products were analyzed with a Hewlett-Packard Model 5750 gas chromatograph equipped with a 3 m by 0.318 cm, 150/200 Poropak Q column and a 1.8 m by 0.318 cm type 13A molecular sieve column. Hydrogen analysis was achieved by injecting into the molecular sieve column operated with a 8.5% hydrogen-in-helium carrier gas. Other gaseous components were analyzed in the Poropak column using a helium carrier gas.

In the procedure used for studying benzaldehyde reduction, 6.0 ml of the MeOH/cyclohexanol solution described above were used as solvent. Aldehyde was added to the reactor via a 5-ml glass syringe, and 1.0 ml of 3 N KOH solution was added by means of a pipette. We were careful to add either the KOH or the aldehyde just before pressurizing with CO to minimize the aldehyde reduction resulting from the noncatalytic Cannizarro reaction. The reactor was then sealed and flushed twice with 600-psi CO before bringing the reactor up to the desired CO pressure. The run was initiated at the time the reactor was immersed in the temperature bath. On completion of the run, the reactor was quenched in cold water and the pressure at 24°C was recorded. The primary analysis was conducted on the liquid solution; however, some runs were also subjected to gas phase analysis. In every experiment, liquid samples were collected in 10-cc sample vials and stored in a refrigerator. In most cases chromatographic analyses were conducted within 6 hours of the termination of the run. The maximum time that any sample was stored before analysis was 72 hours.

To have a basis for comparison, we established a standard run for aldehyde reduction consisting of

0.1 mmole $Rh_6(CO)_{16}$
30 mmole $C_6H_5CHO$
1 ml 3 N KOH
6 ml (MeOH + cyclohexanol)

$P_{CO}$ = 800 psi
T = 125°C
Reaction time = 1 hour

Catalytic activity was measured as a function of turnover frequency [moles product/(mole catalyst) (hour)]. The standard run has a turnover frequency of 105±10. All the parameters investigated were perturbed about this standard and included the effects of catalyst, aldehyde, KOH and water concentration, initial CO pressure, and reaction time. In addition, a few selected runs were also conducted to examine the effects of hydrogen in the gas phase as well as the relative ease with which other aldehydes could be reduced.

## Results

Catalysts and Catalyst Concentration. The most active catalyst for benzaldehyde reduction appears to be rhodium [$Rh_6(CO)_{16}$ precursor], but iron [as $Fe_3(CO)_{12}$] and ruthenium [as $Ru_3(CO)_{12}$] were also examined. The results of these experiments are shown in Table 1. Consistent with earlier results (12), the rhodium catalyst is by far the most active of the metals investigated and the ruthenium catalyst has almost zero activity. The latter is consistent with the fact that ruthenium produces only aldehydes during hydroformylation. Note that a synergistic effect of mixed metals does not appear to be present in aldehyde reduction as contrasted with the noticeable effects observed for the water-gas shift reaction (WGSR) and related reactions (13).

The effect of the concentration of $Rh_6(CO)_{16}$ on the number of turnovers was evaluated by using 0.01 mmole to 0.10 mmole of catalyst, and these results are shown in Figure 1. The turnovers increase with decreasing catalyst precursor. This is indicative of catalysis by cluster fragmentation. The results of our earlier work (12) are also shown in Figure 1. Although the turnovers are higher than those observed here, it should be noted that a higher temperature and a shorter reaction time were used previously. Higher temperature should produce more turnovers, but shorter reaction times would be expected to produce less, depending on the existent reaction orders. This is complicated further by the effect of $CO_2$ production, which tends to consume $OH^-$, and also by the nonisothermal heat-up period (5 min), which is a significant fraction of the 0.5-hr reaction time used previously.

KOH Concentration Studies. The effect of KOH concentration on benzaldehyde reduction was examined, and the results are shown in Figure 2 along with our previous results for ruthenium catalyzed hydroformylation (12).

The Effect of Reactant Concentration. Several experiments were conducted to quantify the effect of reactant concentration on the aldehyde reduction rate. The initial CO pressure was varied

**Table I. Benzaldehyde Reduction with Various Catalysts**

| CATALYST PRECURSOR | NO. OF TURNOVERS[A] |
|---|---|
| $Rh_6(CO)_{16}$ | 105 |
| $Fe_3(CO)_{12}$ | 30 |
| $Ru_3(CO)_{12}$ | 9 |
| $Rh_6(CO)_{16}/Fe_3(CO)_{12}$[B] | 93 |
| $Ru_3(CO)_{12}/Fe_3(CO)_{12}$[B] | 21 |

[A]MOLES ALCOHOL FORMED PER MOLE OF TOTAL CATALYST.
[B]0.05 MMOLES OF EACH.

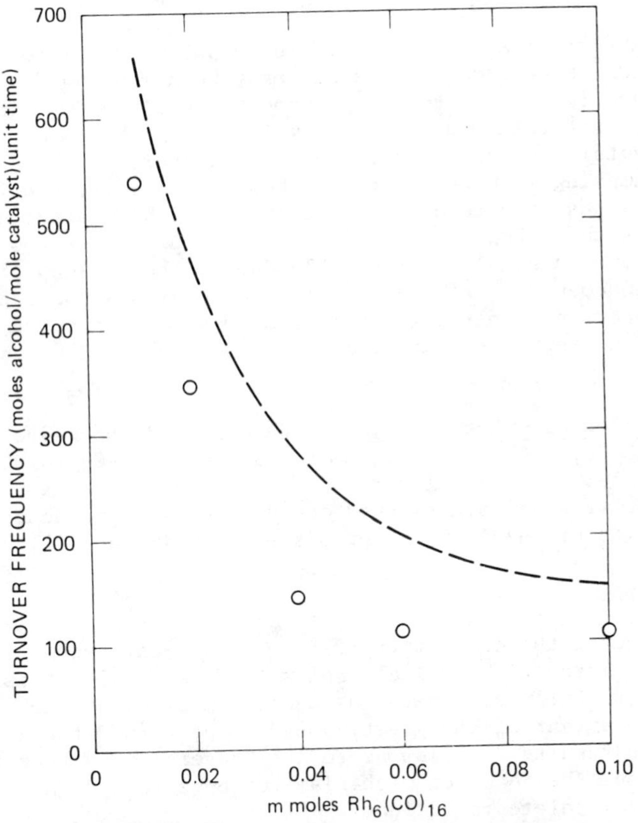


*Figure 1. The effect of catalyst concentration on benzaldehyde reduction: (○) 125°C, 1 h; (– – –) 150°C, 0.5 h (12)*

between 100 and 1200 psi, the $H_2O$ concentration between 55 and 167 mmole, and the $C_6H_5CHO$ concentration between 10 and 40 mmole. Because of solubility problems, the experiments at the highest $H_2O$ and highest $C_6H_5CHO$ concentrations were analyzed after adding 3 ml THF to the mixture recovered at the end of the run.

Figure 3 shows the results of varying the CO pressure. The maximum activity appears to lie near 600 psi for benzaldehyde reduction. Figure 3 is an attempt to elucidate an apparent reaction order with respect to the arithmetically averaged CO pressure. At pressures less than 400 psi, the order is nearly first order. The situation at higher pressures is not clear; however, it is reasonable to speculate that the dominant aspects of the kinetics shift at these pressures. The data suggest the shift is to zero-order dependance.

Studies analyzing the effects of the remaining reactants, $H_2O$ and $C_6H_5CHO$ indicate that the reaction appears to be zero order with respect to both reactants. It is interesting that in previous work we also found similar behavior for $H_2O$ in ruthenium catalyzed hydroformylation (12), as did Ungermann et al. with the WGSR (14).

The Effect of Reaction Time. The problem associated with time-varying $OH^-$ concentrations has already been mentioned. The difficulty associated with the influence of dissolved $CO_2$ can be appreciated by referring to Figure 4, which shows the results of two experiments. In one, samples were taken every hour and in the other sampling occurred every two hours. However, the important factor is that the reactor was recharged with CO after each sample. Note that the effective reaction rate is lower when two hours elapse between samples, presumably due to the buildup of $CO_2$, which consumes $OH^-$. In fact, one experiment was conducted at 94°C for 17 hours and only 27% conversion to alcohol occurred, the same conversion experienced after 3 hours when fresh CO was added hourly.

Reduction of Other Aldehydes. We examined the reduction of anisaldehyde, p-$CH_3OC_6H_4CHO$ and tolualdehyde, p-$CH_3(C_6H_4)CHO$ to examine the effect of electron density on aldehyde reduction. In addition, we also investigated one ketone, acetophenone, $C_6H_5COCH_3$. The results of these experiments are given in Table 2.

## Discussion

Because of the complexity of the rhodium-catalyzed reduction of benzaldehyde to benzyl alcohol with CO and $H_2O$, it is not possible to fully elucidate the mechanism of catalytic reduction given the extent of the kinetic studies performed to date. However, the results do allow us to draw several important conclusions about the reaction mechanism for benzaldehyde hydrogenation and several related reactions.

We recently described (12,15) the use of catalyst concentra-

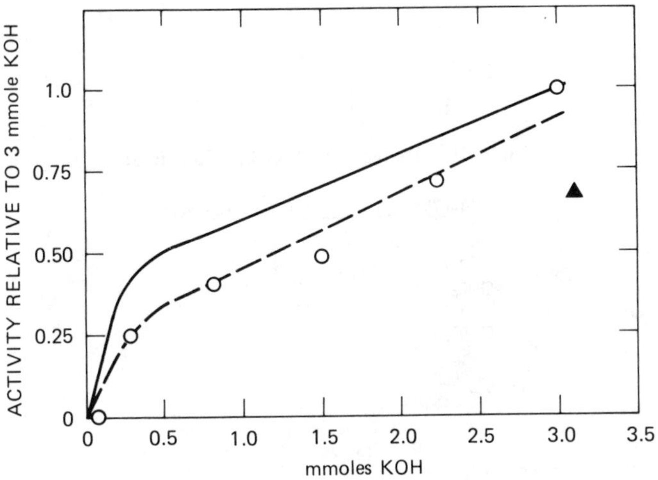


*Figure 2. The Effect of KOH: (-- ○ --) KOH; (▲) 1.56 mmol $K_2CO_3$; (——) hydroformylation (12)*

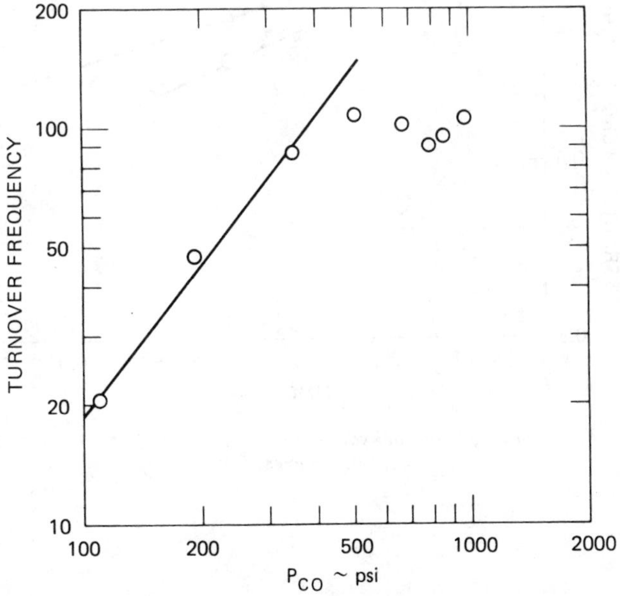


*Figure 3. Benzaldehyde reduction turnovers vs. $P_{CO}$*

**Table II. Reduction of Other Reactants**

| REACTANTS | TURNOVER FREQUENCY |
|---|---|
| $C_6H_5CHO$ | 105 |
| $pCH_3C_6H_4CHO$ | 78 |
| $pCH_3OC_6H_4CHO$ | 63 |
| $C_6H_5COCH_3$ | 24 |

(Standard Conditions, 30 mmoles of Reactants)

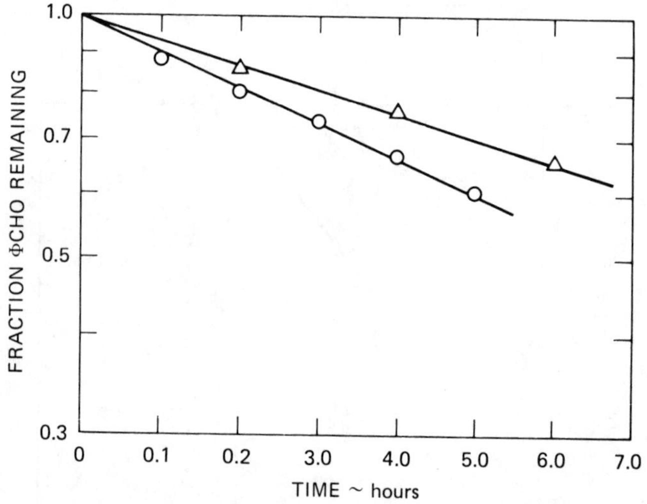


*Figure 4. Time varying benzaldehyde reduction* T = *94°C: (○) purge every 1 h, (△) purge every 2 h*

tion studies to help identify active catalyst species, especially where cluster catalyzed reactions are suspected. In the present work, changes in the amount of rhodium added to the reaction solution have dramatic effects on the turnover frequency. As shown in Figure 1, decreasing the rhodium concentration results in considerable increases in turnover frequencies. These changes are indicative of equilibria involving rhodium cluster complexes that fragment reversibly to smaller cluster complexes or mononuclear complexes wherein the active species are the fragments. In the reaction solutions where benzaldehyde reduction occurs we have observed (12), by IR, both the $Rh_{12}(CO)_{30}^{2-}$ and $Rh_5(CO)_{15}^-$ complexes. Recent work by Chini and coworkers (16) suggests at least one plausible equilibrium:

$$Rh_6(CO)_{16} + OH^- \longrightarrow Rh_6(CO)_{15}H^- + CO_2 \quad (5)$$

$$Rh_6(CO)_{15}H^- + OH^- \longrightarrow Rh_6(CO)_{15}^{2-} + H_2O \quad (6)$$

$$2Rh_6(CO)_{15}H^- \longrightarrow Rh_{12}(CO)_{30}^{2-} + H_2 \quad (7)$$

$$Rh_6(CO)_{15}^{2-} + 2Rh_{12}(CO)_{30}^{2-} + 15CO \rightleftharpoons 6Rh_5(CO)_{15}^- \quad (8)$$

Vidal and Walker (17) have observed that $Rh(CO)_4^-$ is in equilibria with $Rh_5(CO)_{15}^-$. It is likely that $Rh(CO)_4^-$ is also in equilibria with the cluster species in equations (5)-(8); thus, it must also be a part of the benzaldehyde hydrogenation catalyst solution.

The effects of changes in KOH concentration on catalyst activity for benzaldehyde reduction are complex. Figure 2 compares the present work with KOH concentration studies for ruthenium catalyzed hydroformylation:

$$CH_3CH_2CH_2CH{=}CH_2 + H_2O + 2CO \xrightarrow[MeOH/150^\circ]{Ru_3(CO)_{12}/KOH} CH_3CH_2CH_2CH_2CH_2CHO + CO_2 \quad (9)$$

Although the data for $C_6H_5CHO$ reduction lie below the hydroformylation results, it is apparent that the effect of KOH concentration is similar in the two cases.

The initial steep rise can be attributed simply to activation of the catalyst precursor. The amount of base corresponds approximately to one equivalent of KOH for the ruthenium catalyst and two equivalents for the rhodium catalyst. Activation could result from hydroxide attack as in (5) and (6) for rhodium and (10) for ruthenium:

$$Ru_3(CO)_{12} + OH^- \longrightarrow HRu_3(CO)_{11}^- + CO_2 \quad (10)$$

Concentrations of hydroxide beyond those needed for catalyst activation display approximately 0.5 order dependence with regard to turnover frequency. Since these results are for two catalysts performing different functions, it would appear that they correspond to a separate, noncatalytic step in the reaction sequence. The most likely explanation is that the effects are due to $CO_2$-base interactions. This is partially substantiated by the fact that when $K_2CO_3$ was used in place of KOH (filled data point, Figure 2), significant reduction also took place although the activity was lower than that observed with KOH (the same number of K equivalents were used). This drop in activity is due to the difference in initial pH of the carbonate solution (as compared with the KOH solution) which would be governed by dissolved $CO_2$, a variable that changes as the reaction proceeds.

Another noncatalytic step proposed by King et al. (18) in iron carbonyl/base catalysis of the WGSR involves the formation of formate ion; however, we recently observed that formate formation appears to have little importance in the related rhodium catalysis of hydrohydroxymethylation. We plan to perform studies of the $CO + KOH$ and $CO_2 + KOH$ reactions independent of catalysis to more fully appreciate the relationship of these reactions to solution pH and thus the catalytic activity.

The effects of benzaldehyde concentrations on turnover frequency are anomalous. Our results indicate that benzaldehyde hydrogenation turnover frequency is independent of benzaldehyde concentration (an apparent zero-order dependence). However, the data in Table 2 indicate otherwise. If the reaction were independent of aldehyde concentration, the rate data should be independent of the type of aldehyde used. This is especially true with p-tolualdehyde and p-anisaldehyde where the structural changes to the aldehyde (addition of p-methyl or p-methoxy) should influence the reactivity of the aldehyde functionality only through electronic effects. Thus, we are forced to conclude that the aldehyde is involved in the rate determining step even though the concentration study does not support its presence.

Furthermore, the data from Table 2 allow us to draw several valuable conclusions regarding the mechanism of hydride addition to carbon-oxygen double bonds.

There are two possible modes of metal hydride addition to carbon-oxygen double bonds:

$$MH + C_6H_5\text{-}C(=O)\text{-}H \longrightarrow C_6H_5\text{-}C(OH)(H)\text{-}M \qquad (11)$$

$$MH + C_6H_5\text{-}C(=O)\text{-}H \longrightarrow C_6H_5\text{-}CH_2\text{-}OM \qquad (12)$$

In reaction (11) the metal-hydride addition suggests a protonation reaction; whereas, in reaction (12) the addition appears to be a hydride transfer reaction. If the reaction is indeed a hydride transfer reaction then the introduction of p-electron donating substituents, which place more electron density at the carbonyl carbon, (the site of hydride attack) will inhibit hydride addition. The data in Table 2 show that the introduction of p-electron donating substituents reduces the turnover frequency. This is consistent with hydride attack at the benzaldehyde carbonyl carbon, (12).

Further support for addition of the type found in reaction (12) comes from an unusual reaction we observed while investigating hydrogenation of benzaldehyde with $H_2$ using $Rh_6(CO)_{16}$ as catalyst precursor in methanol solution:

$$C_6H_5\text{-}C(=O)H + H_2 \xrightarrow[\text{MeOH}/125°/1h]{Rh_6(CO)_{16}/800\ psi\ H_2} C_6H_5\text{-}CH_2\text{-}OCH_3\ \ 80\% + C_6H_5CH(CH_3)\text{-}OCH_3\ \ 20\% \quad (13)$$

The mechanism for ether formation is quite simple given the recent work of Chini (19) that shows that rhodium clusters can react to give strong acids:

$$Rh_6(CO)_{16} + \text{(traces) } 2H_2O \longrightarrow Rh_{12}(CO)_{30}{}^{2-} + 2H^+ + 2CO_2 + H_2 \quad (14)$$

If some metal hydride species catalyzes the hydrogenation of benzaldehyde:

$$Rh_x(CO)_yH_2{}^- + C_6H_5CHO \longrightarrow Rh_x(CO)_y{}^- + C_6H_5CH_2OH \quad (15)$$

then acid catalyzed etherification will follow:

$$C_6H_5CH_2OH + MeOH \xrightarrow{H^+} C_6H_5CH_2OMe + H_2O \quad (16)$$

The formation of the α-methyl derivative is extremely surprising. We can propose a reasonable mechanism based on Wender's proposed mechanism for cobalt catalyzed CO homologation of methanol (5).

$$CH_3OH + H^+ \rightleftharpoons [CH_3OH_2]^+ \quad (17)$$

$$[CH_3OH_2]^+ + Rh_x(CO)_y{}^- \longrightarrow CH_3\text{-}Rh_x(CO)_y + H_2O \quad (18)$$

$$C_6H_5CHO + CH_3\text{-}Rh_x(CO)_y{}^- \longrightarrow C_6H_5\text{-}CH(CH_3)\text{-}ORh_x(CO)_y{}^- \quad (19)$$

Formation of the methyl-rhodium complex is analogous to the formation of $CH_3\text{-}C(CO)_4$ from $CH_3OH_2^+$ and $Co(CO)_4{}^-$ as proposed by Wender. The difference here is that the nature of the active rhodium species is not known. Under the present conditions, homologation does not occur because CO is not present; however, addition of the methyl-rhodium species to benzaldehyde must occur as shown in (19), metal adds to the oxygen. The product in (19) is then subject to acid catalyzed etherification to obtain the methyl ether.

The formation of metal-oxygen bonds has previously been found to occur for the stoichiometric hydrogenation of CO to methanol with metal hydrides of the early transition metals (20). Moreover, in ruthenium-phosphine catalyzed hydrogenation (with $H_2$) of aldehydes and ketones, metal-oxygen bonded catalytic intermediates have been proposed for the catalytic cycle and in one case isolated (21,22).

It seems extremely likely that similar metal-oxygen bonded

intermediates form in the rhodium-catalyzed reduction of carbon-oxygen double bonds. Thus, it appears that catalytic intermediates such as M-CHR-OH (11) are not viable for carbon-oxygen double bond hydrogenation or methanol synthesis (R=H).

Alternative mechanisms for carbon-oxygen double bond hydrogenation and methanol synthesis are shown in Schemes 4 and 5.

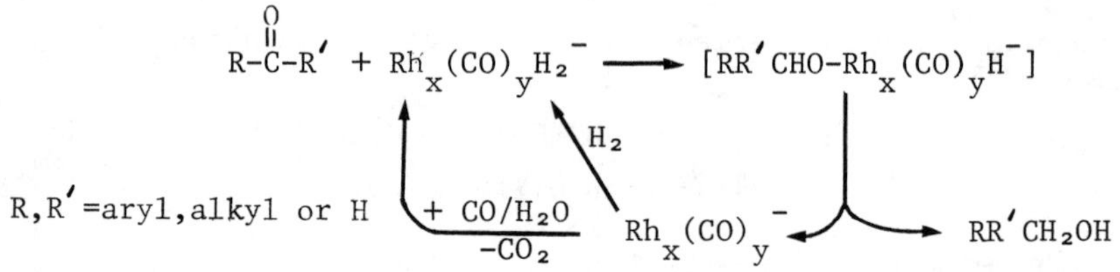
R-C(=O)-R′ + $Rh_x(CO)_yH_2^-$ → [RR′CHO-$Rh_x(CO)_yH^-$]
$H_2$
R,R′=aryl,alkyl or H
+ CO/$H_2O$
$-CO_2$
$Rh_x(CO)_y^-$
RR′$CH_2OH$

Scheme 4

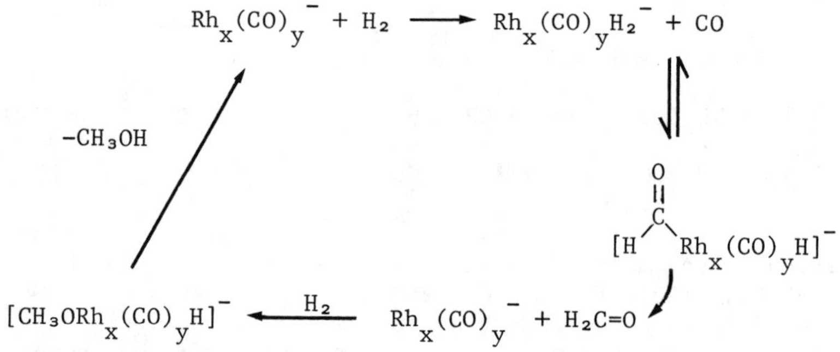
$Rh_x(CO)_y^-$ + $H_2$ → $Rh_x(CO)_yH_2^-$ + CO
$-CH_3OH$
[H-C(=O)-$Rh_x(CO)_yH]^-$
[$CH_3ORh_x(CO)_yH]^-$
$H_2$
$Rh_x(CO)_y^-$ + $H_2C$=O

Scheme 5

Unfortunately, because of the exceptional number of interrelated equilibria between various rhodium clusters and $Rh(CO)_4^-$ it seems unlikely that it will be possible to identify which rhodium species is responsible for the hydrogenation reactions.

## Literature Cited


1. Thomas, M. G., Beier, B. G., and Muetterties, E. L.; J. Am. Chem. Soc. (1976) 98, 1296.
2. Demitras, G. C., Muetterties, E. L. ; J. Am. Chem. Soc. (1977) 99, 2796.
3. Muetterties, E. L.; Pure and Appl. Chem. (1978) 50, 941.
4. Sweet, J. R., Graham, W.A.G.; J. Organomet. Chem. (1979) 173, C9.
5. For a review see Slocum, D. W.; Cat. in Org. Syn. Academic Press, (1980) p. 245.
6. Bradley, J. S.; J. Am. Chem. Soc. (1979) 101, 7419.
7. Rathke, J. W., Feder, H. M.; J. Am. Chem. Soc. (1978) 100, 3623.
8. Pruett, R. L.; Ann. N. Y. Acad. Sci. (1977) 295, 239.
9. Henrici-Olivé, S; Angew. Chem. Int. Ed. Engl. (1976) 15, 136.
10. Muetterties, E. L., Stein, J.; Chem. Rev. (1979) 15, 136.
11. Laine, R. M.; J. Am. Chem. Soc. (1978) 100, 6451.
12. Laine, R. M.; Ann. N.Y. Acad. Sci. (1980) 333, 124.
13. Laine, R. M.; J. Org. Chem. (1980) 45, 3370.
14. Ungermann, C, Landis, V, Moya, S. A., Cohen, H., Walker, H, Pearson, R. G., Rinker, R. G, Ford, P. C.; J. Am. Chem. Soc. (1979) 101, 5922
15. Laine, R. M. Paper presented at the Second Chem. Cong. of the N. American Continent, Aug. 1980, Petr. Div. No. 55.
16. Fumagalli, A., Koetzle, T. F., Takasagawa, F, Chini, P., Martinengo, S., Heaton, B. T.; J. Am. Chem. Soc. (1980) 102, 1740.
17. Vidal, J. L., Walker, W. E.; Inorg. Chem. (1980) 19, 896.
18. King, A. D. Jr., King, R. B., Yang, D. B.; J. Am. Chem. Soc. (1980) 102, 1028.
19. Chini, P., Longoni, G., and Albano, V.; Adv. Organomet. Chem. (1976) 14, 285.
20. Wolozanski, P. T., Threlkel, R. S., Bercaw, J. E., J. Am. Chem. Soc. (1979) 101, 218.
21. Sanchez-Delgado, R. A., de Ochoa, O. L.; Molec. Cat. (1978) 6, 303.
22. Sanchez-Delgado, R. A., de Ochoa, O. L., Suarez, T. Paper presented at the 9th Int. Conf. on Organomet. Chem. P45T.


RECEIVED December 16, 1980.

# 10

# Electrophile-Induced Disproportionation of the Neutral Formyl ($\eta$-$C_5H_5$)Re(NO)($PPh_3$)(CHO)

## Isolation and Properties of the Rhenium Methylidene [($\eta$-$C_5H_5$)Re(NO)($PPh_3$)($CH_2$)]$^+PF_6^-$

J. A. GLADYSZ, WILLIAM A. KIEL, GONG-YU LIN, WAI-KWOK WONG, and WILSON TAM

Department of Chemistry, University of California, Los Angeles, CA 90024

A variety of organic molecules (methane, methanol, higher alkanes and alcohols, glycols, gasoline hydrocarbons) can be obtained from CO/$H_2$ gas mixtures (synthesis gas) in the presence of metallic heterogeneous and homogeneous catalysts [1,2]. Since synthesis gas can be readily produced from coal, and domestic crude oil and natural gas reserves (conventional sources of the aforementioned organic chemicals) are declining, there is intense current interest in CO/$H_2$ chemistry. Research in numerous laboratories is being directed toward the development of milder and/or more selective CO reduction catalysts, and the delineation of CO reduction mechanisms (see other papers contributed to this symposium, and leading references [3-9]).

In considering the formative stages of CO reduction, one is struck by the fact that only a finite number of single carbon catalyst-bound intermediates is possible. Candidate intermediates for which some type of experimental support exists are given in Figure 1 [1,2]. On the basis of available data, it is most probable that more than one distinct mode of CO reduction (and homologation to $C_2$ and higher intermediates) can occur.

It is not at this time practical to probe the finer details of CO reduction using the actual catalysts employed to effect CO/$H_2$ reactions. Reaction conditions are severe and many intermediates are expected to lie in relatively shallow potential energy wells. We therefore initiated a program aimed at synthesizing stable homogeneous transition metal complexes containing ligands corresponding to the catalyst-bound intermediates in Figure 1 [10-18]. Through investigation of their basic chemistry, we have sought to gain insight into possible catalyst reaction pathways.

Considerable challenge is associated with synthesizing stable complexes containing certain of the ligands shown in Figure 1. Although transition metal-CO and transition metal-$CH_3$ complexes have long been known, transition metal complexes containing the other six ligand types in Figure 1 were unknown prior to the early

0097-6156/81/0152-0147$05.00/0
© 1981 American Chemical Society

1970's [1,2]. In all cases, rationalizations can be advanced as to why kinetic instability is anticipated.

Other research groups have actively pursued similar lines of research, and their important contributions (studies by Casey and Graham are particularly relevant: [19, 20, 21]) will be reviewed more thoroughly in our full papers. In this Symposium account, we shall describe the use of cations $[(\eta\text{-}C_5H_5)Re(NO)(CO)_2]^+ BF_4^-$ (1) [22] and $[(\eta\text{-}C_5H_5)Re(NO)(PPh_3)(CO)]^+ BF_4^-$ (2a) as precursors to a number of complexes containing ligands of the types in Figure 1. We chanced upon these systems in the course of prospecting for stable neutral formyl complexes [15]. Earlier, we had found that $[(\eta\text{-}C_5H_5)Mn(NO)(CO)_2]^+ PF_6^-$ reacted with $Li(C_2H_5)_3BH$ to afford the manganese formyl $(\eta\text{-}C_5H_5)Mn(NO)(CO)(CHO)$, which rapidly decomposed at 10 °C [15]. Based upon abundant precedent [12], it seemed probable that rhenium homologs would have greater kinetic stability. We were also influenced by Graham's intriguing 1972 report that $NaBH_4$ reduced 1 to the methyl complex $(\eta\text{-}C_5H_5)Re(NO)(CO)(CH_3)$ (3) [20]. As will be related, the same metal/ligand arrangements which afford kinetically stable neutral formyls have been found to stabilize other reactive ligand types as well.

## Results and Discussion

Rhenium cations 1 and 2a are synthesized by the convenient procedures shown in Figure 2. The use of iodosobenzene ($C_6H_5I^+\text{-}O^-$) in the conversion of 1 to 2a merits note. Substitution of $Ph_3P$ for a CO in 1 could not be effected by standard thermal or photochemical methods. Furthermore, reaction of 1 with $(CH_3)_3N^+\text{-}O^-$ (which is commonly used for the oxidative removal of metal-bound CO) [12] in the presence of $Ph_3P$ did not yield any CO-containing products. Consequently a more selective reagent for the oxidation of ligating CO to $CO_2$ was sought. After surveying several possibilities, it was found that the reaction of 1 in $CH_3CN$ with commercially available iodosobenzene resulted in the smooth formation of $[(\eta\text{-}C_5H_5)Re(NO)(CO)(NCCH_3)]^+ BF_4^-$. As would be expected from a reaction involving attack of iodosobenzene oxygen upon CO, iodobenzene ($C_6H_5I$) was formed in 77% GLC yield. The $[(\eta\text{-}C_5H_5)Re(NO)(CO)(NCCH_3)]^+ BF_4^-$ could be purified or simply refluxed in crude form with $Ph_3P$ in 2-butanone (substitution was slow in refluxing acetone) to afford 2a in 50-65% overall yield.

The reaction of 1 with $Li(C_2H_5)_3BH$ was investigated first [15]. As shown in Figure 3, the relatively stable neutral formyl $(\eta\text{-}C_5H_5)Re(NO)(CO)(CHO)$ (4) formed in quantitative spectroscopic yield. 4 decomposed over several hours at room temperature, and we were unable to isolate 4 in analytically pure form. Similar syntheses and observations were also reported by Casey [19] and Graham [21].

Despite its instability, reactions of 4 with reducing agents were investigated (Figure 4)[15]. Importantly, $BH_3{\cdot}THF$ smoothly reduced formyl 4 to methyl 3. This suggests that 4 (or a $BH_3$

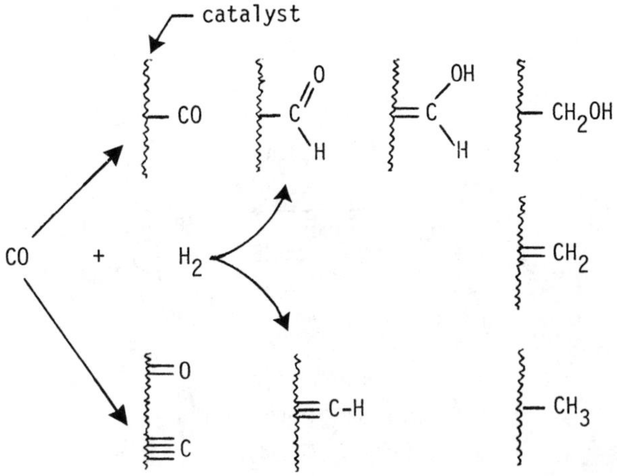


*Figure 1. Possible one-carbon intermediates in CO reduction*

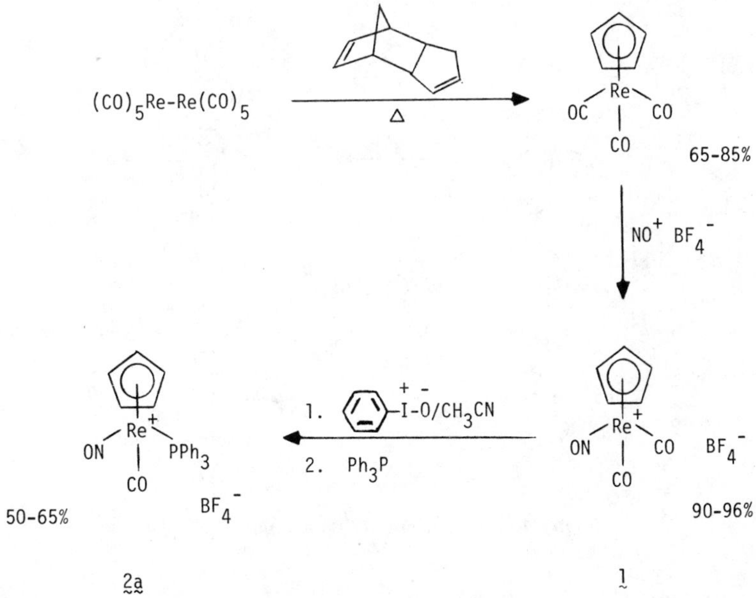


*Figure 2. Synthesis of starting metal carbonyl cations*

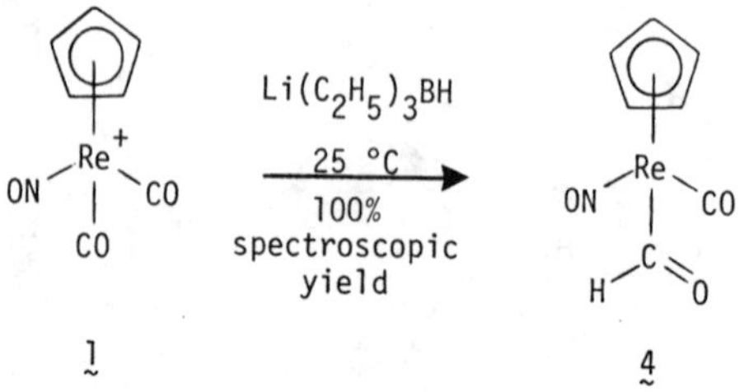


$^1H$ NMR: δ 15.77

$^{13}C$ NMR: 265.9, 200.1, 96.8

(-60 °C, THF-$d_8$)

IR: 1985 s, 1709 s, 1614 s

(THF, $cm^{-1}$)

*Figure 3. A relatively stable neutral formyl complex*

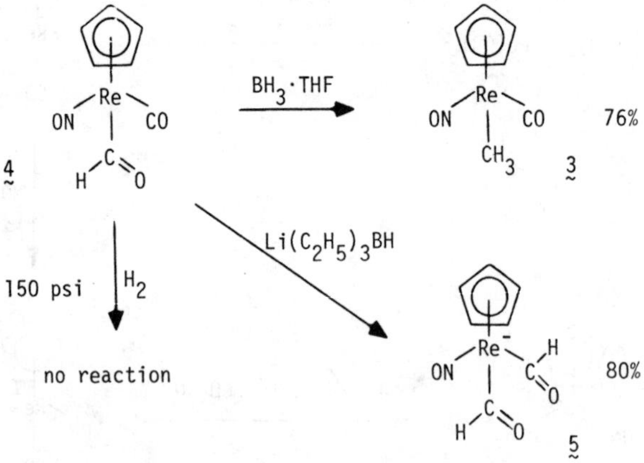


*Figure 4. Reductions of "semi-stable" formyl* **4**

adduct) is an intermediate in Graham's $NaBH_4$ reduction of 1 to 3. Interestingly, reaction of 4 with $Li(C_2H_5)_3BH$ afforded the anionic bis-formyl $Li^+[(\eta\text{-}C_5H_5)Re(NO)(CHO)_2]^-$ (5), derived from attack at the remaining CO of 4. This species had a half life of ca. 2 hr at room temperature. It is significant that the electrophilic reductant $BH_3$ attacks the formyl ligand of 4, whereas the nucleophilic reductant $Li(C_2H_5)_3BH$ attacks the carbonyl ligand of 4. Previously it has been noted that electrophiles such as $Li^+$ bind much more effectively to acyl ligands than carbonyl ligands [13], whereas nucleophiles preferentially attack the carbonyl ligands in metal [25] carbonyl acyls [11,26].

Formyl 4 did not react with 150 psi of $H_2$ at a rate detectably faster than its decomposition, and we considered it unlikely that the other reductions in Figure 4 had an important bearing on the fate of catalyst-bound formyls. Moreover, we sought a crystalline, analytically pure neutral formyl complex whose physical and chemical properties could be subjected to unambiguous definition. Toward this end, the $Ph_3P$-substituted cation 2a would provide a more electron rich rhenium system whose additional phenyl rings might impart greater crystallinity.

Gratifyingly, reaction of 2a with $Li(C_2H_5)_3BH$ afforded the stable (dec pt. ca. 91 °C) formyl $(\eta\text{-}C_5H_5)Re(NO)(PPh_3)(CHO)$ (6), which could be isolated in crystalline analytically pure form (60% yield) after column chromatography (Figure 5)[15]. Later, we found that the reaction of $NaBH_4$ with 2a in $THF/H_2O$ afforded 55-75% yields of 6. This formyl was subjected to an X-ray crystal structure determination [16], a stereoscopic view of which is given in Figure 6. While the characteristic spectroscopic features of formyl complexes have been previously noted [10,11,12,13,19,21], it should be emphasized that the low frequency formyl IR stretch of 6 (1566 $cm^{-1}$ in THF) indicates a substantial resonance contribution from the dipolar carbenoid form 6b. Interestingly, the plane of the formyl ligand virtually eclipses the Re-N-O plane; an identical geometric relationship is observed in the X-ray crystal structures of homologous cationic rhenium alkylidene complexes [27].

Enhanced stability is often detrimental to reactivity, and it came as no surprise that 6 did not react with 150 psi of $H_2$ at 25 °C. When reacted with $BH_3\cdot THF$, 6 was reduced to $(\eta\text{-}C_5H_5)Re(NO)(PPh_3)(CH_3)$ (7) (eq i). Methyl complex 7 could also be obtained by reduction of 2 with $NaBH_4$. However, since the prospects for reduction chemistry relevant to the fate of catalyst-bound formyls seemed bleak, we began to investigate other facets of the chemistry of 6.

Our initial experiment along these lines was a well-precedented attempt to O-methylate the formyl ligand in 6. Considering the seemingly unexciting products shown in Figure 7, I was very grateful that my co-workers were sufficiently curious not to relegate this reaction to the rhenium waste jar. Upon further thought, we speculated that the products 7 and 2b (Figure 7) might constitute formyl reduction and oxidation products, respectively. Therefore

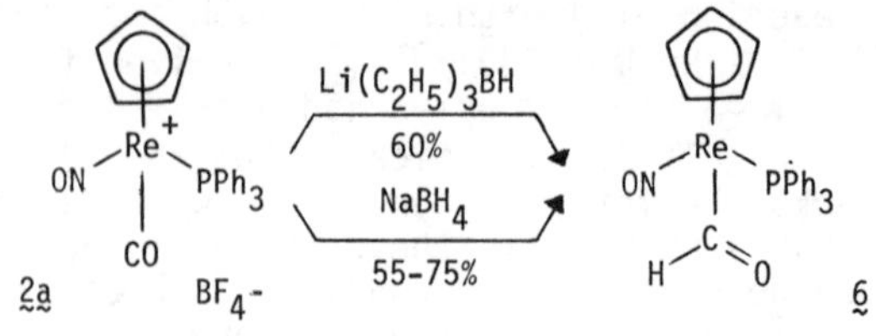


Anal. Calcd: C, 50.34; H, 3.70; N, 2.45, P, 5.41

Found: C, 50.14; H, 3.82; N, 2.39; P, 5.34

IR ($cm^{-1}$, THF): 1663 s, 1566 s;

$^{1}H$ NMR ($C_6D_6$, $\delta$): 17.23

$^{13}C$ NMR (-60 °C, THF-$d_8$): 246.8 ppm.

*Figure 5. An isolable, crystalline neutral formyl complex*

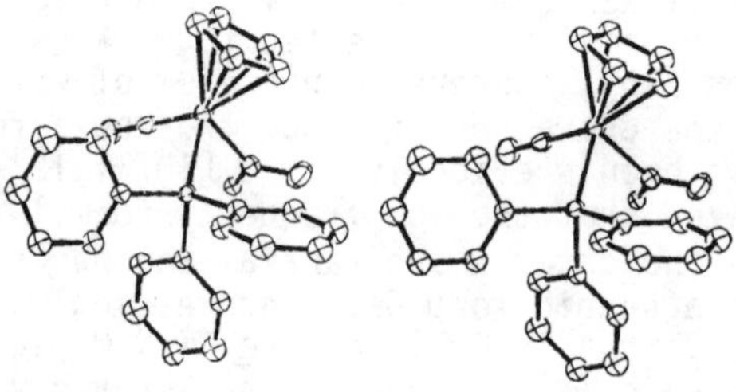

*Figure 6. Stereoscopic view of the x-ray crystal structure of* **6**

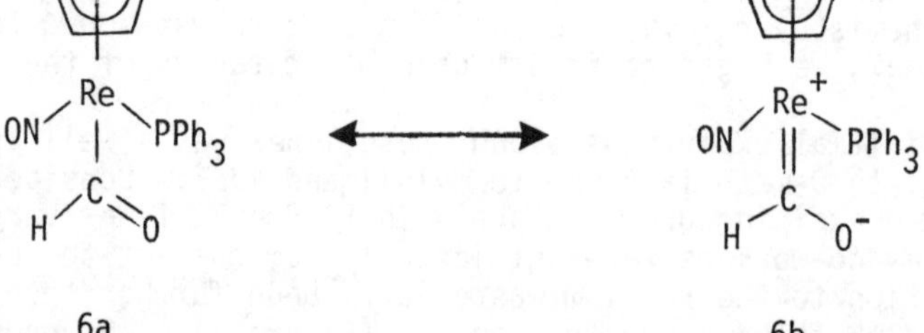


6a

6b

the origin of the methyl ligand was probed by conducting the same reaction with $CD_3SO_3F$ (Figure 8). As shown, the 7 produced was virtually entirely 7-$d_0$. Thus the methyl ligand did not arise from the methylating agent, which strongly suggested that some type of disproportionation was occurring. Remarkably, the formyl ligand was being reduced well below room temperature, and without the addition of an exogenous reducing agent.

Formyl 6 was similarly reacted with electrophiles $(CH_3)_3SiCl$ (Figure 9) and $CF_3CO_2H$ (Figure 10). In both cases, methyl complex 7 and $[(\eta\text{-}C_5H_5)Re(NO)(PPh_3)(CO)]^+$ salts formed in ratios reasonably close to 1:2. The reaction of 6 with $(CH_3)_3SiCl$ also yielded $[(CH_3)_3Si]_2O$ (approximately equimolar with 7), which we postulated to contain oxygen originally from the formyl ligand. $^1H$ NMR monitored reactions of 6 with $CH_3SO_3F$ in $CD_2Cl_2$ showed the presence of similar quantities of dimethyl ether.

In formulating means of unraveling the mechanisms of the reactions in Figures 8-11, we decided to concentrate on the $CH_3SO_3F$ induced disproportionation, since exploratory NMR experiments had shown it to be somewhat slower than the others. We considered the ligand types shown in eq ii (formyl , methoxymethylidene, methoxymethyl, methylidene, methyl) to represent a likely reaction sequence. These have close relationships with several of the potential catalyst-bound intermediates in Figure 1. The $(CH_3)_3SiCl$ and $CF_3CO_2H$ induced disproportionations of 6 might involve $-OSi(CH_3)_3$ and -OH homologs of the $-OCH_3$ containing ligands in eq ii.

A three-stage approach was taken to establish mechanism. First, efforts were directed at the synthesis and isolation of all potential intermediates in the $CH_3SO_3F$ reaction. Secondly, experiments were conducted to test the chemical viability of these intermediates. For instance, since no external reducing agents are added, some of the species in eq ii must be hydride donors, whereas others must be hydride acceptors. Thirdly, after isolating authentic samples of all likely intermediates, it would be possible to convincingly interpret the $^1H$ NMR monitored reaction of 6 with $CH_3SO_3F$, in which numerous transient resonances were observed. In synthesizing the potential intermediates, we worked from right to left through the ligand types shown in eq ii, as detailed in the remainder of this account [17].

Reaction of 3 with $Ph_3C^+PF_6^-$ resulted in the formation of methylidene complex $[(\eta\text{-}C_5H_5)Re(NO)(PPh_3)(CH_2)]^+ PF_6^-$ (8) in 88-100% spectroscopic yields, as shown in Figure 11. Although 8 decomposes in solution slowly at -10 °C and rapidly at 25 °C (the decomposition is second order in 8), it can be isolated as an off-white powder (pure by $^1H$ NMR) when the reaction is worked up at -23 °C. The methylidene $^1H$ and $^{13}C$ NMR chemical shifts are similar to those observed previously for carbene complexes [28]. However, the multiplicity of the $^1H$ NMR spectrum indicates the two methylidene protons to be non-equivalent (Figure 11). Since no coalescence is observed below the decomposition point of 8, a lower limit of $\Delta G^\ddagger$ >15 kcal/mol can be set for the rotational barrier about the rhenium-methylidene bond.

(i)

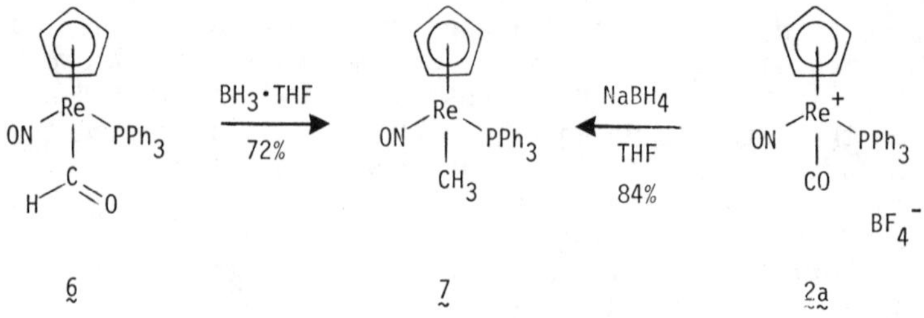


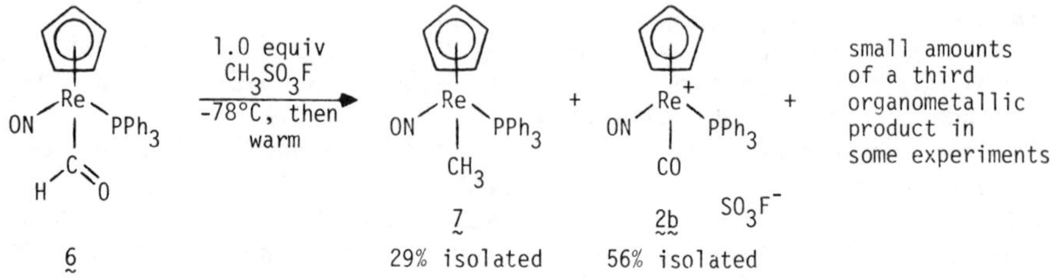


*Figure 7. Reaction of formyl **6** with $CH_3SO_3F$*

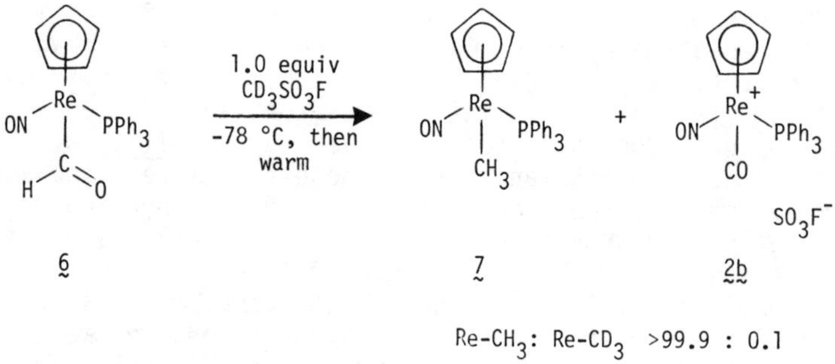


*Figure 8. Origin of Rh-bound methyl*

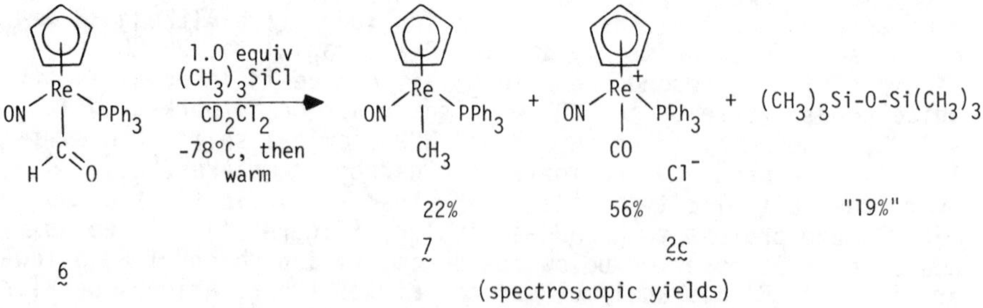


*Figure 9. $(CH_3)_3SiCl$ induced formyl disproportionation*

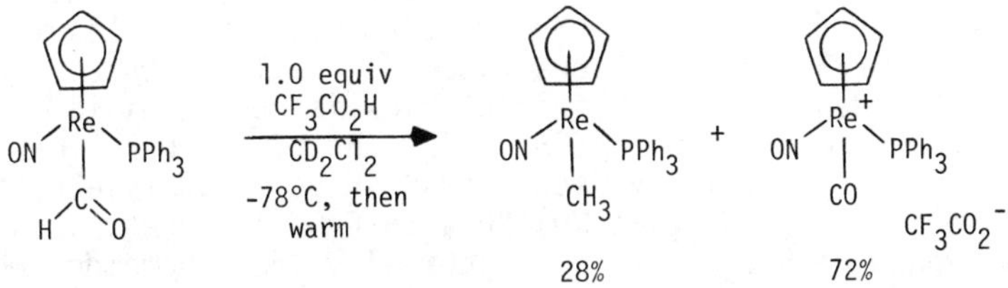

6 7 2d

(spectroscopic yields)

*Figure 10. $CF_3CO_2H$ induced formyl disproportionation*

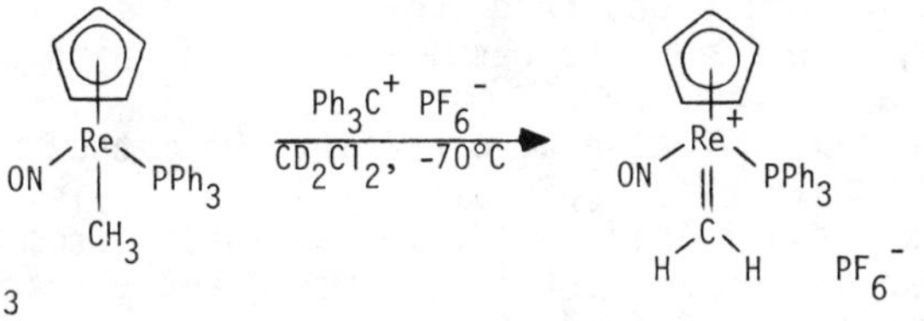

88-100% spectrospic; 85% isolated

3 8

$^1$H NMR: 15.67 (t, $J_{H\text{-}H'} = J_{H\text{-}P} = 4$ Hz)

15.42 (br d, $J_{H\text{-}H'} = 4$, $J_{H'\text{-}P} \leq 1$ Hz)

$^{13}$C NMR, gated decoupled: 290.3 ppm, t, $J_{C\text{-}H} = 151$ Hz

*Figure 11. Synthesis of first detectable electrophilic methylidene complex*

(ii) 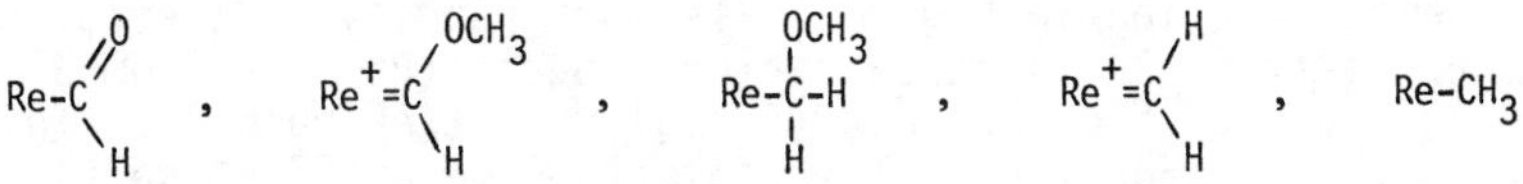

The methylidene complex **8** forms crystalline, analytically pure adducts with pyridine and phosphines, as shown in Figure 12. These reactions establish the methylidene carbon as electrophilic. Relevant to a future mechanistic point, no reaction was observed between **8** and dimethyl ether.

At the time this work was reported, only one other well-characterized non-bridging methylidene complex, Schrock's $(\eta\text{-}C_5H_5)_2Ta(CH_3)(CH_2)$[29], had been described in the literature. However, the methylidene carbon in this complex is nucleophilic, and undergoes ready reaction with $(CH_3)_3SiBr$ and $CD_3I$. Like $(\eta\text{-}C_5H_5)_2Ta(CH_3)(CH_2)$, **8** thermally decomposes (in up to 50% yield) to an olefin complex, $[(\eta\text{-}C_5H_5)Re(NO)(PPh_3)(H_2C{=}CH_2)]^+ PF_6^-$. Since catalyst-bound methylidenes (or higher alkylidene homologs) have been suggested to play important roles in olefin metathesis, olefin cyclopropanation, and Ziegler-Natta polymerization, our studies of **8** are continuing. More recently, additional examples of methylidene complexes have been reported by Brookhart and Flood [30] and Schwartz [31].

At this stage, it must be asked whether or not **8** is a chemically viable intermediate in the formyl disproportions. To be so, it must be able to abstract hydride from other organorhenium species known to be present. Accordingly, when **6** and **8** were mixed in $CD_2Cl_2$ at -70 °C in a $^1H$ NMR monitored reaction, the clean hydride transfer depicted in eq iii occurred immediately (i.e., within the ca. 2-3 minute lag time needed to resume sample spinning and acquire the FT NMR data).

Attention was next directed at preparing the methoxymethyl complex $(\eta\text{-}C_5H_5)Re(NO)(PPh_3)(CH_2OCH_3)$ (**9**). Two high-yield routes were developed, as shown in Figure 13; the synthesis from **8** is slightly more demanding experimentally, since **8** and **9** react rapidly with each other at -70 °C (*vide infra*).

If **9** is to be an intermediate in the $CH_3SO_3F$ induced formyl disproportionation of **6**, it should react with $CH_3SO_3F$ under the conditions of the disproportionation. This would logically lead to dimethyl ether, an observed product, and methylidene **8** ($SO_3F^-$ salt). However, under a variety of conditions, the reaction of **9** with $CH_3SO_3F$ did not yield any detectable **8** ($SO_3F^-$ salt), although the formation of dimethyl ether was always observed. An easier to interpret, "half-methylation" experiment, which resulted in the formation of ca. 1:1:1 ratio of **10a**, **7**, and dimethyl ether, is shown in Figure 14.

The reaction in Figure 14 is more readily understood when it is noted that **10a** is an oxidation product ($H^-$ loss from **9**), whereas **7** is a reduction product (formal $H^-$ attack upon **9**). This suggests that $CH_3SO_3F$ initially converts **9** to **8** ($SO_3F^-$ salt), which then rapidly back reacts with **9** to form the observed products. This can be easily tested by simply reacting **8** and **9** in the $^1H$ NMR monitored reaction shown in Figure 15. Indeed, hydride transfer between **8** and **9** takes place immediately, strongly suggesting that a methylidene intermediate is formed in Figure 14.

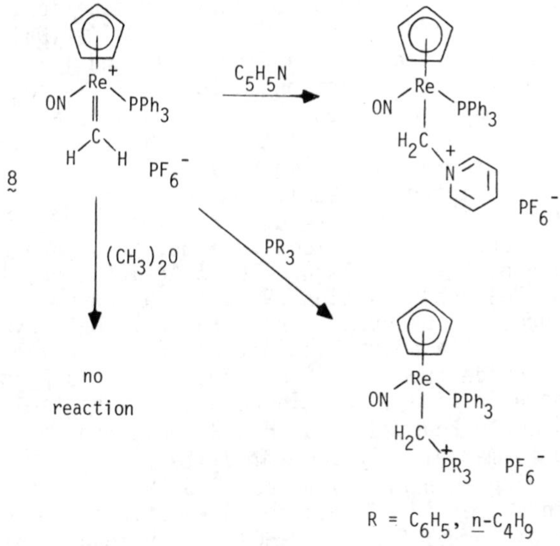


*Figure 12. Formation of methylidene adducts*

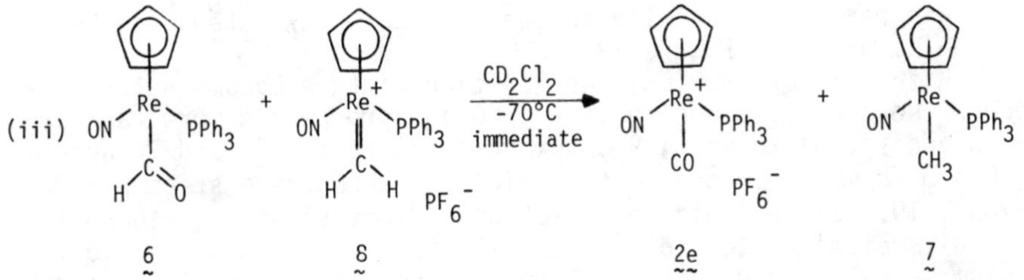


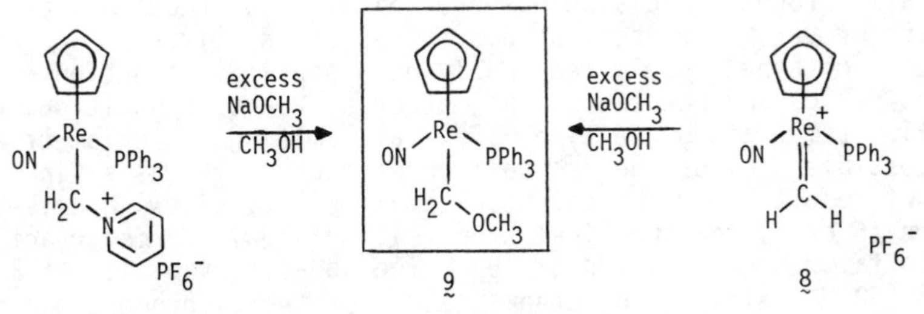


*Figure 13. Synthesis of Rh methoxymethyl complex*

Importantly, the final species under consideration as an intermediate in the $CH_3SO_3F$ induced disproportionation, methoxymethylidene $[(\eta\text{-}C_5H_5)Re(NO)(PPh_3)(CHOCH_3)]^+ SO_3F^-$ (10a), is an isolable product of the reactions in Figures 14 and 15. It is most easily obtained when the reaction in Figure 14 is conducted in toluene, under which conditions 10a precipitates as a toluene solvate. As will be rationalized later, 10a is the "third organometallic product" referred to in Figure 7.

We again attempted to see if 10a could be formed by the methylation of formyl 6. We were unsuccessful, and the most easily interpreted product distributions were obtained when 6 was reacted with 0.5 equiv of $CH_3SO_3F$ (Figure 16). Again, a ca. 1:1 mixture of oxidation (2b) and reduction (9, accompanied by a small amount of "over-reduced" 7) products were obtained. Methoxymethylidene 10a would represent a plausible initial product which might rapidly abstract hydride from starting formyl 6. This hypothesis was tested as shown in Figure 17. Indeed, when isolated 10a and 6 were independently reacted in a $^1H$ NMR monitored reaction at -70 °C, 2b and 9 formed cleanly and immediately.

Since having authentic samples of 8, 9, and 10a enabled us to rigorously interpret $^1H$ NMR monitored reactions, we returned to the reaction of formyl 6 with $CH_3SO_3F$ under conditions similar to those in Figure 7. Accordingly, the addition of 1 equiv of $CH_3$-$SO_3F$ to 6 (0.15 M in $CD_2Cl_2$) at -70 °C resulted in a slow reaction. After warming to -40 °C, formyl 6, 2b, 7, 9, and $CH_3SO_3F$ were present in a 0.5:1.4:0.3:1.0:1.0 ratio; remaining 6 disappeared within 15 minutes. We conclude that at this stage, the disproportionation has passed essentially through the first two steps of the mechanism shown in Figure 18.

With further warming, the reaction mixture became heterogeneous. However, commencing at -10 °C and proceeding more rapidly upon additional warming, 9 disproportionated to 10a and 7, and $(CH_3)_2O$ formed. This transformation corresponds to steps c-e in Figure 19. Since 8 did not react with dimethyl ether (Figure 13), the dissociation step (d) (Figure 18) is likely rapid; however, oxonium salt 11 could conceivably be the species which is reduced by 9 (or 6) to 7.

The mixing of 6 and $CH_3SO_3F$ therefore initiates a complex multistep process involving numerous bimolecular reactions between species of varying concentrations. The precise distribution of products obtained should reasonably be a sensitive function of reactant ratios and concentrations, order of reactant addition, and reaction temperature and time. Several limiting stoichiometries are possible. Also, the sporadic appearance of 10a as a final product results from the potential availability of two hydride donors (6 or 9) for the final step (e). A slowly warmed reaction should favor higher yields of 10a. The non-observability of 10a and 8 ($SO_3F^-$ salt) during steps a and d of the disproportionation is easily understood in terms of the reactions in Figures 14-17.

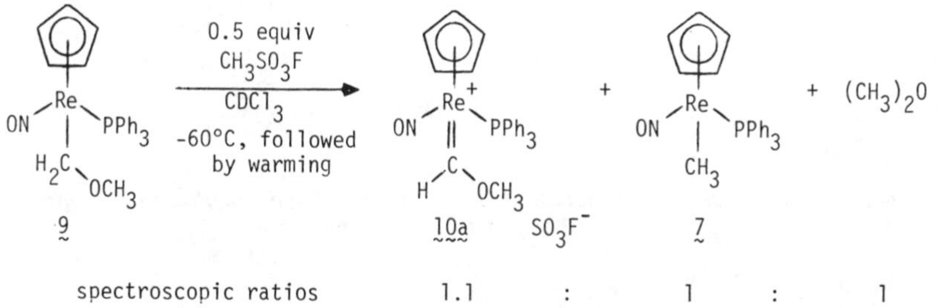


*Figure 14. Half-methylation experiment #1*

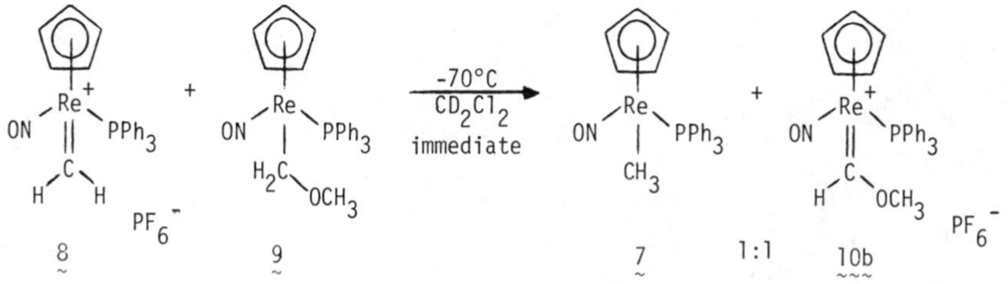


*Figure 15. Reason for the nonobservability of the methylidene complex during the half-methylation experiment*

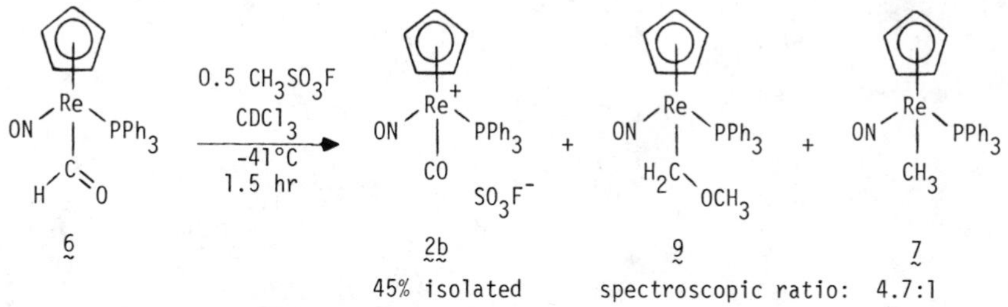


*Figure 16. Half-methylation experiment #2*

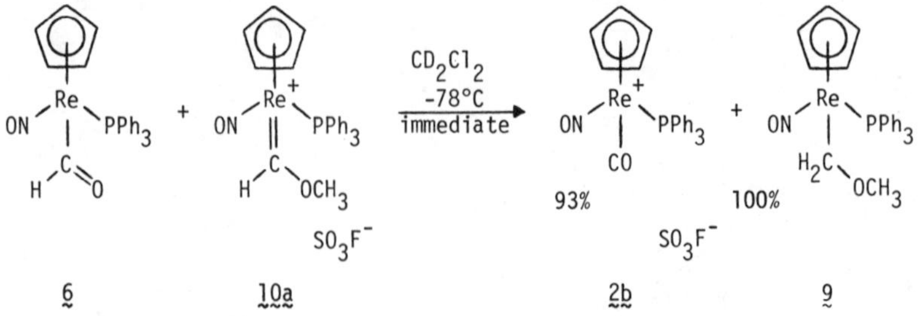

*Figure 17. Reason for the nonobservability of the methoxymethylidene complex during the half methylation experiment*

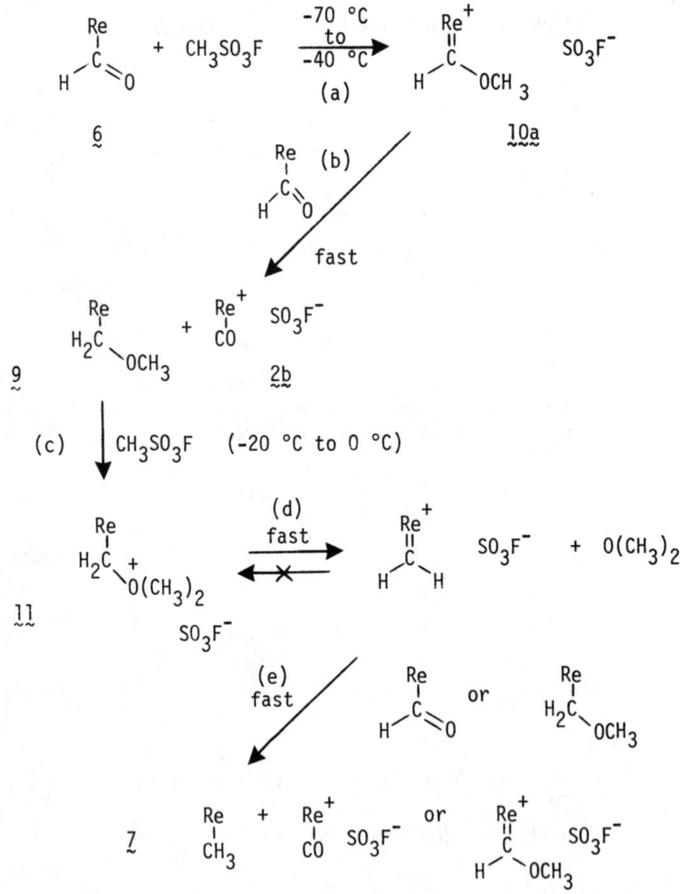

*Figure 18. Mechanism of $CH_3SO_3F$ induced disproportionation of* **6** *(ancillary ligands omitted for clarity)*

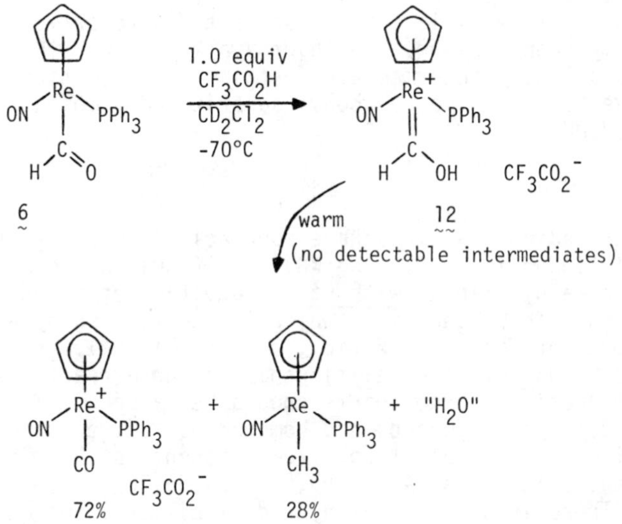


*Figure 19. Detection of hydroxycarbene intermediate in $CF_3CO_2H$ induced disproportionation of* **6**

Available data indicate that $CF_3CO_2H$ and $(CH_3)_3SiCl$ react with 6 by pathways qualitatively similar to the one in Figure 18. Protonation reactions are generally much faster than alkylation reactions. Thus when 6 and $CF_3CO_2H$ are reacted in $CD_2Cl_2$ at -70 °C, a species whose $^1H$ NMR properties indicate it to be $[(\eta\text{-}C_5H_5)Re(NO)(PPh_3)(CHOH)]^+$ $CF_3CO_2^-$ (12, Figure 19) is generated cleanly and quantitatively. Deprotonation to 6 occurs instantly when 12 is reacted with $Li(C_2H_5)_3BH$. The =CHOH ligand (of which 12 is the first detectable complex thereof) is of some historical interest, since it was initially postulated as an intermediate in the Fischer-Tropsch process in 1951 [32]. Upon warming, 12 disproportionates to the product mixture shown in Figure 10 without any detectable intermediates. Referring to a mechanistic scheme for the reaction of 6 and $CF_3CO_2H$ analogous to the one in Figure 18, it can be concluded that step a is rapid relative to step b at -70 °C. However, step a must be reversible, and subsequent disproportionation occurs rapidly upon warming. The hydroxyalkyl $(\eta\text{-}C_5H_5)Re(NO)(PPh_3)(CH_2OH)$ is a likely intermediate in this transformation; Casey and Graham have isolated the carbonyl substituted homolog $(\eta\text{-}C_5H_5)Re(NO)(CO)(CH_2OH)$ [19, 21].

## Overview

The preceding reactions have convincingly demonstrated that the formyl ligand in $(\eta\text{-}C_5H_5)Re(NO)(PPh_3)(CHO)$ can be easily converted to a methyl ligand <u>without</u> the addition of an external reducing agent. This reduction, which is accompanied by a stoichiometric amount of formyl oxidation, occurs well below room temperature. With regard to the relationship of these reactions to catalytic CO reduction, three points should be raised:

(1) Both heterogeneous and homogeneous CO reduction catalyst recipes often contain electrophilic components such as silica supports, metal oxides, and $AlCl_3$ [1,5,33,34,35,36].

(2) There is substantial hydride mobility associated with homogeneous formyl complexes (particularly those which are anionic) [10,11,12,13]. Therefore, the generation of small quantities of catalyst-bound formyls (a step which based upon homogeneous precedent is likely uphill thermodynamically) might be accompanied by a similar electrophile-induced disproportionation.

(3) The reactions of $(\eta\text{-}C_5H_5)Re(NO)(PPh_3)(CHO)$ described were stoichiometric in electrophile "$E^+X^-$." In each case, an "E-O-E" and two $(metal)^+X^-$ species were formed. If an analogous mechanism is to operate on a bona-fide CO reduction catalyst, $H_2$ must be able to convert these end products back to "$E^+X^-$" and $(metal)^0$, concurrently forming $H_2O$. Water is of course a Fischer-Tropsch reaction product [1,32]. While the reduction of oxidized metal species by $H_2$ is commonplace, the suggestion that $H_2$ may regenerate $E^+X^-$ species is more conjectural. Also, as alluded to earlier, there is good evidence that several CO methanation catalysts effect initial CO dissociation to <u>carbide</u> [1,2]; hence we by

no means wish to suggest that electrophilic species may play a role in all CO reduction catalysts.

In important recent work, Shriver has demonstrated that electrophiles can promote the migration of alkyl groups to coordinated CO. Lewis acid adducts of metal acyl complexes are isolated [37]. Thus it is possible that electrophilic species might also facilitate the generation of catalyst-bound formyls.

In conclusion, the use of homogeneous model compounds has enabled the discovery and elucidation of a new formyl reduction mechanism which merits serious consideration as a reaction pathway on certain CO reduction catalysts. Additional studies of the compounds described in this account are actively being pursued.

## Acknowledgement

We thank the Department of Energy for support of this research. We are also grateful for Fellowship support from the Alfred P. Sloan Foundation (JAG) and the Regents of the University of California (WAK, WT).

## References

1. Masters, C. Adv. Organomet. Chem., 1979, 17, 61, and references therein.
2. Muetterties, E.L.; Stein, J. Chem. Rev., 1979, 79, 479, and references therein.
3. Olivé, G.H.-; Olivé, S. Angew. Chem., Int. Ed. Engl., 1979, 18, 77.
4. Pruett, R.L. Ann. N.Y. Acad. Sci., 1977, 295, 239.
5. Fraenkel, D.; Gates, B.C. J. Am. Chem. Soc., 1980, 102, 2478.
6. Clark, G.R.; Headford, C.E.L.; Marsden, K.; Roper, W.R. J. Am. Chem. Soc., 1979, 101, 503.
7. Wong, K.S.; Labinger, J.A. J. Am. Chem. Soc., 1980, 102, 3652.
8. Wong, A.; Harris, M.; Atwood, J.D. J. Am. Chem. Soc., 1980, 102, 4529.
9. Wolczanski, P.T.; Bercaw, J.E. Accts. Chem. Res., 1980, 13, 121.
10. Gladysz, J.A.; Williams, G.M.; Tam, W.; Johnson, D.L. J. Organomet. Chem., 1977, 140, C1.
11. Gladysz, J.A.; Selover, J.C. Tetrahedron Lett., 1978, 319.
12. Gladysz, J.A.; Tam, W. J. Am. Chem. Soc., 1978, 100, 2545.
13. Gladysz, J.A.; Merrifield, J.H. Inorganica Chimica Acta, 1978, 30, L317.
14. Gladysz, J.A.; Selover, J.C.; Strouse, C.E. J. Am. Chem. Soc., 1978, 100, 6766.
15. Tam, W.; Wong, W.-K.; Gladysz, J.A. J. Am. Chem. Soc., 1979, 101, 1589.
16. Wong, W.-K.; Tam, W.; Strouse, C.E.; Gladysz, J.A. J. Chem. Soc., Chem. Commun., 1979, 530.
17. Wong, W.-K.; Tam, W.; Gladysz, J.A. J. Am. Chem. Soc., 1979, 101, 5440.

18. Kiel, W.A.; Lin, G.-Y.; Gladysz, J.A. J. Am. Chem. Soc., 1980, 102, 3299.
19. Casey, C.P.; Andrews, M.A.; McAlister, D.R.; Rinz, J.E. J. Am. Chem. Soc., 1980, 102, 1927.
20. Stewart, R.P.; Okamoto, N.; Graham, W.A.G. J. Organomet. Chem. 1972, 42, C32.
21. Sweet, J.R.; Graham, W.A.G. J. Organomet Chem., 1979, 173, C9.
22. Fischer, E.O.; Strametz, H. Z. Naturforsch. B., 1968, 23, 278.
23. Shvo, Y.; Hazum, E. J. Chem. Soc., Chem. Commun., 1975, 829.
24. Blumer, D.J.; Barnett, K.W.; Brown, T.L. J. Organomet. Chem., 1979, 173, 71.
25. Collman, J.P.; Finke, R.G.; Cawse, J.N.; Brauman, J.I. J. Am. Chem. Soc., 1978, 100, 4766.
26. Casey, C.P.; Bunnell, C.A. J. Am. Chem. Soc., 1976, 98, 436.
27. W.A. Kiel and A.T. Patton, unpublished results, UCLA, 1980.
28. Brookhart, M.; Nelson, G.O. J. Am. Chem. Soc., 1977, 99, 6099.
29. Schrock, R.R.; Sharp, P.R. J. Am. Chem. Soc., 1978, 100, 2389.
30. Brookhart, M.; Tucker, J.R.; Flood, T.C.; Jensen, J. J. Am. Chem. Soc., 1980, 102, 1203.
31. Schwartz, J.; Gell, K.I. J. Organomet. Chem., 1980, 184, C1.
32. Storch, H.H.; Columbic, N.; Anderson, R.B. "The Fischer-Tropsch and Related Syntheses," Wiley, New York, 1951.
33. Demitras, G.C.; Muetterties, E.L. J. Am. Chem. Soc., 1977, 99, 2796.
34. Doesburg, E.B.M.; Orr, S.; Ross, J.H.R.; van Reijen, L.L. J. Chem. Soc., Chem. Commun., 1977, 734.
35. Nijs, H.H.; Jacobs, P.A.; Uytterhoeven, J.B. J. Chem. Soc., Chem. Commun., 1979, 1095.
36. Ichikawa, M. J. Chem. Soc., Chem. Commun., 1978, 566.
37. Butts, S.B.; Strausse, S.H.; Holt, E.M.; Stimson, R.E.; Alcock, N.W.; Shriver, D.F. J. Am. Chem. Soc., 1980, 102, 5093.

RECEIVED December 8, 1980.

# Hydrocarbon Formation on Polymer-Supported $\eta^5$-Cyclopentadienyl Cobalt

LINDA S. BENNER, PATRICK PERKINS, and K. PETER C. VOLLHARDT

Department of Chemistry, University of California, and the Materials and Molecular Research Division, Lawrence Berkeley Laboratory, Berkeley, CA 94720

There has been considerable recent research interest in the activation of carbon monoxide en route to more complex organic molecules. Among the various reactions that have been investigated and/or newly discovered, the transition metal catalyzed reduction of CO to hydrocarbons (Fischer-Tropsch synthesis) has enjoyed particular attention (1-9). Whereas most of the successful efforts in this area have been directed toward the development of heterogeneous catalysts, there are relatively few homogeneous systems. Among these, two are based on clusters (10,11) and others are stoichiometric in metal (12-17). In this report we detail the synthesis and catalytic chemistry of polystyrene (Ⓟ) supported $\eta^5$-cyclopentadienyldicarbonyl cobalt, $CpCo(CO)_2$. This material is active in the hydrogenation of CO to saturated linear hydrocarbons and appears to retain its "homogeneous", mononuclear character during the course of its catalysis.

Since cobalt on kieselguhr in one of the original Fischer-Tropsch catalysts (1-9), it appeared attractive to investigate the catalytic activity of cobalt complexes immobilized on polystyrene. Although there are many supported cobalt-based Fischer-Tropsch catalysts known (see, for example, references 18-21), no polystyrene-bound systems had been reported. During the course of our work 18% (22,60,61) and 20% (23) crosslinked analogs of Ⓟ $CpCo(CO)_2$ were shown to exhibit limited catalytic activity but no CO reduction. A preliminary disclosure of our work has appeared (24).

## Synthesis of Polystyrene Supported Catalysts

Ⓟ $CH_2CpCo(CO)_2$ **3** and Ⓟ $CpCo(CO)_2$ **5** were prepared utilizing the procedures of Grubbs et al. for the syntheses of polystyrene-bound cyclopentadiene (25) and Rausch and Genetti for the synthesis of $CpCo(CO)_2$ (26). Thus, for **3**, commercially available Ⓟ $CH_2Cl$ (1% DVB, microporous, 1.48 mmol Cl/g. resin) was treated with excess NaCp to form Ⓟ $CH_2CpH$ **2**. This was then exposed to $Co_2(CO)_8$ to form desired compound **3** (0.3-0.5 mmol Co/g. resin,

0097-6156/81/0152-0165$05.25/0
© 1981 American Chemical Society

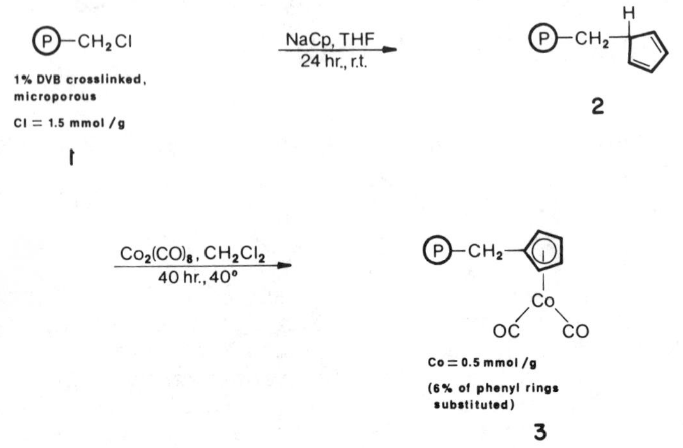
P—CH2Cl
1% DVB crosslinked, microporous
Cl = 1.5 mmol /g
1
NaCp, THF
24 hr., r.t.
P—CH2
H
2
Co2(CO)8, CH2Cl2
40 hr., 40°
P—CH2
Co
OC
CO
Co = 0.5 mmol /g
(6% of phenyl rings substituted)
3

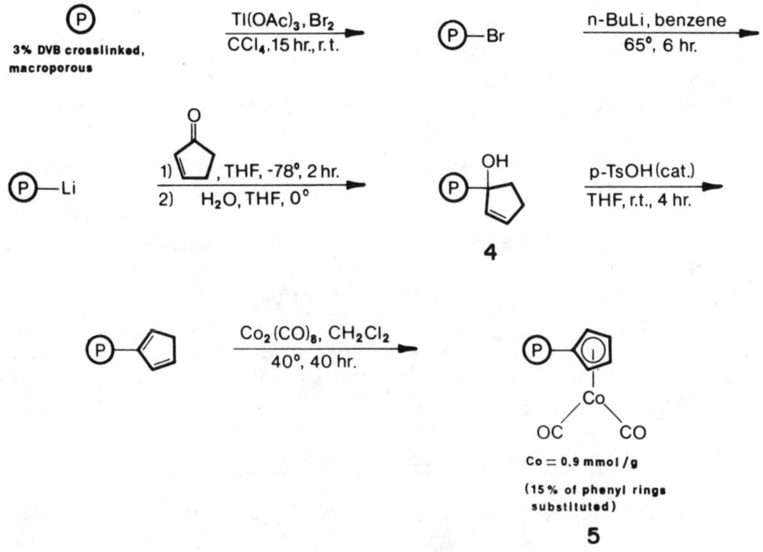
P
3% DVB crosslinked, macroporous
Tl(OAc)3, Br2
CCl4, 15 hr., r.t.
P—Br
n-BuLi, benzene
65°, 6 hr.
P—Li
O
1) , THF, -78°, 2 hr.
2) H2O, THF, 0°
OH
P
4
p-TsOH (cat.)
THF, r.t., 4 hr.
P
Co2(CO)8, CH2Cl2
40°, 40 hr.
P
Co
OC
CO
Co = 0.9 mmol /g
(15% of phenyl rings substituted)
5

ca. 6% of phenyl rings substituted); Soxhlet extraction (benzene or $CH_2Cl_2$) was used in an attempt to remove non-attached species. Resin 3 showed the characteristic IR absorptions of an $M(CO)_2$ complex: 2012 and 1954 $cm^{-1}$ (KBr); compare with $CpCo(CO)_2$: 2033 and 1972 $cm^{-1}$ (cyclohexane); 2017 and 1954 $cm^{-1}$ (acetone) (27).

A problem associated with this procedure is the difficulty in removing excess reagents from the microporous resin. The chloride content was fairly high (0.25 mmol/g., ca 15% of original) in Ⓟ$CH_2CpH$ 2; as no chloromethyl absorbance was seen in the IR, this implied that NaCl was trapped in the resin. Elemental analysis (C, 88.90%; H, 7.47%; Cl, 0.90%; total, 97.27%) suggested the presence of other impurities, which appeared to persist even after extensive extraction with solvent (THF-ethanol).

Assuming that the hydrophobic and nonionic nature of the polystyrene matrix prevented infiltration by solvent, an alternative synthesis was devised starting with macroporous 3% cross-linked polystyrene. This resin was first washed (28) with: $CH_2Cl_2$, THF, THF saturated with lithium aluminum hydride, THF, twice with 1M HCl (97°C), 1M KOH (75°C), 1M HCl (75°C), three times with $H_2O$ (75°C), DMF (40°C), twice with $H_2O$, 1M HCl (80°C), $H_2O$ (80°C), methanol, methanol/$CH_2Cl_2$ (1/1), methanol/$CH_2Cl_2$ (1/3), and $CH_2Cl_2$. After drying in vacuo (70°C, 24 hr.), the resin was brominated using 1.05 equivalents bromine with $Tl(OAc)_3$ as a catalyst (29). Cream-white ⓅBr showed 99% substitution (43.44% Br, 5.44 mmol/g.) at the para ring position (IR). If anhydrous conditions were not used in the bromination, a heterogeneous mixture of beads ranging from red-brown to white in color was obtained.

Lithiation was achieved using two portions of n-butyllithium (n-BuLi), three equivalents each, at 65°C in benzene under nitrogen atmosphere (29). The brown ⓅLi was not isolated due to its extreme reactivity; bromine analysis on the final product, Ⓟ$CpCo(CO)_2$ 5, however, proved that the lithiation at this step was essentially complete (0.1 mmol residual Br/g. resin). Upon addition of 2-cyclopentenone in THF at -78°C (1.01 equivalents, syringe pump addition over 2h), the brown color changed gradually to beige. Warming to room temperature followed by quenching with ice-cold $H_2O$ yielded cyclopentenol derivative 4 (IR (KBr), 3425 and 1047 $cm^{-1}$). Some unsubstituted phenyl rings and presumably 3-(polystyryl)cyclopentanone (IR (KBr), 1730 $cm^{-1}$, result of conjugate addition of ⓅLi to 2-cyclopentenone) were noted. Addition of excess enone gave no increase in ring substitution; faster addition or higher reaction temperature (0°C) yielded a higher proportion of cyclopentanone formation (IR).

Vacuum dehydration of 4 was unsuccessful (0.005 torr, temperatures from 65-190°C, 24 hr.) in contrast to other reports (25,30). However, mild treatment (31) with p-toluenesulfonic acid (THF, 25°C, 4h) produced the desired ⓅCpH (IR (KBr), 675 $cm^{-1}$, loss of bands due to 4). Higher temperature (60°C) or longer reaction times caused discoloration (black resin).

Reaction of Ⓟ CpH with excess $Co_2(CO)_8$ ($CH_2Cl_2$, 40°C, 40-48 h) yielded Ⓟ $CpCo(CO)_2$ **5**, (0.8-1.0 mmol Co/g., IR (KBr), 2012 and 1953 $cm^{-1}$). Residual $Co_2(CO)_8$ was removed via Soxhlet extraction with benzene or $CH_2Cl_2$. Interestingly, reaction of **4** with $Co_2(CO)_8$ yielded **5** directly and as efficiently. Apparently, dehydration of **4** occurred under these conditions. The "overall yield" (based on cobalt incorporation) for this synthetic sequence is 15%.

The relatively low percentage of ring substitution can be attributed to several side reactions: 1,4-addition of Ⓟ Li to 2-cyclopentenone, incomplete dehydration of **4** as evidenced by the presence of a small hydroxyl absorption (3425 $cm^{-1}$) in the IR spectrum of **5**, and reduction of the polymer-bound cyclopentadiene in its reaction with $Co_2(CO)_8$ (26,27,32).

Resin **3** is light orange in color when dry; **5** is tan. Both turn dark brown in a swelling solvent. Exposure of either resin to air slowly results in a grey-green color due to oxidation, but 28% of the resin-bound $CpCo(CO)_2$ is left after one month's exposure (IR).

## Thermal and Photochemical Decarbonylation

A brief investigation of the potential thermal and photolytical chemistry of resins **3** and **5** was undertaken. It was prompted by a report of the observation of site-site isolation on irradiation of 18% crosslinked **3** (22). This is in contrast to soluble $CpCo(CO)_2$ which forms di- and trinuclear clusters under the same conditions (33,34). Similar apparent prevention of the formation of higher nuclear species was noted in the decarbonylations of Ⓟ $CH_2Fe(CO)_2H$ (22) and Ⓟ $CH_2CpM(CO)_3H$ (M = Cr, Mo, W) (35), and in the catalytic activity of Ⓟ $CH_2CpTiCp$ (25). However, these species are all supported by a relatively highly crosslinked polymer, and there was ample indication that lesser crosslinking enables "bimolecular" reactions of active centers (36,37,38,39).

Irradiation of Ⓟ $CH_2CpCo(CO)_2$ **3** (1% DVB, 6% ring substitution, brown color) was carried out through Pyrex (-20 to 25°C, toluene). A red-brown cast was apparent on the resin after about 15 minutes; continued irradiation resulted in a gradual change of the resin's color (green-black). A gradual loss in the terminal CO absorptions due to Ⓟ $CH_2CpCo(CO)_2$ was seen by IR analysis (2012 and 1954 $cm^{-1}$), accompanied by a rise and decline of two bands in the bridging CO region (1790 and 1770 $cm^{-1}$). After cessation of irradiation (4h), only 6% of the resin-bound $CpCo(CO)_2$ remained; no other carbonyl-containing species were present. In analogy to $CpCo(CO)_2$, it appears that **3** forms (Ⓟ $CH_2CpCoCO)_2$ **6** (1770 $cm^{-1}$) and (Ⓟ $CH_2CpCo)_2(CO)_3$ **7** (1790 $cm^{-1}$) upon irradiation. Upon thermal decarbonylation (190°C, vacuum or 128°C, n-octane), **3** turned bright green in a few minutes; however, IR analysis indicated the absence of new carbonyl-containing species.

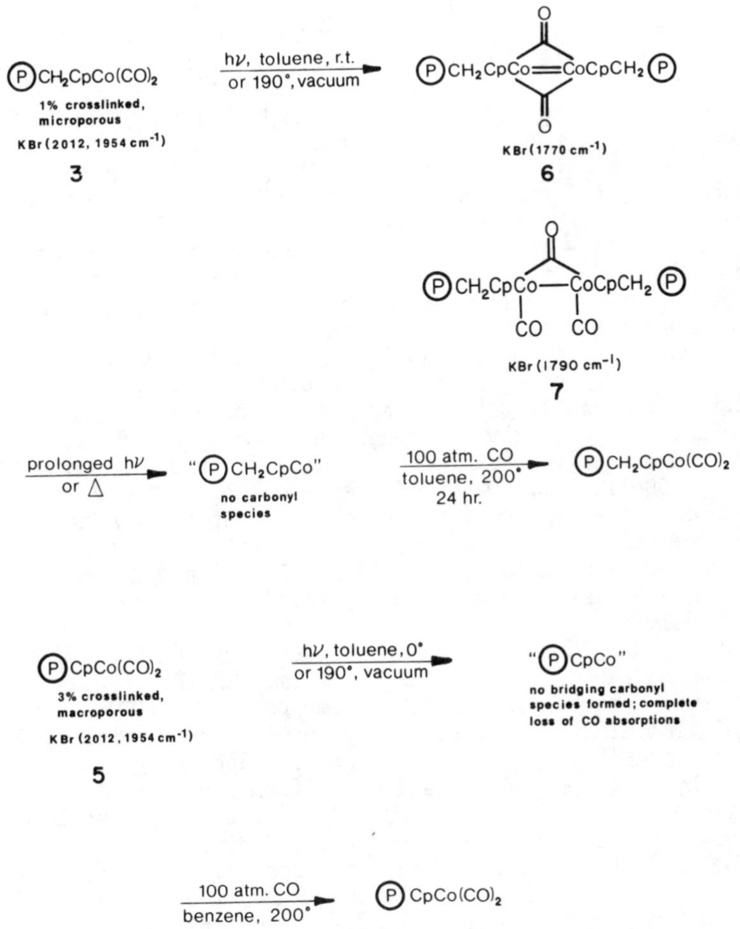
Ⓟ CH₂CpCo(CO)₂
1% crosslinked, microporous
KBr (2012, 1954 cm⁻¹)
3
hν, toluene, r.t. or 190°, vacuum
Ⓟ CH₂CpCo=CoCpCH₂ Ⓟ
KBr (1770 cm⁻¹)
6
Ⓟ CH₂CpCo—CoCpCH₂ Ⓟ
CO CO
KBr (1790 cm⁻¹)
7
prolonged hν or Δ
"Ⓟ CH₂CpCo"
no carbonyl species
100 atm. CO toluene, 200° 24 hr.
Ⓟ CH₂CpCo(CO)₂
Ⓟ CpCo(CO)₂
3% crosslinked, macroporous
KBr (2012, 1954 cm⁻¹)
5
hν, toluene, 0° or 190°, vacuum
"Ⓟ CpCo"
no bridging carbonyl species formed; complete loss of CO absorptions
100 atm. CO benzene, 200°
Ⓟ CpCo(CO)₂

Continued heating resulted in darkening of the resin, formation of a 1770 $cm^{-1}$ absorption for 6 in the IR, development of a broad absorption covering 1700-1300 $cm^{-1}$, and decrease of the absorptions due to Ⓟ $CH_2CpCo(CO)_2$. Evidently, 7 was not formed under these conditions. No other absorptions were noted in either the thermal or photolytic decarbonylation. It can thus be concluded that the 1% DVB matrix with 6% substitution does not maintain site-site isolation.

Decarbonylation of Ⓟ $CpCo(CO)_2$ 5 (3% DVB, 15% ring substitution) via photolysis produced no color change, but a gradual disappearance of the terminal IR carbonyl absorptions. No other CO bands were noted. Vacuum thermolysis (0.005 torr, 110-190°C) resulted in slow color changes from tan to bright green to bronze to steel gray. During this time, the only changes noted in the IR spectrum were the slow, complete disappearance of the CO absorptions due to 5 and the development of a broad absorption from 1700-1300 $cm^{-1}$. Again no species such as 6 and 7 were observable. Thus, resin 5 is either capable of preventing the aggregation of the coordinatively unsaturated intermediate metal centers, or the species resulting from such aggregation are kinetically unstable under the reaction conditions.

Uptake of CO (1 atm. CO, 50°C, 12 h, or 9.5 atm. CO, room temp., 40 h) by the decarbonylated resins was sluggish; under conditions comparable to the complete regeneration of $CpCo(CO)_2$ from clusters (33), only 20% regeneration of 3 and 5 was detected by IR. Quantitative recarbonylation (IR spectrum) was achieved by using 100-110 atm. CO, 200°C, benzene solvent for 24 h. No new carbonyl bearing species were detected at intermediate stages of this reaction. The strenuous conditions are necessary, as a reaction for 18 h yielded only 92% reconversion. Minimal cobalt loss from the resin had occurred by this drastic treatment (elemental analysis).

The data, although inconclusive, suggest the possibility that mononuclear cobalt species are formed in the irradiative and thermal decarbonylation of polymer-bound $CpCo(CO)_2$. These may gain coordinative saturation by interaction with the polymer backbone, possibly by π-donation from a neighboring phenyl group (22) or by some sort of oxidative addition process into a phenyl-hydrogen or benzyl-hydrogen bond. Two speculative possibilities are shown in 8 and 9. Such structures, particularly 8, should they be present in the decarbonylated resins, might show interesting activity in the catalytic chemistry of carbon monoxide and other unsaturated small molecules. Indeed (in addition to hydrocarbon formation to be described subsequently) some, albeit limited, activity was observed in the hydroformylation reaction, included in this account, and in the catalytic trimerization of alkynes, reported elsewhere (24). Interestingly, although the latter activity was also observed by others when using a 20% crosslinked Ⓟ $CH_2CpCo(CO)_2$, the former was not (23). On the other hand, hydroformylation activity was detected with silica gel supported $CpCo(CO)_2$ (40), free $CpCo(CO)_2$ (41,60), and 18% Ⓟ$CH_2CpCo(CO)_2$ (60,61).

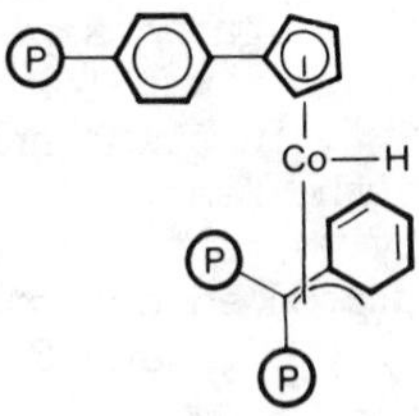


**8**

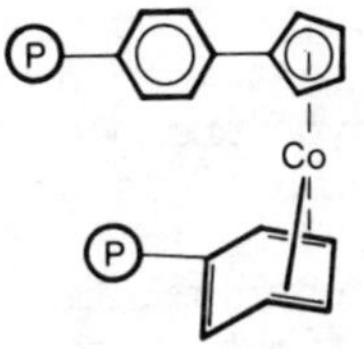


**9**

## Hydroformylation of 1-Pentene in the Presence of (P) $CpCo(CO)_2$

There are several polymer supported transition metal hydroformylation catalysts (42). Most are attached by phosphine ligation and suffer from catalyst leaching. There are no $\eta^5$-cyclopentadienyl half sandwich systems despite the potentially, clearly advantageous presence of the relatively strong Cp-metal bond (43,44). Resin 5 was used in the following brief study in which the potential of polystyrene-supported $CpCo(CO)_2$ to function as a hydroformylation catalyst was tested.

The reaction was conducted at 270 psig., 1/1 $H_2/CO$, 140°C, 90 h, using purified n-octane as the solvent and 1-pentene as the substrate. During this time, a slow pressure drop ensued. Analysis of the solution by gas chromatography yielded: hexanal (13%), 2-methylpentanal (11%), pentane (3%), 2-pentene (21%); the balance was unreacted 1-pentene (52%). The normal to branched ratio was poor (1.1/1), indicating the failure of the resin to exert significant steric control. Olefin isomerization is evidently competitive with product formation (ratio of aldehydes to 2-pentene was 1.14/1); this would contribute to a low n/b ratio. No alcohols were observed, and no gaseous products (e.g. methane) were noted. The turnover number was low (1.68 mmol aldehyde produced per mmol Co per day), in accordance with the mild conditions used. In contrast to the reaction of 5 with alkynes (24) no loss of catalytic activity was observed on prolonged hydroformylation. This result prompted an extension of this investigation to the potential activity of 3 and 5 in the Fischer-Tropsch reaction.

## Fischer-Tropsch Catalysis by Polystyrene Supported $CpCo(CO)_2$

The conditions chosen for this investigation (3/1 $H_2/CO$, 100 psig. (6.8 atm), 140°C) are analogous to those typically encountered in Fischer-Tropsch catalysis by heterogeneous cobalt (175-300°C, 5-30 atm.) (1). The reaction was conducted in a medium pressure glass and stainless steel reactor (static conditions), heated by an external oil bath at 190°C. The temperature inside the reactor averaged at ca. 140°C. (P) $CpCo(CO)_2$ 5 was suspended in purified n-octane to swell the resin and allow access to catalyst sites in the interior of the beads; no detectable reaction took place if the solvent was excluded. Upon reaching the reaction temperature, a slow pressure drop ensued; an induction period could not be ascertained. Analysis of the solution phase after 100 h by gas chromatography showed the existence of methane and higher hydrocarbons up to $C_{25}H_{52}$ as shown in Figure 1. In addition to $H_2$ and CO, methane was predominant in the gas phase (gas chromatography, mass spectra, IR). Smaller

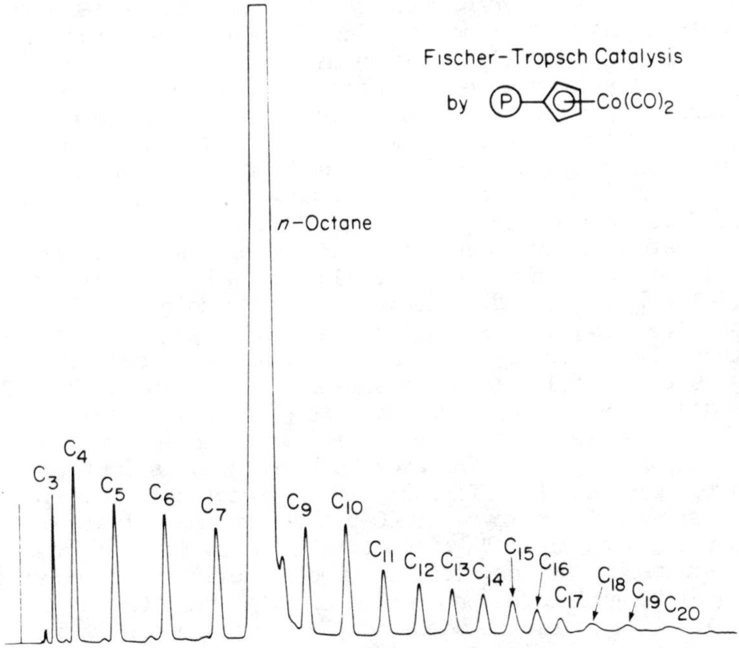
Fischer-Tropsch Catalysis
by (P)—Co(CO)2
n-Octane
C3
C4
C5
C6
C7
C9
C10
C11
C12
C13
C14
C15
C16
C17
C18
C19
C20

*Figure 1.*

amounts of ethane, propane, butane, water, and $CO_2$ were also detected in the gas phase. The used catalyst had suffered a 50% loss of the absorbance intensity of the terminal CO stretches due to Ⓟ $CpCo(CO)_2$ 5; no additional IR bands were observed. This, however, did not result in a loss of catalytic activity, as recycling the catalyst with a fresh charge of n-octane and $H_2/CO$ led to a pressure drop at the same rate. The solution after reaction was colorless and (after removal of the catalyst beads) catalytically inactive; thus the catalytic activity clearly resided within the resin.

As a check to confirm that no extraneous non-polymer-attached catalytic species were present, the following experiment was performed. Polystyrene without attached cyclopentadiene was exposed to $Co_2(CO)_8$, extracted using a Soxhlet extractor and dried in vacuo in exactly the same manner as was used to synthesize 5. When used under the above Fischer-Tropsch reaction conditions, these treated, white polystyrene beads did not discolor, release any detectable species into solution, cause a $CO/H_2$ pressure drop, or result in the formation of any detectable amounts of methane. These observations argue against the presence of small amounts of occluded $Co_2(CO)_8$ or $Co_4(CO)_{12}$ which could conceivably have been active or precursors to active species. It should be noted that the above clusters were reported to be essentially inactive under Fischer-Tropsch conditions (140°C, toluene, 1.5 atm., 3/1 $H_2/CO$, three days) leading to mere traces of methane (11). The lack of products under our conditions also indicates that, at least in the absence of resin-bound $CpCo(CO)_2$ or its derivatives, the polystyrene support did not degrade.

The extremely low turnover rate (0.011 mmol CO/mmol Co/h, production of 0.003 mmol $CH_4$/mmol Co/h average), stimulated attempts to increase the efficiency of the reaction. Indeed, pretreatment involving removal of CO via vacuum thermolysis of Ⓟ $CpCo(CO)_2$ 5 to yield "Ⓟ CpCo", followed by its use in a Fischer-Tropsch reaction, led to improved activity (turnover of 0.130 mmol CO/mmol Co/h, production of 0.053 mmol $CH_4$/mmol Co/h average). It appears that decarbonylation is necessary for formation of the catalytically active species.

Resin 3 was found to be more than 100 times less active than 5, hence all subsequent experiments were carried out using 5.

The number of turnovers (10) in hydrocarbon production without apparent decrease in activity proved that the reaction was indeed catalytic. The IR spectrum of the recovered resin showed small absorptions due to Ⓟ $CpCo(CO)_2$ 5 due to some recarbonylation of "Ⓟ CpCo". In addition, a broad distinct band at 1887 $cm^{-1}$ was seen. The identity of the species exhibiting this carbonyl band is still a mystery; in particular, the band position does not match that reported for any of the $CpCo(CO)_2$-derived di- and trinuclear carbonyls (vide supra). It is tempting to associate this band with some catalytic intermediate, such as the polymer-bound analogues of $CpCo(H)_2(CO)$ and $CpCo(H)(Ph)(CO)$, but this is pure speculation.

Recarbonylation of the used resin was successful at 1430 psig. CO (97.3 atm.) and 200°C; previous experiments with "Ⓟ CpCo" (vide supra) suggested the necessity of such drastic conditions. The intensity of the Ⓟ $CpCo(CO)_2$ absorptions returned to the original value, and an analysis of the resin showed no cobalt loss (elemental analysis).

The Fischer-Tropsch reaction with "Ⓟ CpCo" was repeated using deuterium; GC/MS analysis after five turnovers confirmed the existence of per-deuterated hydrocarbons in solution. In addition, all the major fragment ions were of even mass, indicating that virtually no undeuterated or partially deuterated products were formed (all odd masses <15% of the base peak). Some $CD_3H$ was present (mass spectrum) (approximately 10% of the amount of $CD_4$). The IR spectrum of the resin from this reaction showed some regeneration of the attached $CpCo(CO)_2$ groups, the band at 1887 $cm^{-1}$, and a weak band at 2168 $cm^{-1}$, assigned to a C-D stretch. It appears that $CD_3H$ arises via insertion of cobalt into a phenyl or benzyl C-H bond, exchange of the hydride with the deuterium gas to form a cobalt deuteride, and reductive elimination to form a monodeuterated phenyl ring or benzyl position ($\nu_{C-D} \sim$ 2140-2200 $cm^{-1}$); reaction of the HD (or cobalt hydride) so produced would yield $CD_3H$. Both the average rate of CO consumption (0.064 mmol/mmol Co/h) and average rate of $CD_4$ production (0.040 mmol/mmol Co/h) were lower than the rates for the $H_2$/CO reaction (0.130 and 0.053 mmol/mmol Co/h, respectively) implying a possible deuterium isotope effect on the reaction.

High $H_2$/CO ratios in the Fischer-Tropsch reaction on metal surfaces result in the production of methane in high selectivity. In line with these observations, pure hydrogen over Ⓟ $CpCo(CO)_2$ **5** (0.291 mmol Co) was used in one experiment (75 psig $H_2$ at 25°C, 68h, 140°C). This reaction produced a mixture of hydrocarbon gases (GC/MS): methane (120 μmol), ethane (1.7 μmol), propane (0.17 μmol), isobutane (ca. 0.001 μmol), and n-butane (0.09 μmol); 21% of the carbon of the original resin-bound CO was accounted for in these products. Free CO was found (26% of the original present on the resin (determined by GC); 46% of the CO was still bound to the resin (IR). When exposed to 3/1 $H_2$/CO, this pretreated resin showed Fischer-Tropsch activity similar to decarbonylated **5**. Thus, higher $H_2$/CO ratios favor methanation, as expected, and unlike many heterogeneous catalysts, there is no advantage in hydrogen pretreatment.

Further control experiments were performed to rule out the possibility of hydrocarbon production from sources other than CO. To this end, "Ⓟ CpCo" with 2.2% residual Ⓟ $CpCo(CO)_2$ (0.015 mmol CO) was exposed to the same partial pressure of $H_2$ as used above in the absence of CO under the experimental conditions. The reaction produced 0.014 mmol $CH_4$ total; no other hydrocarbon products were detected. It is clear that only CO, not the polystyrene support, attached catalyst, or solvent, is the carbon source for the hydrocarbon products. Further evidence in support of this statement is obtained from the mass balance on carbon in all the reactions described; starting carbon inventory based on CO and ending carbon inventory based on hydrocarbons and unreacted CO always agree to within 10%. If a contribution from another source were present (degradation of the polymer, hydrocracking of the n-octane) in significant amounts, there should be a gross discrepancy. Finally, an experiment using $^{13}CO/H_2$ and 5 furnished completely $^{13}C$-labeled hydrocarbons (g.c. mass spectrometry up to $C_{25}H_{52}$).

The ability of Ⓟ$CpCo(CO)_2$ 5 or a derived species to catalyze the Fischer-Tropsch reaction suggested that some conditions might be found whereby soluble $CpCo(CO)_2$ might exhibit the same activity. To these ends, $CpCo(CO)_2$ was subjected to a variety of conditions: $H_2$, $H_2/CO$, 150°C, 190°C, n-octane solvent, toluene solvent, with or without added polystyrene, 2-5 days reaction time, in various permutations. Essentially complete decomposition of the complex was noted for all cases. Methane (amounting to 5-10% of the CO originally present) was produced, especially under forcing conditions (higher temperature, higher partial pressure of $H_2$, no added CO). Small amounts of higher hydrocarbons were detected, but the major products were cyclopentadiene, cyclopentene, and cyclopentane (1.5/88.7/9.8). A shiny coating on the walls of the glass reaction vessel (assumed to be cobalt metal) was present after reaction. Significantly, no methane or higher hydrocarbons were observed until the onset of decomposition, usually 24 h after the start of the reaction. It can be concluded that $CpCo(CO)_2$ does not possess Fischer-Tropsch activity, in accordance with the suggestions of others (10).

The observation of hydrogenated cyclopentadiene products obtained in the homogeneous reaction of $CpCo(CO)_2$ has an interesting bearing on the polymer-supported case. Were a significant amount of decomposition of "ⓅCpCo" to take place by the same mechanism, one would not expect quantitative regeneration of the $\eta^5$-linkage and, in continuation, quantitative recarbonylation to form 5 as observed. Thus, it appears unlikely that polymer supported (but unbound) cobalt clusters (45), crystallites, or atoms are responsible for the observed activity. Moreover, although polystyrene seems to prevent irreversible aggregation of metal clusters (37,46) and metal carbonyls on alumina may be reconstituted with CO after thermal or oxidative decarbonylation (47), a mechanism that involves cobalt atoms or aggregates would require the formation of soluble species subject to leaching. Of course, it is difficult to completely rule out trace amounts of such species as being responsible for catalytic action. One would have assumed, however, that such compounds should have been formed in the attempt to use polystyrene and $CpCo(CO)_2$ or $Co_2(CO)_8$ as Fischer-Tropsch catalysts, where no activity was found. In addition, widely differing degrees of activity of 5

Table I. Product Distribution using Ⓟ $CpCo(CO)_2$ 5 as a Fischer-Tropsch Catalyst

| n-alkane | weight (mg.) | weight fraction | mmol | mol% | mmol CO consumed to form product | mol % CO consumed to form product |
|---|---|---|---|---|---|---|
| $C_1$ | 4.006 | .321 | .2504 | 73.4 | .2504 | 29.7 |
| $C_2$ | .024 | .002 | .0008 | .2 | .0016 | .2 |
| $C_3$ | .274 | .022 | .0062 | 1.8 | .0187 | 2.2 |
| $C_4$ | .694 | .056 | .0120 | 3.5 | .0479 | 5.7 |
| $C_5$ | .768 | .062 | .0107 | 3.4 | .0533 | 6.3 |
| $C_6$ | 1.968 | .158 | .0229 | 6.7 | .1373 | 16.3 |
| $C_7$ | 1.320 | .106 | .0132 | 3.9 | .0924 | 11.0 |
| $C_9$ | .818 | .065 | .0064 | 1.9 | .0575 | 6.8 |
| $C_{10}$ | .084 | .007 | .0006 | .2 | .0059 | .7 |
| $C_{11}$ | .448 | .035 | .0029 | .8 | .0316 | 3.8 |
| $C_{12}$ | .422 | .034 | .0025 | .7 | .0298 | 3.5 |
| $C_{13}$ | .410 | .033 | .0022 | .6 | .0290 | 3.4 |
| $C_{14}$ | .278 | .022 | .0014 | .4 | .0200 | 2.4 |
| $C_{15}$ | .190 | .015 | .0009 | .3 | .0134 | 1.6 |
| $C_{16}$ | .150 | .012 | .0007 | .2 | .0106 | 1.3 |
| $C_{17}$ | .134 | .011 | .0006 | .2 | .0095 | 1.1 |
| $C_{18}$ | .158 | .013 | .0006 | .2 | .0112 | 1.3 |
| $C_{19}$ | .208 | .017 | .0008 | .2 | .0147 | 1.7 |
| $C_{20}$ | .110 | .008 | .0004 | .1 | .0078 | .9 |

might have been expected depending on pretreatment of catalyst and recycle time, an effect that was not observed.

The product distribution from the Fischer-Tropsch reaction on 5 is shown in Table I. It is similar but not identical to that obtained over other cobalt catalysts (18-21,48,49). The relatively low amount of methane production (73 mol%) when compared with other metals and the abnormally low amount of ethane are typical (6). The distribution of hydrocarbons over other cobalt catalysts has been found to fit the Schulz-Flory equation [indicative of a polymerization-type process (6)]. The Schulz-Flory equation in logarithmic form is

$$\log \frac{M_p}{p} = \log(\ln^2\alpha) + p \log\alpha,$$

where $M_p$ = the weight fraction of the compound containing p carbon atoms

$\alpha$ = the chain growth probability factor

defined as $$\alpha = \frac{r_p}{r_p + r_t}$$

$r_p$ = rate of propagation

$r_t$ = rate of termination

A plot of $\log(M_p/p)$ vs. p yields $\alpha$ from either the slope or the intercept (see Figure 2); the "goodness of fit" is indicated by the relative agreement of $\alpha$ obtained from the slope or the intercept. For Ⓟ$CpCo(CO)_2$, the data yield $\alpha$ = 0.81 from the slope and 0.84 from the intercept; considering the fact that a blank contribution from the n-octane solvent had to be subtracted out of $C_5$- $C_{10}$, the agreement is good. Typical values of $\alpha$ are 0.80-0.87 (6).

This catalyst seems to have a better selectivity towards normal paraffins than other catalysts but is not shape selective (19,21): Isobutane/n-butane = 0.008, isopentane/n-pentane = 0.029 (gas chromatography and GC/MS); usual values are 0.1 and higher. Some compounds giving the correct masses for propene, butenes, pentenes, and hexenes were found; however, they were present in even lesser quantities than the branched paraffins.

The actual structure of the active catalyst in the above reactions is a matter of speculation. The evidence, however, points to the presence of a homogeneous but immobilized Fischer-Tropsch catalyst. Since soluble $CpCo(CO)_2$ does not possess Fischer-Tropsch activity, this activity is a unique feature of the polymer-bound system. The finding that 5 is regenerated quantitatively upon exposure of the active Fischer-Tropsch catalyst resin to CO implies that the $\eta^5$-cyclopentadienylcobalt bond remains intact throughout the Fischer-Tropsch reaction. Similar

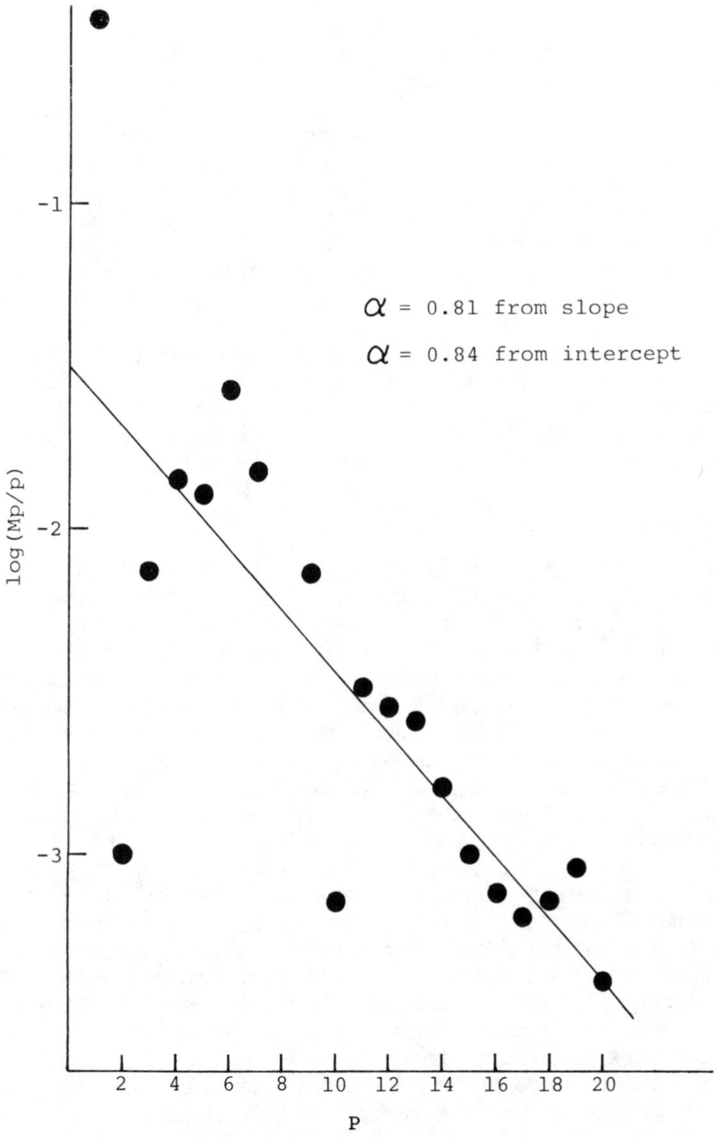
α = 0.81 from slope
α = 0.84 from intercept
log(Mp/p)
-1
-2
-3
2
4
6
8
10
12
14
16
18
20
P

*Figure 2.*

conclusions have been reached recently by other workers in connection with work carried out on 18% crosslinked polystyrene (60,61). This, along with the observation that 5 apparently enjoys some site-site isolation, disfavors the notion that (bound or unbound) clusters are responsible for catalytic action (50).

The Fischer-Tropsch activity of resin 5 and the unique reaction conditions have important consequences. The use of a reaction solvent raises the possibility of controlling heat removal in this appreciably exothermic process. The apparent homogeneous nature of the catalytic species suggests that other soluble Fischer-Tropsch catalysts may be forthcoming. Finally, Ⓟ CpCo($CO)_2$ 5 possesses catalytic activity not found in soluble CpCo($CO)_2$; this demonstrates that attachment to a polymer support not only may induce changes in catalytic activity of a transition metal complex, but also might give rise to completely new activity (51,52,53).

The above considerations led to the expectation that it might be possible to construct soluble analogs of 5 in which an appropriate ligand mimicked the environment imparted by the polystyrene matrix in the immediate vicinity of the metal. The results of these experiments are reported in the final section.

## Homogeneous Analogs of Ⓟ CpCo($CO)_2$ 5

Since substitution of the Cp-ligand in CpCo($CO)_2$ by the polystyrene chain appeared to be the origin of the special stability of the CpCo-moiety in the presence of hydrogen, it was reasoned that similar substitution by phenyl group carrying alkyl chains might have the same effect. Should these models undergo oxidative addition to benzylic (as in 8) or phenylic C-H bonds this might be readily ascertainable by structural or H-D exchange studies. Several model compounds of varying alkyl-chain length and incorporating secondary and tertiary benzylic hydrogens were synthesized in one step according to the scheme shown. The phenyl substituted alkyl halides are available by literature procedures (54,55). Complexation to cobalt occurred most efficiently in the presence of a slight excess $Co_2(CO)_8$ (26) and cyclohexene. The latter alkene evidently serves to quench any intermediate $HCo(CO)_4$ (32) protecting the now valuable cyclopentadiene ligand from reduction (56). Interestingly, both mono- and bis-alkylated cyclopentadienyl complexes were isolated in an approximate ratio of 6:1 by column chromatography as red-brown oils. The latter were assumed to have the 1,3-disubstitution pattern. Both sets of compounds proved to be considerably more air and thermally sensitive than the parent system. Whereas irradiation in sealed tubes did not result in any observable change, simple heating produced insoluble clusters (33). In an effort to suppress bimolecular reactions two of the complexes [R = $C_6H_5(CH_2)_3$, $C_6H_5CHCH_3(CH_2)_2$] were subjected to flash vacuum pyrolysis at low contact times (300-450°C, $2x10^{-4}$torr) (57). The resulting

$$\mathrm{RBr + NaCp} \xrightarrow[\mathrm{THF,\ 0^\circ}]{\mathrm{N_2,\ 3\,hr}} \xrightarrow[\mathrm{N_2,\ 24\,hr,\ r.t.\ cyclohexene}]{\mathrm{Co_2(CO)_8,\ 1.2\ eq}} \underset{\sim 60\%}{\mathrm{RCpCo(CO)_2}} + \underset{\sim 20\%}{\mathrm{1,3\text{-}R_2CpCo(CO)_2}}$$

$$\mathrm{R = C_6H_5(CH_2)}_m \quad \text{or} \quad \mathrm{R = C_6H_5CHCH_3(CH_2)}_n$$

$$m = 3\text{-}5 \qquad\qquad n = 2,4$$

product mixtures contained starting material, free protonated ligand RCpH, and cobaltocenes $(RCp)_2Co$, and the pyrolysis tube was covered by a cobalt mirror. Several of the model compounds were then exposed to Fischer-Tropsch conditions similar to the ones employed in the hydrocarbon synthesis using resin 5. No activity was found until decomposition set in, producing traces of methane. Pressurization with $CO/D_2$, however, furnished some deuterated RCpH and recovered deuterated complex. Extensive deuteration of the latter was achieved under conditions which left the structure of the complex relatively intact (200°C, 4d, $D_2/CO$ = 10/1, 130 psig). In the case of $RCpCo(CO)_2$ $[R=(CH_2)_4CH(CH_3)-C_6H_5]$ the incorporation of up to 10 deuterium atoms could be ascertained by mass spectrometry. Surprisingly, as indicated by the fragmentation pattern, very little (if any) of the deuterium label is found on the aromatic ring. The proton NMR spectrum reveals that most of the H-D exchange has occurred on the Cp ligand, its ordinarily complex AA'BB' pattern having been simplified to two broad singlet absorptions of relative combined intensity of 0.5. The aryl (5H) and aliphatic (12H) regions appear unchanged. That deuterium uptake is preferred on the RCp ligand is also suggested by the relative intensity of the $RCp\text{-}d_4Co(CO)_2$ molecular ion peak which is the signal of maximum height in the peak envelop of deuterated parent ion. It is possible that H-D exchange occurs through a cobalt hydride intermediate in an intramolecular manner as recently also postulated for a $\eta^5$-cyclopentadienyl iron hydride (58). Further exchange along the alkyl chain might occur via insertion of cobalt into ligand C-H bonds, behavior similar to that observed with zirconocene derivatives (59). It seems likely that deuterium incorporation into resin 5 occurs through a similar mechanism. The instability of the model compounds again highlights the special stability associated with the polymer supported system. Perhaps choice of a differently substituted homogeneous analog of 5 will provide a structure endowed with the catalytic capabilities of the resin bound system. This is the subject of continuing work.

## Acknowledgement

This work was supported by the Division of Chemical Sciences, Office of Basic Energy Sciences, U.S. Department of Energy under Contract No. W-7405 Eng-48. We thank Dr. H. Heinemann and Professors A. T. Bell, R. G. Bergman, and G. A. Somorjai for stimuulating comments. K.P.C.V. is an Alfred P. Sloan Fellow (1976-1980) and a Camille and Henry Dreyfus Teacher-Scholar (1978-1983).

## Literature Cited

1. Storch, H., Golumbic, N., and Anderson, R., "The Fischer-Tropsch and Related Syntheses", Wiley, New York, N.Y., 1951.
2. Nefedov, B. K. and Eidus, Y. T., Russ. Chem. Rev., 1965, 34, 272.

3. Eidus, Y. T., Russ. Chem. Rev., 1967, 36, 338.
4. Pichler, H. and Schulz, H., Chem. Ing. Tech., 1970, 42, 1162.
5. Vannice, M. A., Catal. Rev.-Sci. Eng., 1976, 14, 153.
6. Henrici-Olivé, G. and Olivé, S., Angew. Chem., 1976, 88, 144; Angew. Chem., Int. Ed. Engl., 1976, 15, 136.
7. Schulz, H., Erdoel, Kohle, Erdgas, Petrochem. Brennst.-Chem., 1977, 30, 123.
8. Falbe, J., "Chemierohstoffe aus Kohle", Georg Thieme, Stuttgart, West Germany, 1977.
9. Masters, C., Adv. Organomet. Chem., 1979, 17, 61.
10. Thomas, M. G., Beier, B. F., and Muetterties, E. L., J. Am. Chem. Soc., 1976, 98, 1297.
11. Demitras, G. C. and Muetterties, E. L., J. Am. Chem. Soc. 1977, 99, 2796.
12. Huffman, J. C., Stone, J. G., Krusell, W. C., and Caulton, K. G., J. Am. Chem. Soc., 1977, 99, 5829.
13. Labinger, J. A., Wong, K. S., and Scheidt, W. R., J. Am. Chem. Soc., 1978, 100, 3254.
14. Wong, K. S. and Labinger, J. A., J. Am. Chem. Soc., 1980, 102, 3652.
15. Masters, C., van der Woude, C., and van Doorn, J. A., J. Am. Chem. Soc., 1979, 101, 1633.
16. Whitmire, K., and Shriver, D. F., J. Am. Chem. Soc., 1980, 102, 1456.
17. Wong, A., Harris, M., and Atwood, J. D., J. Am. Chem. Soc., 1980, 102, 4529.
18. Blanchard, M. and Bonnet, R., Bull. Soc. Chim. Fr., 1977, 7.
19. Vanhove, D., Makambo, P., and Blanchard, M., J. Chem. Soc., Chem. Commun., 1979, 605.
20. Blanchard, M., Vanhove, D., and Derouault, A., J. Chem. Res. (S), 1979, 404.
21. Fraenkel, D. and Gates, B. C., J. Am. Chem. Soc., 1980, 102, 2478.
22. Gubitosa, G., Boldt, M., and Brintzinger, H. H., J. Am. Chem. Soc., 1977, 99, 5174.
23. Chang, B.-H., Grubbs, R. H., and Brubaker, C. H., J. Organomet. Chem., 1979, 172, 81.
24. Perkins, P. and Vollhardt, K.P.C., J. Am. Chem. Soc. 1979, 101, 3985.
25. Bonds, W. D., Brubaker, C. H., Chandrasekaran, E. S., Gibbons, C., Grubbs, R. H., and Kroll, L. C., J. Am. Chem. Soc., 1975, 97, 2128.
26. Rausch, M. D. and Genetti, R. A., J. Org. Chem., 1970, 35, 3888.
27. "Gmelins Handbuch der anorganischen Chemie", Supplement to the 8th edition, Vol. 5, part I, 1973.
28. Relles, H. M. and Schluenz, R. W., J. Am. Chem. Soc., 1974, 96, 6469.

29. Farrall, M. J. and Fréchet, J. M. J., J. Org. Chem., 1976, 41, 3877.
30. Riemschneider, R. and Nerin, R., Monatsh. Chem., 1961, 91, 829.
31. Miller, R. D., J. Chem. Soc., Chem. Commun., 1976, 277.
32. Sternberg, H. W. and Wender, I., Chem. Soc. Spec. Publ., 1959, 13, 35.
33. Vollhardt, K.P.C., Bercaw, J. E., and Bergman, R. G., J. Organomet. Chem., 1975, 97, 283.
34. Lee, W. S. and Brintzinger, H. H., J. Organomet. Chem., 1977, 127, 87.
35. Gubitosa, G. and Brintzinger, H. H., J. Organomet. Chem., 1977, 140, 187.
36. Regen, S. L., J. Am. Chem. Soc., 1975, 97, 3108.
37. Collman, J. P., Hegedus, L. S., Cooke, M. P., Norton, J.R., Dolcetti, G., and Marquardt, D. N., J. Am. Chem. Soc., 1972, 94, 1790.
38. Grubbs, R. H., Gibbons, C., Kroll, L. C., Bonds, W. D., and Brubaker, C. H., J. Am. Chem. Soc., 1973, 95, 2373.
39. Rebek, J. and Trend, J. E., J. Am. Chem. Soc., 1979, 101, 737.
40. Wild, F.R.W.P., Gubitosa, G., and Brintzinger, H. H., J. Organomet. Chem., 1978, 148, 73.
41. Craven, W. J., Wiese, E., and Wiese, H. K., U.S. Patent 3026344 (1958/62), C.A., 1962, 57, 7312.
42. Hartley, F. R. and Vezey, P. N., Adv. Organomet. Chem., 1977, 15, 189.
43. Connor, J. A., Top. Curr. Chem., 1977, 71, 71.
44. Dyagileva, L. M., Mar'in, V. P., Tsyganova, E. I., and Razuvaev, G. A., J. Organomet. Chem., 1979, 175, 63.
45. Smith, A. K., Theolier, A., Basset, J. M., Ugo, R. Commereuc, D., and Chauvin, Y., J. Am. Chem. Soc., 1978, 100, 2590.
46. Invatate, K., Dasgupta, S. R., Schneider, R. L., Smith, G. C., and Watters, K. L., Inorg. Chim. Acta, 1975, 15, 191.
47. Brenner, A. and Burwell, R. L., J. Am. Chem. Soc., 1975, 97, 2565.
48. Vannice, M. A., J. Catal., 1975, 37, 449.
49. Vannice, M. A., J. Catal., 1977, 50, 228.
50. Muetterties, E. L., Bull. Soc. Chim. Belg., 1975, 84, 959.
51. Leznoff, C. C., Acc. Chem. Res., 1978, 11, 328.
52. Manecke, G. and Stork, W., Angew. Chem., 1978, 90, 691; Angew. Chem., Int. Ed. Engl., 1978, 17, 657.
53. Crowley, J. I. and Rapoport, H., Acc. Chem. Res., 1976, 9, 135.
54. Friedman, L. and Shani, A., J. Am. Chem. Soc., 1974, 96, 7101.
55. Noller, C. R. and Dinsmore, R., Org. Syn. Coll. Vol., 1943, 2, 358.
56. Vollhardt, K.P.C. and Winter, M. J., unpublished.

57. Fritch, J. R. and Vollhardt, K.P.C., Angew. Chem., 1979, 91, 439; Angew. Chem., Int. Ed. Engl., 1979, 18, 409.
58. Davies, S. G., Felkin, H., and Watts, O., J. Chem. Soc., Chem. Commun., 1980, 159.
59. Bercaw, J. E., A.C.S. Adv. Chem. Ser., 1978, 167, 136.
60. Gubitosa, G. and Brintzinger, H. H., Coll. Int. CNRS, 1977, 281, 173.
61. Boldt, M., Gubitosa, G., and Brintzinger, H. H., German Patent 2727245, 1980.

RECEIVED December 8, 1980.

12

# Chain-Length Control in the Conversion of Syngas over Carbonyl Compounds Anchored into a Zeolite Matrix

D. BALLIVET–TKATCHENKO, N. D. CHAU, H. MOZZANEGA, M. C. ROUX, and I. TKATCHENKO

Institut de Recherches sur la Catalyse, 2 avenue Einstein, F- 69626 Villeurbanne Cedex, France

It is now superflous to point out the renewed interest for the Fischer-Tropsch (F-T) synthesis (1) i.e. the conversion of $CO+H_2$ mixtures into a broad range of products including alkanes, alkenes, alcohols. Recent reviews (2,3,4,5) emphasized the central problem in F-T synthesis: selectivity or more precisely chain-length control.

Recent patents and publications reported the use of modified "classical" F-T catalysts or new supported metal catalysts. Ruhrchemie disclosed (6) iron catalysts modified by additives like Ti, Mn and Mo, which are claimed to produce ca 50% of $C_2$-$C_4$ alkenes. Ichikawa, by using rhodium carbonyl compounds deposited on various supports, was able to produce selectively (7) $C_2$-$C_4$ alcohols and hydrocarbons. Nijs et al. reported (8) that Ru(III) ions exchanged into Y-zeolites reduced with hydrogen lead to selective catalysts for the synthesis of hydrocarbons in the $C_1$-$C_{10}$ range. Similarly, Fraenkel and Gates have shown (9) that catalysts prepared by reduction with cadmium of Co(II) ions exchanged into A- and Y-zeolites may produce propylene as the only hydrocarbon product under well-defined conditions. Blanchard et al., by using di-cobalt octacarbonyl supported on alumina, have observed (10) that good selectivities for $C_2$-$C_6$ hydrocarbons could be obtained when the support presents a mean pore size of 5nm.

The aim of this work is to prepare better defined catalytic systems by combining components with well characterized structures and properties such as zeolites and transition metal molecular complexes. Among these complexes, clusters are potential candidates for selective F-T catalysts since neighboring atoms with unique topologic and electronic features may help "hydro-oligomerization" of carbon monoxide. However, until now only very low activities have been achieved (11). Zeolites are well defined, crystalline alumino-silicates (12) which may offer stabilization of metal particles (13) and shape selectivity (14) owing to their geometrical frame properties. From the point of view of reactivity and catalysis, hope is still high that such adducts will be efficient as/or more efficient than classical heterogeneous

0097-6156/81/0152-0187$05.00/0
© 1981 American Chemical Society

catalysts. The interaction between the support and the cluster and the redox behaviour of these partners are important factors which will influence the type of catalyst formed. Earlier work from this Laboratory (15,16) has shown that the adsorption of certain transition metal carbonyls into an HY-zeolite framework and appropriate thermal desorption lead either to ions (Mo, Fe) or to metal (Re, Ru). However the oxidation reaction could be prevented by the use of non-acidic zeolites like the Na-Y type(17).

We report here results related to the catalytic behaviour of dodecacarbonyl-tri-iron and tri-ruthenium, bis(cyclopentadienyl-dicarbonyliron) and octacarbonyl-di-cobalt deposited on Y-zeolites under F-T conditions. The influence of the nature of the zeolite and of the metal, the dispersion of the metal and the reaction conditions upon activity and products distribution were investigated.

## Experimental.

Materials. The NaY faujasite was supplied by Linde Co (SK 40 Sieves). A conventional exchange with $NH_4Cl$ provides a $NH_4Y$ sample (unit cell composition : $(NH_4)_{46}Na_{10}Al_{56}Si_{136}O_{384}$). Heating this sample for 15h in oxygen and 3h in vacuo ($10^{-5}$torr) at 350°C leads to the hydrogen form HY.

The silica-alumina was supplied by Ketjen with a 13% alumina content. It was vacuum-treated at 450°C ($10^{-5}$torr) for 15hrs before anchoring the cluster.

Tri-iron dodecacarbonyl was prepared according to (18);bis (cyclopentadienyldicarbonyliron) was prepared according to (19); tri-ruthenium dodecacarbonyl and di-cobalt octacarbonyl were supplied by Strem Chemicals.

The $Fe_3(CO)_{12}$-Y adducts are prepared under argon atmosphere with the HY and NaY zeolites previously heated under vacuum at 350°C. The impregnation of the support is performed either from pentane solution at 25°C or from dry mixing of the carbonyl and the zeolite to avoid any complication from the solvent. In this last preparation the sample stands in vacuo for 24h at 60°C in order to favour the sublimation of the carbonyl into the pores of the zeolite.

The $Ru_3(CO)_{12}$-Y adducts are similarly prepared both through impregnation (solvent: cyclohexane) or dry mixing (heating at 90°C under vacuum.

The $|CpFe(CO)_2|_2$-Y adducts are prepared in the same way but with heating at 40°C under vacuum when the dry mixing procedure is used.

The $Co_2(CO)_8$-NaY adducts are prepared by the dry mixing procedure.

In any instance, the amount of metal anchored is determined by chemical analysis. The loadings correspond to 6-12 metal atoms per unit cell.

Catalytic experiments. The runs are performed in a 300mL static reactor (Autoclave Engineers Model AE 300) for 15hrs under an initial 20 bar pressure with a sample weight leading to 0.4 mg-atom of metal. Neither the unloaded zeolites nor the molecular clusters are active in CO hydrogenation under our experimental conditions.

The products are analysed by gas chromatography usually on five different columns in order to detect CO, $H_2$, $CO_2$, $H_2O$ and $C_1$-$C_n$ hydrocarbons (alkanes, alkenes) and alcohols. The mass balance for carbon, based on CO consumption, generally lies within 85-95%.

## Results and Discussion.

It should be emphasized that the results were intended to demonstrate qualitative trends rather than quantitative kinetic data with these typical catalysts. Moreover the high CO conversion levels achieved in the present work will not be the limiting factor to observation of side reactions and long-chain hydrocarbons.

Tri-iron dodecacarbonyl zeolites adducts. The adducts $Fe_3(CO)_{12}$-HY and $Fe_3(CO)_{12}$-NaY were used as starting materials. The catalytic runs are performed with these adducts or with the materials recovered from their total decarbonylation at 200°C under vacuum.

The $Fe_3(CO)_{12}$-HY adduct exhibits no catalytic activity in the temperature range studied (200-300°C). During the study of the decarbonylation under vacuum, several stoichiometric reactions take place as evidenced by mass spectrometry and infrared spectroscopy (16). The oxidation of Fe(0) into Fe(II) species by the zeolite protons, the water-gas shift reaction and the hydrogenation of $CO_2$ account to some extent for the formation of $H_2$, $Fe^{2+}$, $CO_2$, $H_2O$, $CH_4$ and higher hydrocarbons. The sample thus obtained, i.e. $Fe^{2+}$-HY, is as inactive in the catalytic syngas conversion as are standard $Fe^{2+}$-NaY and $Fe^{3+}$-NaY exchanged zeolites. Therefore $Fe^{2+,3+}$-Y zeolites in contrast to $Ru^{n+}$-Y zeolites (8) are not catalyst precursors. It is worth mentioning that molecular hydrogen is unable to reduce $Fe^{2+}$-Y zeolite even under drastic conditions (20).

Conversly, the $Fe_3(CO)_{12}$-NaY adduct is active for syngas conversion. A non-decomposed sample exhibits a significant activity at 230°C whereas the catalytic efficiency for the decarbonylated one already appears at 200°C. Infrared experiments show an increase in the stability of the $Fe_3(CO)_{12}$ units upon thermal treatment under CO atmosphere so that total carbon monoxide evolution only takes place at 230°C thus suggesting that the catalyst is certainly not $Fe_3(CO)_{12}$. This cluster has to be transformed into higher nuclearity species which bind less strongly with carbon monoxide upon CO re-adsorption (17).

Effects of the inlet $H_2/CO$ ratio and reaction temperature on $H_2+CO$ conversion and products selectivity were studied at constant initial pressure and reaction time with non-decomposed samples of $Fe_3(CO)_{12}$-NaY adduct (4%Fe).

In all experiments, $H_2O$, $CO_2$ alkanes and alkenes (up to $C_{12}$) are produced. Linear alcohols are also detected to minor amounts.

At increasing reaction temperatures (230-350°C) the product selectivity is shifted towards $C_1$-$C_4$. The alkene to alkane ratio declines at higher reaction temperatures whereas the branched to linear alkane ratio increases as well as $CO_2$ formation. These observations are entirely consistent with the behaviour of classical F-T catalysts (Table 1).

A reaction temperature of 250°C was used to study the other reaction parameters. A decrease in the $H_2/CO$ ratio increases the consumption of CO whereas the percent converted decreases. Conversly the $H_2$ conversion increases but its consumption remains constant. In fact this means that higher molecular weight products are formed under low hydrogen partial pressure : as indicated in Table 2, quantitative analyses of the products show indeed a decrease in $C_1$-$C_3$ yield and a concomitant increase in $C_3^+$ hydrocarbons. The selectivity for $CO_2$ remains constant which apparently indicates that $CO_2$ is at least in the case of iron a primary product in the F-T synthesis. We have checked that the iron/zeolite catalyst activities for hydrogenation and the water-gas-shift are not significant with the precursor $Fe_3(CO)_{12}$/NaY. The alkene to alkane ratio greatly varies with (i) $H_2/CO$ and (ii) the chain length. An increase in hydrogen partial pressure increases the alkane production as alkene hydrogenation is a secondary reaction which takes place with F-T catalyst and, in this work, with the zeolite system. For a steady syngas inlet, $C_2^=/C_2$ ratio is consistently much lower than the $C_3^=/C_3$ one. $C_4^=/C_4$ ratio is complicated by the existence of the butene isomers and isobutene (Table 2).

Although ethylene is more readily hydrogenated than the other alkenes, it has been reported (21) to participate to the formation of higher molecular weight hydrocarbons under F-T conditions. This is observed too for $Fe_3(CO)_{12}$-NaY catalysts. If the F-T synthesis is performed with ethylene as a co-reactant, significant changes in selectivity are found for $CO_2$, $C_3$ and for the i-$C_4$/n-$C_4$ ratio. The variations for $CH_4$ and $C_3^=/C_3$ ratio can also be attibuted to the $H_2/CO$ ratio modification (3.5/1 instead of 4/1) if one takes into account the hydrogen consumed for ethylene hydrogenation (Table 3). It appears that as $CO_2$ seems to be a primary product (vide supra) the decrease for its selectivity in the presence of ethylene suggests that CO consumption now occurs partly through a reaction leading to more hydrocarbons, especially $C_3$, and no $CO_2$. Such a reaction pathway could involve the insertion of CO into a metal-ethyl bond as already well documented in coordination chemistry and homogeneous catalysis or the insertion of a surface carbene

Table 1. Effect of the reaction temperature on CO and $H_2$ conversions, and product selectivities (expressed as mole percent of CO converted into the desired product).
Experimental conditions: catalyst = $Fe_3(CO)_{12}$-NaY (4%Fe); $H_2/CO$ = 4/1; initial pressure = 20 bar; reaction time = 15hrs.

| T°C | $H_2$ conv. % | CO conv. % | $CO_2$ % | $CH_4$ % | $C_2$-$C_4$ % | $C_4^+$ % |
|---|---|---|---|---|---|---|
| 200 | 8.7 | 36 | 5.5 | 10.3 | 14.2 | 70 |
| 240 | 40 | 82 | 7.6 | 16 | 19 | 57.4 |
| 255 | 43 | 85 | 13 | 19 | 24 | 44 |
| 295 | 29 | 74 | 20 | 26 | 34 | 20 |
| 365 | 37 | 88 | 19 | 19 | 27 | 35 |

Table 2. Effect of the $H_2/CO$ ratio on CO and $H_2$ conversions and product selectivities (expressed as mole percent of CO converted into the desired product).
Experimental conditions: catalyst = $Fe_3(CO)_{12}$-NaY (4%Fe); initial pressure = 20 bar; reaction temperature = 250°C; reaction time = 15hrs.

| $H_2/CO$ | $H_2$ conv. % | CO conv. % | $CO_2$ % | $CH_4$ % | $C_2$ % | $C_3$ % | $C_2^=/C_2$ % | $C_3^=/C_3$ % | $C_4^=/C_4$ % |
|---|---|---|---|---|---|---|---|---|---|
| 4/1 | 39 | 73 | 10 | 20.6 | 9 | 9.9 | - | 0.8 | - |
| 2/1 | 57 | 64 | 9.5 | 7.4 | 4.4 | 7.7 | 4.5 | 45 | 16 |
| 1/1 | 66 | 48 | 10.4 | 5.2 | 2.8 | 5.3 | 7.1 | 47 | 19.5 |

into the adsorbed ethylene (22). However, if ethylene can participate to the F-T reactions the presence of CO is essential for chain growth as an ethylene + $H_2$ feed mainly gives ethane in the present reaction conditions.

The formation of $C_n$ hydrocarbons from syngas involve a chain-growth mechanism (23). The molecular weight distribution will depend on (i) the propagation to transfer rate ratio and (ii) the side reactions. From the reaction mechanism standpoint, these secondary reactions (e.g. hydrogenation, isomerization and hydrogenolysis) are masking the primary growth process and make difficult such an approach. A non-decomposed $Fe_3(CO)_{12}$-NaY adduct (4%Fe) provides a typical molecular weight distribution which is reported on Figure 1, curve 1. An exponential decrease of the mole percent of CO consumed to form $C_n$ is observed from $C_1$ to $C_9$ with a consistently lower value for $C_2$. Hydrocarbons higher than $C_9$ are only present in trace amounts. This distribution is independent of the $H_2$/CO ratio and of the reaction temperature. Only the slope of the straight line sligthly changes indicating a variation in the value of the chain-growth probability. In order to assess the peculiar distribution reported in Figure 1, curve 1, another Fe-NaY catalyst precursor was prepared in such a way that metallic particles on the external surface of the zeolite are obtained. This is performed by heating the $Fe_3(CO)_{12}$-NaY adduct (4%Fe) in vacuum from 25 to 250°C within 1 hr and further evacuation at 250°C for 15hrs. Particles of 20-30nm in diameter are thus obtained. Higher CO, $H_2$ conversions and $CO_2$ yield (13%) are found with this Fe-NaY sample. The chain-length distribution (Figure 1, curve 2) shows that a drastic change occurs in the $C_6$-$C_{10}$ domain. It can be concluded from these experiments that the $Fe_3(CO)_{12}$-NaY precursor induces a peculiar selectivity for the hydrocarbon chain-length. As the $Fe_3(CO)_{12}$ are located in the supercages and since large iron particles are not detected by X-ray techniques after the catalytic runs, this hydrocarbon distribution can be achieved (i) by encapsulation (i.e. stabilization) of small iron particles and (ii) by a cage effect since the $C_9$ length fits the supercage diameter. Molecular weight distribution as depicted in Figure 1, curve 1 must be related with the presence of small metal particles. This relationship is demonstrated for ruthenium catalysts (vide infra). Parallel studies by Jacobs (8), Gates (9), Blanchard (10) and their co-workers point out the same effect of the porosity of the inorganic matrix on the upper limit of the hydrocarbon chain-length. However a clear-cut between points (i) and (ii) is difficult to assess since the particle size of the catalyst can be limited by the pore diameter in which it is entrapped.

As already mentioned, the acidity of the HY zeolite precludes its use as a support for $Fe_3(CO)_{12}$, then iron particles. The same behaviour is observed for $Fe_3(CO)_{12}$-silica-alumina system. This material, when decomposed at 200°C is not an efficient catalyst for F-T synthesis and only $C_1$-$C_4$ products are

Table 3. Effect of ethylene as a co-reactant on CO, $H_2$ conversions, $C_3^=$ formation and $CO_2$, $C_1$-$C_4$ selectivities (expressed as mole percent of CO converted into the desired product). Experimental conditions: catalyst = $Fe_3(CO)_{12}$-NaY (4%Fe); initial pressure = 20 bar; $C_2^=/H_2/CO$ = 1/4/1; reaction temperature = 250°C; reaction time = 15hrs.

| Co-reactants | $H_2$ conv. % | CO conv. % | $CO_2$ % | $CH_4$ % | $C_3$ % | $C_3^=/C_3$ % | $iC_4/nC_4$ % |
|---|---|---|---|---|---|---|---|
| $H_2$ + CO | 39 | 73 | 10 | 20.6 | 9.9 | 0.8 | 4.8 |
| $C_2^=$ + $H_2$ + CO | 48 | 73 | 7.6 | 11 | 13.2 | 16.7 | 1.6 |

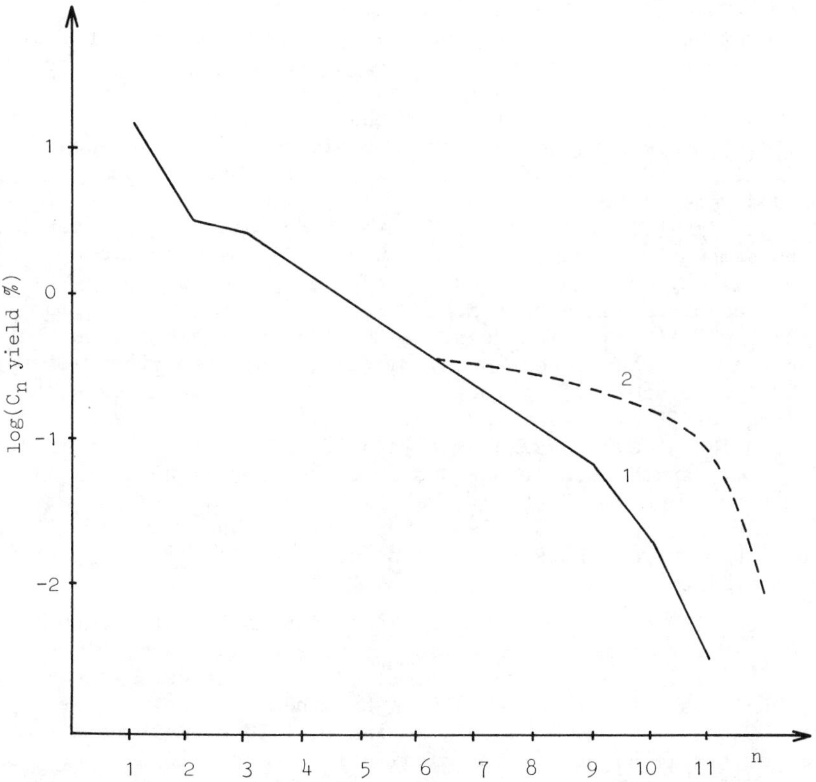


*Figure 1. A Schulz-Flory-type representation for catalysts derived from Curve 1, in situ decomposition of the $Fe_3(CO)_{12}$-NaY adduct; Curve 2, ex situ decomposition of the $Fe_3(CO)_{12}$-NaY adduct leading to large Fe crystallites (see text): initial pressure, 20 bar; $H_2/CO = 4/1$; reaction temperature, 250°C*

obtained (23). However, the chain-length selectivity can be altered by the addition of an acidic support, e.g. HY zeolite, to the $Fe_3(CO)_{12}$-NaY zeolite in the autoclave. A drastic change is observed in the $C_6$-$C_{10}$ range (Figure 2, curve 2). This behaviour is explained by the occurence of secondary reactions like isomerization and cracking owing to the acidity of the HY zeolite. Experiments performed under hydrogen pressure with HY zeolite alone show that n-octane is transformed into $C_2$ (traces), $C_3$, $C_4$ (predominant, i-$C_4$/n-$C_4$=5), $C_5$ and $C_6$ whereas n-butane is transformed into $C_2$ (traces), $C_3$,i-$C_4$ $C_5$ and $C_6$ but, in this latter case, the activity is quite low. The presence of the HY zeolite leads to an increase in the $C_4$-$C_5$ fraction with a concomitant decrease of higher hydrocarbons. Thus, this experiment further supports the role of NaY entrapped iron particles as active sites for the F-T synthesis.

*Bis(cyclopentadienyldicarbonyliron)-zeolites adducts*. The adducts $|CpFe(CO)_2|_2$-HY and $|CpFe(CO)_2|_2$-NaY were used as starting materials. They are not decomposed before the catalytic run. Infrared spectra of these materials show that the integrity of the molecular complex is retained; it is expected that owing to its size, which is smaller than that of $Fe_3(CO)_{12}$, $|CpFe(CO)_2|_2$ lies inside the cavities. Infrared, UV-VIS and ESR data will provide relevant information on the fate of this compound under thermal treatment (24).

The $|CpFe(CO)_2|_2$-HY adduct exhibits no catalytic activity in the temperature range studied (200-300°C). However cyclopentene and cyclopentane are detected through GC monitoring of the gas-phase. A redox reaction $Fe(0)/H^+$ is still occurring as for the $Fe_3(CO)_{12}$-HY adduct; the evolved hydrogen allows the reduction of the cyclopentadienyl ligands. This behaviour already provides evidence for the location of the Fe(I) complexes within the large cavities of the zeolite.

The $|CpFe(CO)_2|_2$-NaY adduct is active for syngas conversion. Under the standard conditions the CO conversion is quite similar to that observed for $Fe_3(CO)_{12}$-NaY (Table 4). However hydrogen conversion is higher and this is reflected in the chain-length distribution which shows a better selectivity for light hydrocarbons (Figure 3).

*Dicobalt octacarbonyl-NaY zeolite adducts*.Only the adducts $Co_2(CO)_8$-NaY were used as starting materials. No decomposition trough thermal treatment was attempted before the catalytic runs.

These adducts are more active than the iron ones in the conversion of syngas. At 250°C, a higher yield of methane is observed (Table 4) and carbon dioxide is produced in smaller amounts. Inspection of Table 5 summarizing the influence of the $H_2$/CO ratio on products selectivity also indicates a higher production of saturated hydrocarbons. This behavior is typical for cobalt catalysts in F-T synthesis (2,25). The chain-length distribution is similar to that observed for catalysts derived

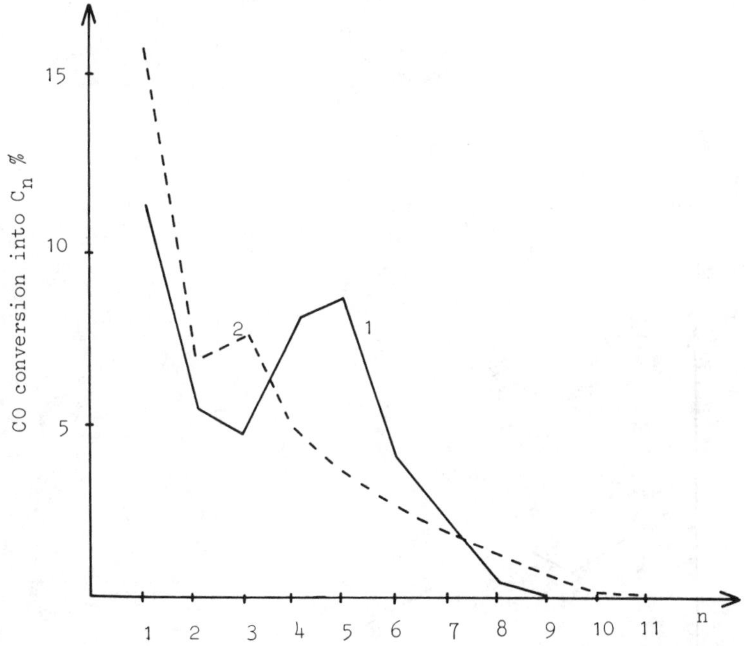


*Figure 2. The effect of added HY zeolite on the hydrocarbon distribution; Curve 1, $Fe_3(CO)_{12}$-NaY + HY; Curve 2, $Fe_3(CO)_{12}$-NaY: initial pressure, 20 bar; $H_2/CO = 4/1$; reaction temperature, 250°C*

Table 4. Comparative performances of the $Fe_3(CO)_{12}$, $|CpFe(CO)_2|_2$ and $Co_2(CO)_8$-NaY systems ($C_1$ selectivities expressed as mole percent of CO converted into the desired product).
Experimental conditions: initial pressure = 20 bar; $H_2/CO$ = 4/1; reaction temperature = 250°C; reaction time = 15hrs.

| Metal precursor | $H_2$ conv. % | CO conv. % | $CO_2$ % | $CH_4$ % |
|---|---|---|---|---|
| $Fe_3(CO)_{12}$ | 39 | 73 | 10 | 20.6 |
| $\|CpFe(CO)_2\|_2$ | 50 | 75 | 10 | 22 |
| $Co_2(CO)_8$ | | 100 | 0.8 | 42.5 |

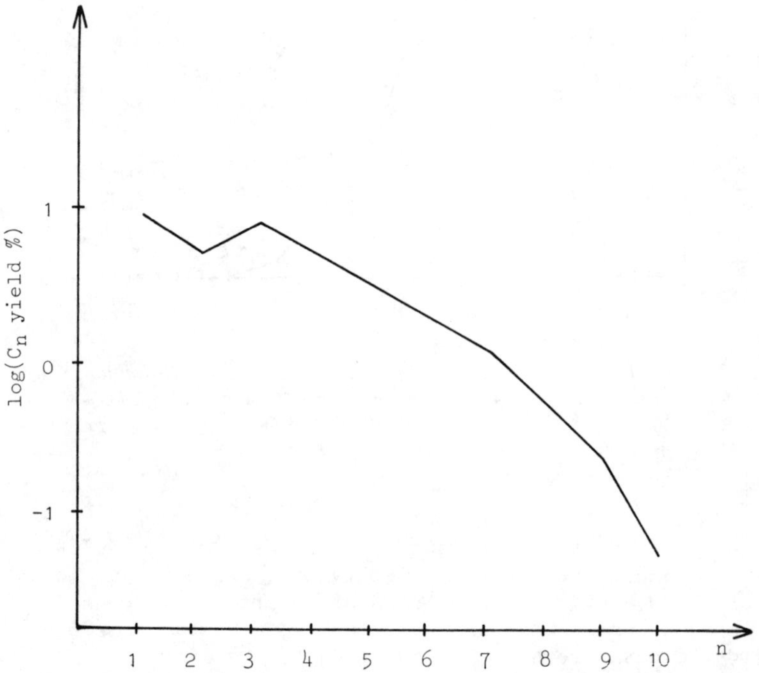


*Figure 3. A Schulz-Flory-type plot for catalyst derived from $|CpFe(CO)_2|_2$-NaY adduct: initial pressure, 20 bar; $H_2CO$ = 4/1; reaction temperature, 250°C*

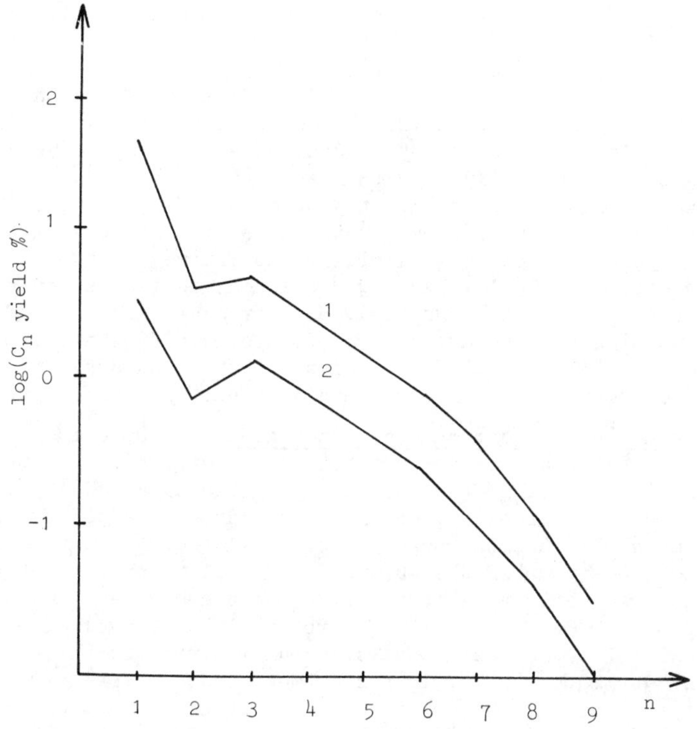


*Figure 4. A Schulz-Flory-type plot for catalyst derived from $Co_2(CO)_8$-NaY adduct: Curve 1, $H_2/CO = 4/1$; Curve 2, $H_2/CO = 1/1$; initial pressure, 20 bar; reaction temperature, 250°C*

Table 5. Effect of the $H_2/CO$ ratio on CO and $H_2$ conversions and product selectivities (expressed as mole percent of CO converted into the desired product). Experimental conditions : catalyst = $Co_2(CO)_8$-NaY (4%Fe); initial pressure = 20 bar; reaction temperature = 250°C; reaction time = 15hrs.

| $H_2/CO$ | $H_2$ conv. % | CO conv. % | $CO_2$ % | $CH_4$ % | $C_2$ % | $C_3$ % | $C_3^+$ % | $C_2^=/C_2$ % | $C_3^=/C_3$ % |
|---|---|---|---|---|---|---|---|---|---|
| 4/1 | | 100 | 0.8 | 42.5 | 8 | 15 | 33.7 | - | - |
| 1/1 | 59.5 | 34 | 6.1 | 11.3 | 4.4 | 12 | 66 | 5.8 | 13.5 |

from $Fe_3(CO)_{12}$-NaY adducts, i.e. ethylene is again produced in smaller amounts (Figure 3) and chain-length is practically limited to $C_9$-$C_{10}$.

However, comparison of Figures 1 and 3 indicates a greater decrease in hydrocarbon content at the $C_6$-$C_7$ level for cobalt catalysts than for iron catalysts. These results agree well with the already reported greater hydrogenation and hydrogenolysis abilities of cobalt catalysts with respect to iron ones (2). Thus besides the intrinsic shape selectivity of the Y zeolite, the nature of the metal and presumably the size of the metal particles are also important factors in limiting the chain-growth process.

Interestingly, Co(II) exchanged NaY zeolite is also operative under the F-T conditions (26). However, large amounts of methane and higher hydrocarbons are produced. This behaviour is reminiscent of "classical" F-T catalysts (2,25) and suggests the occurence of cobalt crystallites outside the Y zeolite cavities.

Triruthenium dodecarbonyl-zeolites adducts. Ruthenium-Y zeolites have already been reported as catalysts for selective F-T synthesis. Starting from $Ru_3(CO)_{12}$ instead of Ru(II) complexes (8) on HY zeolite, catalysts could be obtained provided that the materials are thoroughly decarbonylated. A study of the thermal decomposition of the $Ru_3(CO)_{12}$-HY zeolite shows that up to 200°C, three carbonyl groups per $Ru_3$ unit evolve and a plateau occurs up to 320°C. The remaining carbonyl ligands evolve between 320 and 420°C (15). Syngas conversion only occurs with samples pretreated at 320 or 400°C. This treatment lead to small ruthenium particles of *ca* 1.5-2nm as indicated by electron microscopy (17). If the thermal decomposition from 60 to 320°C is operated in vacuum within 1 hr and is followed by further evacuation at 320°C, particles of 10 to 100 nm are obtained. These catalysts are more active than the corresponding iron ones, *i.e.* carbon monoxide hydrocondensation starts below 200°C. All the catalytic runs were performed at 200°C as methane is selectively produced at 250°C. Table 6 and Figure 5 show the dramatic difference in behaviour between the small particles- and large particles- containing catalysts. A high yield of methane is obtained for Ru-Y 10 nm. The selectivity of the Ru-Y 1.5 nm catalysts is similar to that observed for the Fe-NaY-HY bifunctional catalyst (Figures 5 and 2). However the comparison can only be qualitative since the reaction temperatures are significantly different.

After the catalytic runs no modification of mean particle size is observed for this last system. Conversly, $Ru_3(CO)_{12}$ deposited on silica-alumina is readily decomposed at 200°C to metallic particles of 1 nm mean size which are also catalysts for the F-T synthesis. The catalytic activity at 200°C is *ca* one tenth of the Y zeolite supported ones and methane is practically the only hydrocarbon formed. Electron microscopy examination of the catalyst after reaction reveals a drastic sintering of the

Table 6. Effect of the ruthenium particle size on CO and $H_2$ conversions and product selectivities (expressed as mole percent of CO converted into the desired product). Experimental conditions: initial pressure = 20 bar; $H_2/CO$ = 4/1; reaction temperature = 200°C; reaction time = 15hrs.

| Catalyst and mean particle size | $H_2$ conv. % | CO conv. % | $CO_2$ % | $C_1$ % | $C_2$ % | $C_3$ % | $C_3^+$ % |
|---|---|---|---|---|---|---|---|
| Ru-Y 10 nm | 66 | 100 | 0 | 41 | 10.7 | 6.9 | 41.4 |
| Ru-Y 1.5 nm | 38 | 37 | 3 | 23 | 2.6 | 14.6 | 56.8 |

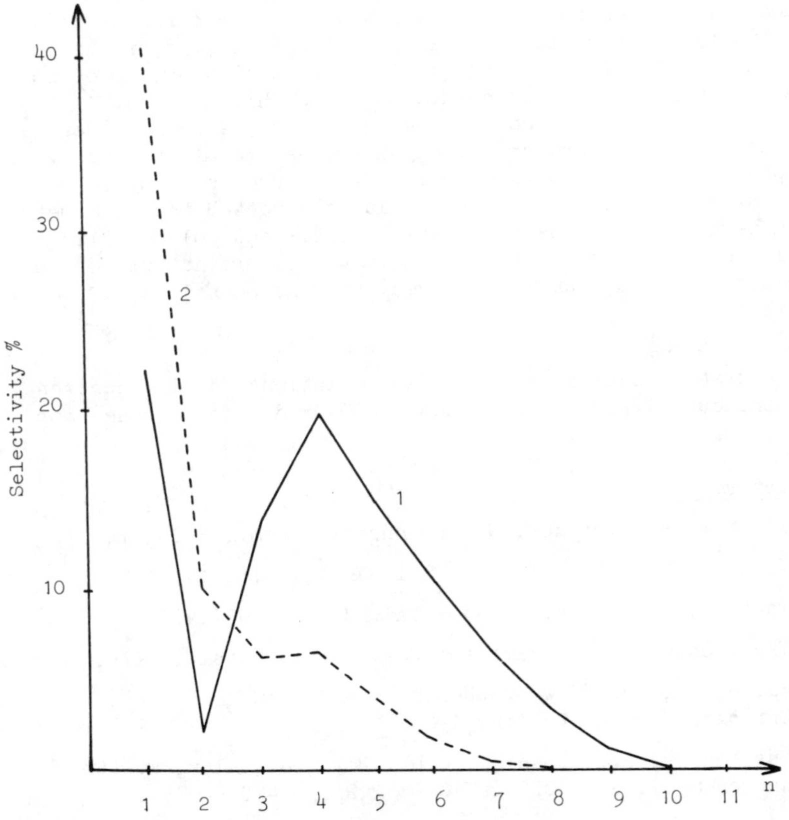


*Figure 5. The influence of the Ru particles size on the hydrocarbon selectivities: Curve 1, distribution for a mean-particle size of 1.5 nm; Curve 2, distribution for a mean-particle size of 10 nm; initial pressure, 20 bar; $H_2/CO = 4/1$; reaction temperature, 200°C*

ruthenium aggregates: well crystallized hexagonal plates of 10 to 100 nm are now present. This result points out the stabilizing effect of the zeolite supercages upon sintering of the small aggregates and serves as an indirect evidence for the occurence of small aggregates in the case of the (air-sensitive) iron and cobalt catalysts.

## Conclusion.

Fischer-Tropsch synthesis could be "tailored" by the use of iron, cobalt and ruthenium carbonyl complexes deposited on faujasite Y-type zeolite as starting materials for the preparation of catalysts. Short chain hydrocarbons, i.e. in the $C_1$-$C_9$ range are obtained. It appears that the formation and the stabilization of small metallic aggregates into the zeolite supercage are the prerequisite to induce a chain length limitation in the hydrocondensation of carbon monoxide. However, the control of this selectivity through either a definite particle size of the metal or a shape selectivity of the zeolite is still a matter of speculation. Further work is needed to solve this dilemna.

In a more general context, metal carbonyls on zeolites can be a unique way to prepare highly dispersed metal catalysts. In the present work, this is especially the case for iron as no other mild methods are operative. It is expected that the method could be applied to the preparation of bi- and polymetallic catalysts even though the starting material are not bi- or polymetallic clusters, but more conveniently homometallic clusters.

## Acknowledgements.

We are greatly indebted to the Centre National de la Recherche Scientifique (ATP grand 2333 and GRECO Oxydes de Carbone) for support of this work.

## Literature Cited.

1. Fischer, H., Tropsch, H., Brennstoff Chem., 1926, 7, 97.
2. Vannice, M.A., Catal. Rev.-Sci. Eng., 1976, 14, 153.
3. Masters, C., Advan. Organometal. Chem., 1979, 17, 61.
4. Tkatchenko, I., Fundamental Res. Homog. Catal., 1979, 3, 119.
5. Falbe, J., Ed. "New Syntheses with Carbon Monoxide", Springer Verlag; Berlin, 1980.
6. Büssemeier, B., Frohning, C.D., Horn, G., Kluy, W. (to Ruhrchemie), Ger. OLS 2 518 964, 2 536 488 (1975).
7. Ichikawa, M., J.C.S. Chem. Comm., 1978, 567.
8. Nijs, H.H., Jacobs, P.A., Uytterhoeven, J.V., J.C.S. Chem. Comm., 1979, 180, 1095.

9. Fraenkel, D., Gates, B.C., J. Am. Chem. Soc., 1980, 102, 2478.

10. Vanhove, D., Makambo, P., Blanchard, M., J. Chem. Soc., Chem. Comm., 1979, 605.

11. Demitras, G.C., Muetterties, E.L., J. Am. Chem. Soc., 1977, 99, 2796.

12. Rabo, J.A., Ed., "Zeolite Chemistry and Catalysis", the American Chemical Society, Washington, D.C., 1976; Katzer, J.R., Ed. "Molecular Sieves II", the American Chemical Society, Washington, D.C., 1977.

13. Gallezot, P., Catal. Rev.-Sci. Eng., 1979, 20, 121.

14. Csicsery, S.M., "Zeolite Chemistry and Catalysis", Rabo,J.A., Ed., the American Chemical Society, Washington, D.C., 1976, p.680.

15. Gallezot, P., Coudurier, G., Primet, M., Imelik, B., "Molecular Sieves II", Katzer, J.R., Ed., the American Chemical Society, Washington, D.C., 1977, p.144.

16. Ballivet-Tkatchenko, D., Coudurier, G., Inorg. Chem., 1979, 18, 558.

17. Ballivet-Tkatchenko, D., Coudurier, G., Tkatchenko,I., Figueiredo, C., Preprints Div. Pet. Chem. Am. Chem. Soc. 1980.

18. Mc Farlane, W. Wilkinson, G., Inorg. Synth. 1966, 8, 181.

19. King, R.B., Stone, F.G.A., Inorg. Synth., 1963, 7, 110.

20. Huang, Y.-Y., Anderson, J.R., J. Catal., 1975, 40, 143.

21. Pichler, H. Schulz, H., Erdöl u.Kohle-Erdas Petrochemie, 1970, 23, 651.

22. Sumner, Jr., C.E., Riley, P.E., Davis, R.E., Pettit, R., J. Am. Chem. Soc., 1980, 102, 1752.

23. Schulz, H., el Deen, A.Z., Fuel Process. Technol., 1977, 1, 45.

24. Ballivet-Tkatchenko, D., Coudurier, G., Mozzanega, H, Tkatchenko, I., Fundamental Res. Homog. Catal., 1979, 3, 257.

25. Ballivet-Tkatchenko, D., Coudurier, G., Praliaud, H., Chau, N.D., work in progress.

26. Pichler, H., Hector, A., Encyclopedia of Chemical Technology, Wiley New-York, 2nd Ed., 1964, vol. 4, p. 446.

27. Ballivet-Tkatchenko, D., Praliaud, H., unpublished results.

RECEIVED December 8, 1980.

# 13

# Hydrogenation of Carbon Monoxide on Alumina-Supported Metals

## A Tunneling Spectroscopy Study

R. M. KROEKER—I.B.M. Research Laboratory, 5600 Cottle Road, San Jose, CA 95193

P. K. HANSMA—Department of Physics, University of California, Santa Barbara, CA 93106

W. C. KASKA—Department of Chemistry, University of California, Santa Barbara, CA 93106

The hydrogenation of CO on alumina supported metals has been the subject of many research efforts. This report is about the application of a relatively new technique, tunneling spectroscopy, to the problem of identifying the reaction intermediates that are formed on the catalyst surface. Tunneling spectroscopy is a vibrational spectroscopy that presents information about the system being studied in much the same way as do infrared and raman spectroscopy. The selection rules of tunneling are very weak; the spectrum observed consists of both the infrared and raman allowed vibrations. This similarity between techniques invites comparison of the results obtained in this work with those of previous infrared studies, such as those done by G.Blyholder, et al.(1), and R.A.Dalla Betta, et al.(2). From such comparisons it is possible to say that all three techniques find the observation of submonolayers of hydrocarbons to be technically demanding. No experimental method is yet available that allows the complete determination of all reaction pathways on a catalyst surface. As a first step toward this ultimate goal we report the observation of the hydrogenation of CO chemisorbed on alumina-supported metal particles with tunneling spectroscopy.

Inelastic Electron Tunneling Spectroscopy (IETS) measures the energies, and thus the frequencies, of the normal modes of vibration of molecules that are incorporated in order near the insulator of a metal-insulator-metal tunnel junction. In this work all junctions are made with an aluminum-aluminum oxide-dopant-lead structure, where the dopant consists of small metal particles (with 10-40 Angstrom diameters) that are exposed to CO and hydrogen. The insulator plus dopant thickness is approximately 30 Angstroms, thin enough to allow electrons to tunnel from the aluminum to the lead electrode whenever a voltage (aluminum negative) is applied. The maximum energy of a tunneling electron above the fermi energy of the lead electrode is the energy it gains from the applied voltage, eV. Experimentally it is found that these tunneling electrons can excite the vibrations of molecules in or near the insulating barrier. The dominant criterion

0097-6156/81/0152-0203$05.00/0
© 1981 American Chemical Society

is simply that the electron must have an energy greater or equal to the vibrational energy, hν, of the molecule excited. This gives rise to a threshold voltage, V = hν/e, that can be observed as a conductance increase. This increase is then measured and displayed in a form comparable to the absorbence in an infrared experiment. A more detailed description of the experimental techniques of tunneling can be found in the review literature (3).

## Rhodium

The top trace in Figure 1 shows the spectrum obtained for CO chemisorbed on rhodium particles. This differential spectrum is the result of subtracting the signals from two junctions; one prepared with rhodium particles, and another without rhodium particles. The resulting differential spectrum clearly shows vibrations due to the chemisorbed CO at 413, 465, 600, 1721, and 1942 $cm^{-1}$. The exact frequencies measured depend on the particle size, the temperature of the junction during formation, the degree of surface coverage, and the amount of other gases (such as water) chemisorbed on the surface. Detailed studies of these small shifts (typically a few percent) have not been completed. A recent study of such shifts due to temperature has been published by P.R.Antoniewicz, et al. using infrared spectroscopy (4). Structure due to the aluminum oxide, aluminum, and lead electrodes is greatly suppressed. The CO is chemisorbed as three different species on the rhodium surface (5,6,7). Three species have also been observed with infrared spectroscopy on similar systems. The identification of the three types of chemisorbed CO by infrared workers (8,9,10,11) as a gem dicarbonyl, $Rh(CO)_2$, a linear species, RhCO, and a multiply bonded species, $Rh_xCO$, agrees with the tunneling identifications derived from isotopic shifts. In tunneling spectra the dicarbonyl species is best seen by observing a low frequency bending mode at 413 $cm^{-1}$, the linear species can be identified by a low frequency bending mode at 465 $cm^{-1}$, and the multiply bonded species can be identified by the presence of the CO stretching vibration at 1721 $cm^{-1}$. The broad band at 600 $cm^{-1}$ contains the rhodium-carbon stretching vibrations for all three species, and the CO stretching mode at 1942 $cm^{-1}$ contains contributions from both the linear and dicarbonyl species.

The lower trace in Figure 1 shows the results of heating the tunnel junctions (complete with a lead top electrode) in a high pressure cell with hydrogen. It is seen that the CO reacts with the hydrogen to produce hydrocarbons on the rhodium particles. Studies with isotopes and comparison of mode positions with model compounds identify the dominant hydrocarbon as an ethylidene species (12). The importance of this observation is obviously not that CO and hydrogen react on rhodium to produce hydrocarbons, but that they will do so in a tunneling junction in a way so that the reaction can be observed. The hydrocarbon is seen as it forms from the chemisorbed monolayer of CO (verified by isotopes). As

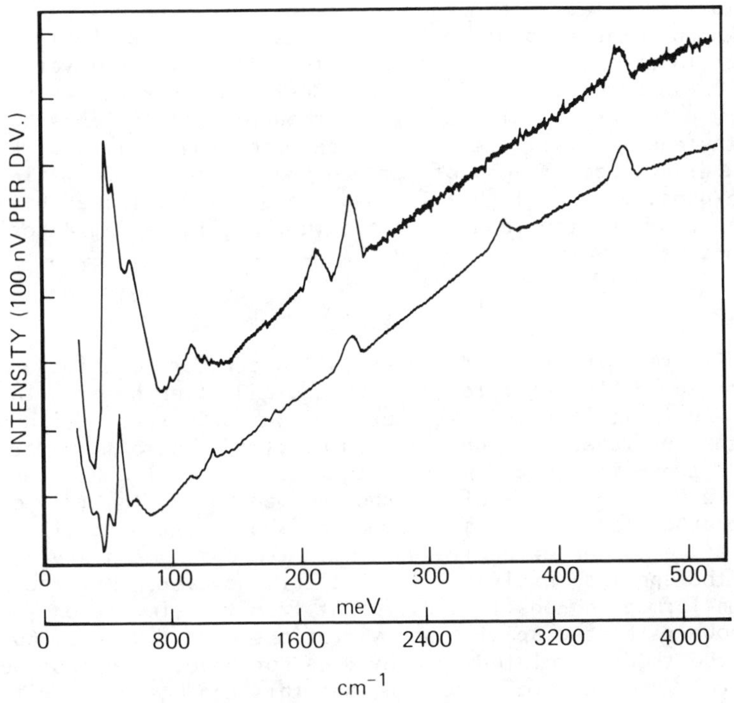


*Figure 1. Differential spectra of CO chemisorbed on alumina-supported Rh particles before and after heating to 420 K in hydrogen. One of the three species of chemisorbed CO remains after heating and can be identified by a bending mode at 478 $cm^{-1}$, a stretching mode at 586 $cm^{-1}$, and a stretching mode at 1937 $cm^{-1}$ as a linear CO species. The other CO species react and/or desorb while producing hydrocarbons on the Rh particles. The dominant species formed has been identified as an ethylidene species.*

it forms without the presence of gas phase CO, and no surface species containing oxygen is observed, it seems unlikely to be formed by CO insertion. A possible mechanism would be the polymerization of $CH_2$ groups from dissociated CO that had been hydrogenated. At any rate, the ethylidene species is a relatively stable surface species when CO is hydrogenated on rhodium particles <u>that are incorporated in tunneling junctions</u>. At present, it is not known if this same species is formed on supported rhodium particles in a more conventional reaction cell. It is not known because no other technique has developed the technology needed to observe the formation of the first submonolayer of hydrocarbons on supported metals. A similar species has been observed to form on single crystal rhodium from the chemisorption of ethylene (13,14). It is to be expected that future work will allow direct comparisons of surface species formed by different techniques, but until this is possible questions raised by the presence of the top lead electrode in these tunneling experiments can not be answered.

## Cobalt

Figure 2 shows the differential spectra of CO chemisorbed on supported cobalt particles both before and after heating in hydrogen. Again it is seen that the chemisorbed CO reacts with hydrogen that diffuses through the lead electrode to form hydrocarbons in the tunneling junction. For the case of cobalt much less is known about the nature of the chemisorbed CO and the type of hydrocarbon formed. This information should become available as soon as we do the extensive work with isotopes necessary to identify the species involved. It is clear, however, that the hydrocarbon formed on cobalt is different from that formed on rhodium. The cobalt related species has vibrations near 1600 and 760 $cm^{-1}$ that the rhodium related species does not have. The mode near 1600 $cm^{-1}$ should involve oxygen, and this can be tested with isotopes. As mentioned above, the rhodium species does not contain oxygen. The mode near 760 $cm^{-1}$ possibly is due to $CH_2$, a group that also does not appear in the rhodium species. Thus even before this species is identified, it is clear that the reaction pathway for hydrogenation of CO on cobalt is distinct from that on rhodium. This is, of course, no surprise; it has long been known that each metal has its own product distribution and reaction kinetics when used as a catalyst. What is noteworthy is that tunneling spectroscopy has been able to model supported catalysts well enough to reflect this difference between different metals. This suggests that whatever the effect of the lead electrode is on these reactions, there is information to be had from comparisons between the reactions of different metals under the same conditions.

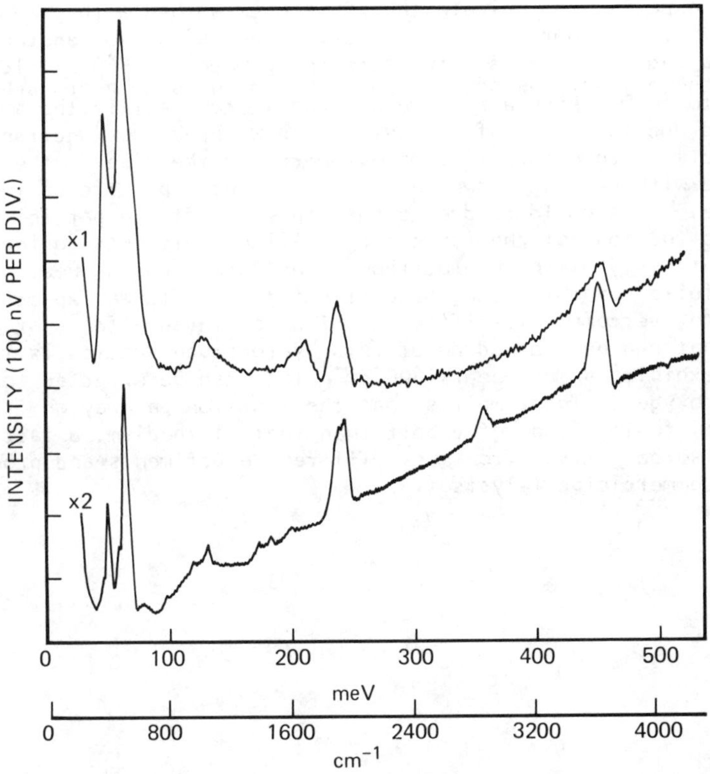


*Figure 2. Differential spectra of CO chemisorbed on alumina-supported Co particles both before and after heating in hydrogen to 415 K. The chemisorbed CO is seen to react and form hydrocarbons in the tunnel junction. This hydrocarbon species is distinct from that formed on Rh as seen by vibrational modes near 1600 $cm^{-1}$ and 760 $cm^{-1}$.*

## Iron

Figure 3 shows the results obtained to date on the hydrogenation of CO chemisorbed on supported iron particles. Iron is a difficult metal to work with in tunneling junctions due to its magnetic properties. We have developed a technique (15) that allows us to obtain spectra of highly dispersed iron with chemisorbed CO, as shown in the lower trace in the figure. Isotopic shift experiments indicate the CO is chemisorbed with predominantly linear character. The spectrum contains two bending modes at 436 and 519 $cm^{-1}$, and two stretching modes at 569 and 1856 $cm^{-1}$. When these junctions are heated in hydrogen, some hydrocarbon is seen to form. At present, we are unable to identify the species formed due to a lack of intensity. As we heat the junctions the particles become magnetic, as evidenced by the rise of the background with heating shown in the middle and upper traces in the figure. This could be due to the sintering of the particles or due to the loss of chemisorbed CO. All our attempts to increase the intensity of the hydrocarbon modes formed during heating have been foiled by this background structure. While we expect to be able to overcome this difficulty with continued effort, at present all that can be learned about this hydrocarbon species is that it also exhibits a mode near 1600 $cm^{-1}$ that can be expected to involve oxygen. This implies that the reaction pathway on iron more closely follows that of cobalt than that of rhodium, a fact that can also be derived from many different experiments and processes with commercial catalysts (16).

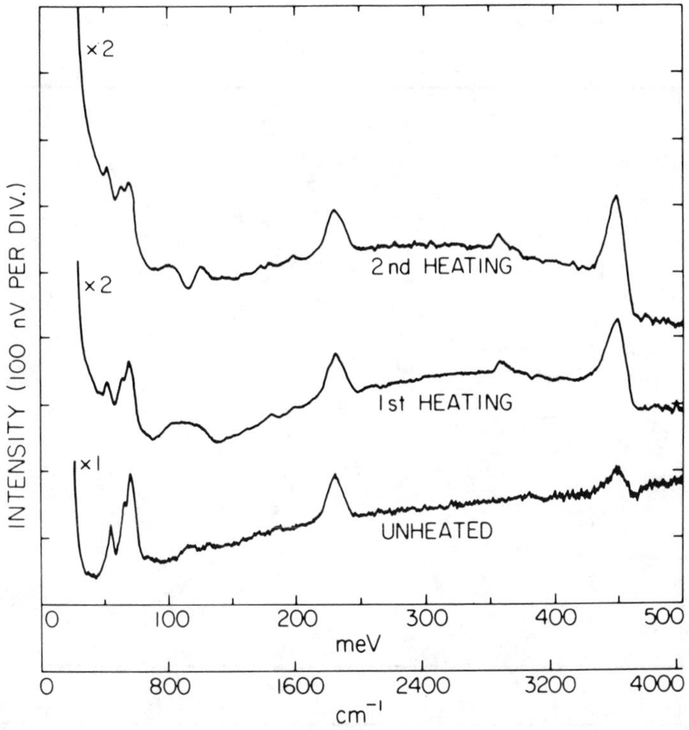


*Figure 3. Differential spectra of CO chemisorbed on alumina-supported Fe particles shown before and after two heatings in hydrogen to 420 K. Some CO reacts to form hydrocarbons on the Fe particles. The rising background seen at low frequencies indicates the formation of magnetic particles, either through sintering or the desorption of CO. The formation of OH in the junction upon heating does not correlate with the formation of a C–O bond nor with the formation of the C–H bonds.*

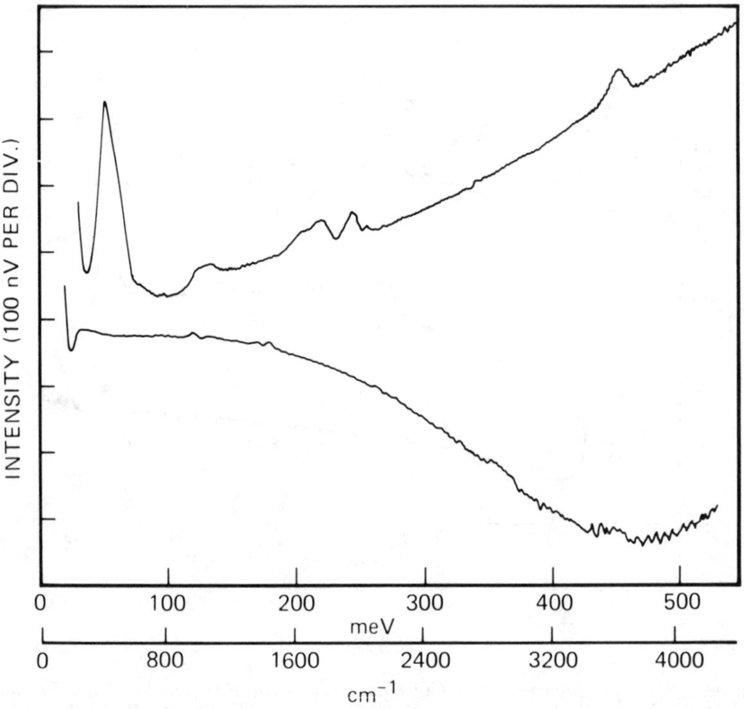


*Figure 4. Differential spectra of CO chemisorbed on alumina-supported Ni particles both before and after heating to 425 K. Very little surface hydrocarbon is seen to form on the Ni particles. This lack of surface hydrocarbon reflects the selectivity of such catalysts for methanation over Fisher–Tropsch synthesis.*

## Nickel

Nickel is known for its selectivity for methanation. This selectivity for methane implies a lack of carbon-carbon bonding on the metal surface when compared to cobalt or rhodium. Figure 4 shows our results for the hydrogenation of chemisorbed CO on supported nickel particles. The CO is chemisorbed on the nickel in at least four different ways (17). Upon heating in hydrogen the CO reacts and/or desorbs forming very little surface hydrocarbon. We feel this lack of surface hydrocarbon reflects the selectivity of alumina supported nickel particles for the formation of methane. The modes that are seen to form are again too weak for identification, as was the case for iron. We expect that future work with nickel will improve the observed signal to noise ratio significantly. It should be remembered that tunneling spectroscopy is less than twenty years old; its application to studies of the activation of CO is less than five years old. With this first work with supported metal particles we hope to have demonstrated some of the potential tunneling spectroscopy for the modelling of adsorption and reaction on catalist surfaces.

## Acknowledgments

We thank the Office of Naval Research and the Division of Materials Research of the National Science Foundation (Grant DMR 79-25430) for partial support of this work.

## Literature Cited

1. Blyholder,G.; Neff,L.D., J. Phys. Chem., 1962, 66, 1664-1667.
2. Dalla Betta,R.A.; Shelef,M., J. Catal., 1977, 48, 111-119.
3. Hansma,P.K., Phys. Rep., 1977, 30C, 145-206.
4. Antoniewicz,P.R.; Cavanagh,R.R.; Yates,J.T.Jr., J. Chem. Phys., 1980, 73, 3456-3459.
5. Hansma,P.K.; Kaska,W.C.; Laine,R.M., J. Amer. Chem. Soc., 1976 98, 6064-6065.
6. Kroeker,R.M.; Kaska,W.C.; Hansma,P.K., J. Catal., 1979, 57, 72-79.
7. Kroeker,R.M.; Kaska,W.C.; Hansma,P.K., J. Catal., 1980, 63, 487-490.
8. Yang,A.C.; Garland,C.W., J. Phys. Chem., 1957, 61, 1504-1512.
9. Arai,H.; Tomanaga,H., J. Catal., 1976, 43, 131-142.
10. Yao,H.C.; Rothschild,W.G., J. Chem. Phys., 1978, 68, 4774-4780.
11. Yates,J.T.; Duncan,T.M.; Worley,S.D.; Vaughan,R.W., J. Chem. Phys., 1979, 70, 1219-1224.
12. Kroeker,R.M.; Kaska,W.C.; Hansma,P.K., J. Catal., 1980, 61, 87-95.
13. Ibach,H.; Hopster,H.; Sexton,B., Appl. Sur. Sci., 1977, 1, 1-24.

14. Kesmodel,L.L.; Dubois,L.H.; Somerjai,G.A., Chem. Phys. Lett., 1978, 56, 267-271.
15. Kroeker,R.M.; Kaska,W.C.; Hansma,P.K., J. Chem. Phys., 1980, 72, 4845-4852.
16. Mills,G.A.; Steffgen,F.W., Catal. Rev., 1973, 8, 159-210.
17. Kroeker,R.M.; Kaska,W.C.; Hansma,P.K., J. Chem. Phys., 1980, in press.

RECEIVED December 19, 1980.

14

# Hydrogenation of Carbon Monoxide to Methanol and Ethylene Glycol by Homogeneous Ruthenium Catalysts

B. DUANE DOMBEK

Union Carbide Corporation, South Charleston, WV 25303

Hydrogenation of the carbon monoxide molecule is a reaction of both inherent chemical interest and practical significance, since processes based on it could become a major source of chemicals and fuels in the future. Heterogeneously catalyzed processes for synthesis gas conversion are already practiced commercially (1) (methanol synthesis and the Fischer-Tropsch process, for example), but homogeneous catalysts have not yet arrived at this stage. The major potential for homogeneously catalyzed processes appears to be in highly selective production of useful chemicals and intermediates from synthesis gas. The goals of our research have been to attempt to understand the chemical steps involved in formation of one of these useful chemicals - ethylene glycol - from synthesis gas, and to use this knowledge in designing more effective catalytic systems.

Most homogeneous catalysts for hydrogenation of carbon monoxide have been discovered and studied under extremely high pressures. For example, the first demonstration that organic products (including ethylene glycol and glycerine) could be obtained from $H_2/CO$ by homogeneous catalysis was performed with cobalt complexes under pressures from 1500 to 5000 atm (2). Subsequently, rhodium complexes were found to be catalytically active at elevated pressures (3), and continued research on this system has given improved results at lower pressures (4). Many metal carbonyl complexes have recently been reported to hydrogenate CO at pressures substantially above 1000 atm (5,6,7,8,9).

Although related reactions have also been done under low pressures, very low rates of product formation are observed (8,10,11). We have found, however, that a ruthenium carbonyl catalyst is quite active for converting $H_2/CO$ to methanol under moderate pressures (below 340 atm). More significantly, we also discovered that an ethylene glycol product could be obtained from this catalyst by use of carboxylic acid promoters or solvents (12). This remarkable and intriguing promoter effect deserved, we felt, further mechanistic investigation

0097-6156/81/0152-0213$05.00/0
© 1981 American Chemical Society

because of its potential implications for influencing catalyst selectivity toward two-carbon and longer-chain products.

## Results

Reaction of acetic acid solutions of $Ru_3(CO)_{12}$ with mixtures of CO and $H_2$ under pressure produces substantial amounts of methyl acetate and smaller quantities of ethylene glycol diacetate, as shown in Table I. Other products observed in these reactions are traces of glycerine triacetate and small amounts of ethyl acetate. (The ethanol is apparently derived largely from acetic acid by catalytic hydrogenation, since reactions in propionic acid solvent yield similar quantities of propyl propionate and only traces of ethyl propionate.)

Infrared spectra of reaction solutions immediately after depressurization show that the ruthenium is mainly in the form of $Ru(CO)_5$, and only traces of $Ru_3(CO)_{12}$ are present. However upon standing at ambient conditions the solutions precipitate $Ru_3(CO)_{12}$ in nearly quantitative yields. Infrared spectra under reaction conditions (400 atm of 1:1 $H_2/CO$, 200°C) also correspond to the spectrum of $Ru(CO)_5$; no acetate or cluster complexes are observed. However, there is evidence for the presence of small amounts of $Ru_3(CO)_{12}$ under somewhat lower pressures (ca. 200 atm). Many other ruthenium complexes were used as catalyst precursors, and were found to be converted to the same ruthenium products under reaction conditions. For example, $H_4Ru_4(CO)_{12}$ (13), $[Ru(CO)_2(CH_3CO_2)_2]_n$ (14), $Ru_6C(CO)_{17}$ (15), $H_3Ru_3(CO)_9(CCH_3)$ (16), and $Ru(acac)_3$ all gave equal rates to organic products and were recovered as $Ru_3(CO)_{12}$.

No evidence of ruthenium metal formation was found in catalytic reactions until temperatures above about 265°C (at 340 atm) were reached. The presence of Ru metal in such runs could be easily characterized by its visual appearance on glass liners and by the formation of hydrocarbon products (8,17). The actual catalyst involved in methyl and glycol acetate formation is therefore almost certainly a soluble ruthenium species. In addition, the observation of predominantly a mononuclear complex under reaction conditions in combination with a first-order reaction rate dependence on ruthenium concentration (e.g., see reactions 1 and 3 in Table I) strongly suggests that the catalytically active species is mononuclear.

Reactions in solvents other than carboxylic acids (e.g., ethers, alcohols, esters, hydrocarbons, etc.) under conditions given in Table I do not produce detectable amounts of ethylene glycol (less than about 0.02 mmole). (Experiments were carried out to demonstrate that glycol would survive had it been produced in some of these solvents.) However, methanol yields nearly equivalent to those obtained in acetic acid are found in some of these solvents (cf. reactions 1, 10 and 11 in Table I). Reactions in acetic acid diluted with these solvents also give

Table I. Hydrogenation of carbon monoxide with ruthenium catalysts. All reactions performed in a glass-lined rocker bomb with 2.35 mmol Ru (charged as $Ru_3CO)_{12}$), at 230°C under 340 atm 1:1 $H_2$/CO for 2 h, unless noted otherwise.

| reaction | solvent | mmol $CH_3O$- | mmol $-OCH_2CH_2O$- | notes |
|---|---|---|---|---|
| 1 | 50 mL acetic acid | 52.2 | 1.37 | |
| 2 | 50 mL propionic acid | 61.0 | 1.03 | |
| 3 | 50 mL acetic acid | 14.9 | 0.41 | a |
| 4 | 50 mL acetic acid | 139. | 1.58 | b |
| 5 | 40 mL acetic acid, 10 mL $H_2O$ | 66.8 | 0.75 | |
| 6 | 40 mL acetic acid, 10 mL c-$C_6H_{12}$ | 39.7 | 0.82 | |
| 7 | 40 mL acetic acid, 10 mL THF | 45.9 | 1.03 | |
| 8 | 40 mL acetic acid, 10g $H_3PO_4$ | 48.2 | 0.21 | |
| 9 | 50 mL THF | 19.4 | - | |
| 10 | 50 mL ethyl acetate | 33.1 | - | |
| 11 | 50 mL ethanol | 55.6 | - | |
| 12 | 50 mL ethanol | 109. | - | b |
| 13 | 30 g $C_6Cl_5OH$ | n.d. | - | |

[a] 0.70 mmol Ru charged.

[b] Reaction at 260°C.

similar methanol yields, but the amount of ethylene glycol produced is less than is formed in undiluted acetic acid solvent (see Table I). For example, Fig. 1 shows the effect of increasingly diluting the acetic acid solvent with water and methyl acetate; there is an approximate dependence of glycol rate on the second power of acetic acid concentration.

A number of carboxylic acids other than acetic were investigated as solvents or promoters. All of these acids which were stable to reaction conditions were found to be effective in promoting glycol ester production (e.g., propionic, pivalic, benzoic, etc.). However, other Brønsted acids of non-carboxylic nature were not found to be effective promoters. Thus pentachlorophenol, although it has a $pK_a$ value (4.82) very close to that of acetic acid (4.76), is not a comparable promoter (Table I, reaction 13). Likewise, phosphoric acid ($pK_1$=2.15) is not an effective solvent or co-solvent with acetic acid (Table I, reaction 8). Experiments with lower concentrations of these acids in sulfolane solvent also showed that carboxylic acids are unique in promoting glycol formation. The promoter function of carboxylic acids thus appears not to be dependent (only) upon their acidity, but on some other chemical or structural property.

The glycol-producing reaction exhibits an interesting dependence on the partial pressure of carbon monoxide. As shown in Fig. 2 the dependence of rate on CO pressure is large at low CO partial pressure, but approaches zero-order at higher CO levels. A more constant dependence on $H_2$ partial pressure is observed, but a non-integral relationship between first- and second-order (*ca.* 1.3) is found. The formation of the methanol product in these reactions exhibits nearly the same behavior with respect to CO and $H_2$ partial pressures, and only minor changes in product distribution are observed on changing the gas composition or pressure.

## Discussion

The nearly identical dependences of methanol and glycol product rates on ruthenium concentration, $H_2$ partial pressure, and CO partial pressure perhaps suggest that the same catalyst is involved in forming both products, and that the branching point occurs relatively late in the process. Essentially the only difference in the empirical rate equations for the two products is the zero-order dependence on acetic acid concentration for methanol production *vs*. a high dependence on acid concentration for glycol formation. Presumably the acid is only involved in the glycol-forming reaction after the branching point. It is possible to postulate a mechanistic sequence consistent with these assumptions and with all of the experimental observations, as shown in Scheme 1. Reaction steps outlined in this scheme are combinations of elementary steps,

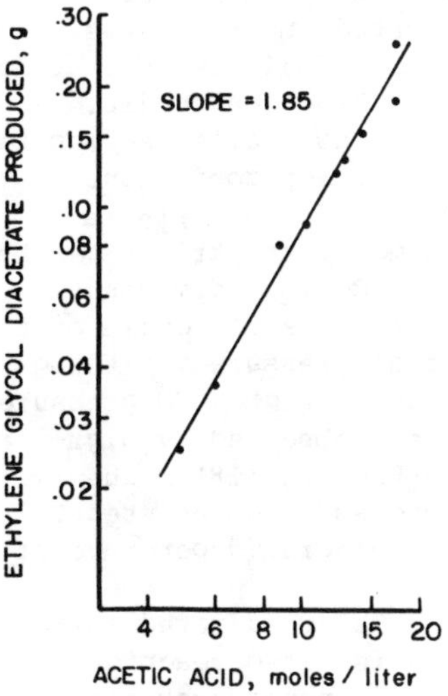


Journal of the American Chemical Society

*Figure 1. Log–log plot of ethylene glycol diacetate yield vs. acetic acid concentration when diluted with varying amounts of methyl acetate and $H_2O$; conditions are specified in Table I (12)*

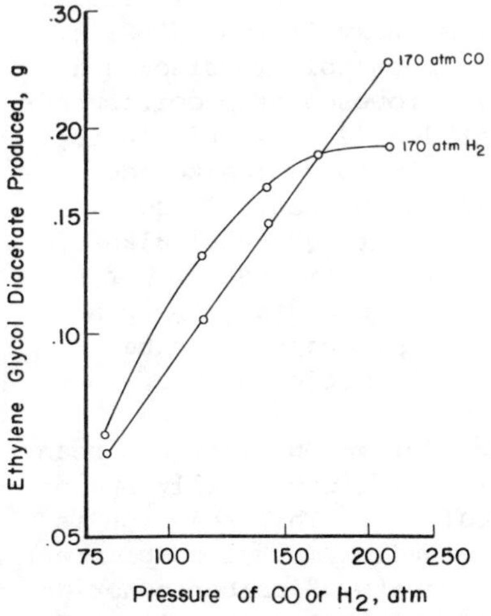


Journal of the American Chemical Society

*Figure 2. Ethylene glycol diacetate yield as a function of varying CO and $H_2$ partial pressures, at constant 170 atm $H_2$ and CO partial pressures, respectively; other conditions are listed in Table I. (12)*

and reagents specified are those consumed by the entire step. Rate-determining processes may be elementary steps which are not explicitly shown; coordinately unsaturated intermediates, for example, are not included. Double arrows indicate that at least partial reversibility is expected under reaction conditions.

Information on step 1 in this scheme is available from high-pressure infrared measurements under reaction conditions. At CO partial pressures above about 150 atm no $Ru_3(CO)_{12}$ is observed, but at lower CO pressures some of the trimer can be detected. This equilibrium between $Ru_3(CO)_{12}$ and mononuclear, catalytically active species may therefore be the cause of the CO dependence found under low CO partial pressures, although zero-order CO dependence is observed under higher CO pressures.

Reaction of $Ru(CO)_5$ with $H_2$ has been observed by high-pressure IR spectroscopy to produce $H_2Ru(CO)_4$ (18). The involvement of $H_2$ in an equilibrium process such as step 2 could be the root of the observed non-integral dependence of reaction rate on $H_2$ pressure.

Intramolecular hydride migration to a coordinated carbonyl ligand has not yet been observed as an isolated reaction, but the involvement of this process in step 3 seems probable. Reductive elimination of the resultant formyl ligand could yield coordinated formaldehyde (step 4), as previously proposed in a mechanism for the Fischer-Tropsch reaction (19). This formaldehyde ligand could presumably insert into a Ru-H bond in two directions; either a metal-oxygen bond ($Ru\text{-}OCH_3$) or a metal-carbon bond ($Ru\text{-}CH_2OH$) could be formed. Because this catalytic system is highly specific for methanol formation in the absence of carboxylic acids, a methoxy ligand rather than hydroxymethyl is presumed to be the methanol precursor under these conditions, via step 5. This proposed step corresponds to the known hydrogenation of aldehydes (20,21) and, in the reverse sense, to the conversion of alcohols or alkoxides (22,23,24) to aldehydes by ruthenium complexes. These processes are all believed to proceed through metal alkoxide complexes rather than $\alpha$-hydroxyalkyl intermediates. If a hydroxymethyl complex were indeed an intermediate in this system, at least traces of two-carbon products would be expected to be produced by migratory insertion of a carbonyl ligand -- a step which is surely very facile under the conditions employed for catalysis. For example, it has been reported (25) that formaldehyde is stoichiometrically hydroformylated to glycolaldehyde by $HCo(CO)_4$. This reaction is presumed to involve an intermediate hydroxymethyl complex which is carbonylated even at $0^oC$ under one atm of carbon monoxide.

Since it is experimentally observed that carboxylic acids are required to promote glycol production by this system, and since acid concentration appears in the empirical rate equation for glycol production with a substantial exponent (ca. 1.8), the formation of a metal-carbon bonded intermediate (step 6) may involve an acid "dimer". The character of such an aggregate is not certain - it could be a hydrogen-bonded acid dimer, a protonated acid molecule, or a small equilibrium concentration of acid anhydride, for example. Whatever the form of this complex it could act as an effective acylating agent, attacking the oxygen atom of a coordinated formaldehyde intermediate and promoting metal-carbon bond formation. Esterification of alcohols by acetic acid, for example, is observed to be second-order in acid concentration, and an acid dimer is believed to be involved (26). (Ruthenium-catalyzed $H_2$/CO reactions performed in acetic anhydride also produce glycol diacetate, but much of the anhydride is hydrogenated, yielding ethyl acetate and acetic acid.) A stable osmium complex containing coordinated formaldehyde has been shown to undergo an electrophilic addition entirely analogous to that proposed for step 6 (27). Reaction with the alkylating agent $CF_3SO_3CH_3$ results in methylation of the formaldehyde oxygen atom and formation of a metal-carbon bonded methoxymethyl ligand (eq. 1).

$$Os(h^2\text{-}CH_2O)(CO)_2(PPh_3)_2 \cdot H_2O \xrightarrow{CF_3SO_3CH_3} [Os(CH_2OCH_3)(H_2O)(CO)_2(PPh_3)_2]^+ \quad (1)$$

The ruthenium acyloxymethyl complex produced by step 6 of Scheme 1 could, of course, eliminate the methyl ester product, but it also has the possibility of leading to a two-carbon product via alkyl group migration to coordinated CO (eq. 2).

$$CH_3O\overset{O}{\overset{\|}{C}}R \longleftarrow OC\text{-}\overset{H}{\overset{|}{Ru}}\text{-}CH_2O\overset{O}{\overset{\|}{C}}R \longrightarrow \overset{H}{\overset{|}{Ru}}\text{-}\overset{O}{\overset{\|}{C}}CH_2O\overset{O}{\overset{\|}{C}}R \quad (2)$$

The chemistry of such an acyloxymethyl ligand has been investigated in a closely related manganese model system (28). The complex $(CO)_5Mn\text{-}CH_2OC(O)Bu^t$ reacts with $H_2$ under very mild conditions (100 psi, 75°C) to give good yields of ethylene glycol ester $HOCH_2CH_2OC(O)Bu^t$ (Scheme 2). Only traces of methyl ester $CH_3OC(O)Bu^t$ are observed, but under other conditions this can become the major product. The glycol product apparently arises from hydrogenation of a glycolaldehyde ester intermediate, $HC(O)CH_2OC(O)Bu^t$, since this aldehyde is observed in early stages of the reaction, but disappears rapidly as the alcohol product is formed. This model system demonstrates that an acyloxymethyl ligand can be converted to a glycol or methanol product under very mild conditions.

The formation of a longer-chain product, glycerine triacetate, in $Ru_3(CO)_{12}$-acetic acid catalytic reactions can also be accounted for by Scheme 1. A glycolaldehyde ester intermediate would presumably be largely hydrogenated to a

*Scheme 1. Postulated mechanism of methanol and ethylene glycol ester formation by Ru catalysts*

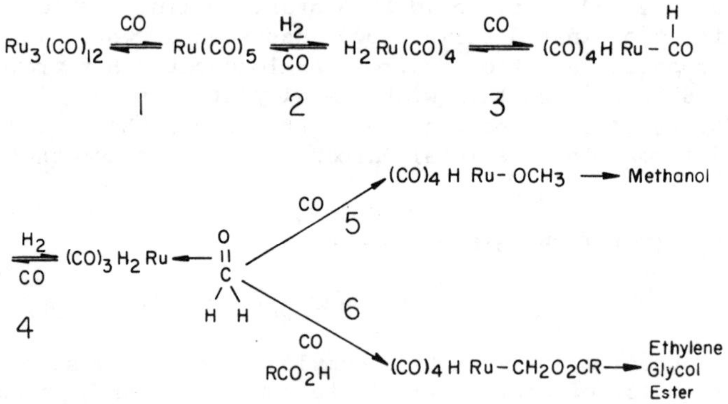


*Scheme 2. Reactions of a Mn acyloxymethyl complex with $H_2$ (R = C(O)Bu$^t$)*

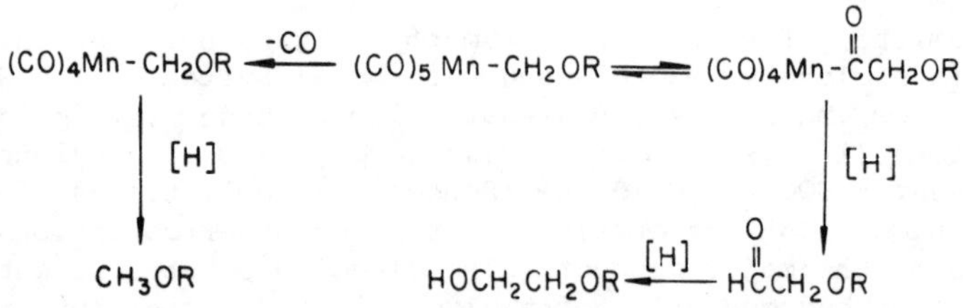

glycol product by a process similar to step 5. However, hydroformylation of some of the aldehyde intermediate by the process of step 6 would lead to the glycerine product.

The experimental results are consistent with the rate-determining step for methanol formation being $H_2$ addition in step 4, and that for glycol formation being step 6 after involvement of CO and acid. Later steps could possibly be rate-limiting instead, and other mechanistic schemes could perhaps be written which are consistent with the observations. However, all of the steps in Scheme 1 are reasonably well precedented in studies of Ru chemistry or closely related systems with the exception of step 3, hydride migration to CO. (Indeed, first-order homogeneous reactions such as this are perhaps the best existing evidence for intramolecular hydride migration to a carbonyl ligand.) The scheme provides one explanation for the unusual promoter effect observed to be specific to carboxylic acids. This phenomenon is not a normal solvent or acidity effect, but appears to involve the introduction of a reagent which can intercept a catalytic intermediate and significantly alter the course of its reaction.

Acknowledgments

The author is grateful to Dr. Leonard Kaplan and Dr. George O'Connor for support and many helpful discussions, to Mr. T. D. Myers and Mr. R. B. James for experimental assistance, and to Union Carbide Corporation for permission to publish this research.

Abstract

Solutions of $Ru_3(CO)_{12}$ in carboxylic acids are active catalysts for hydrogenation of carbon monoxide at low pressures (below 340 atm). Methanol is the major product (obtained as its ester), and smaller amounts of ethylene glycol diester are also formed. At 340 atm and 260°C a combined rate to these products of $8.3 \times 10^{-3}$ turnovers $s^{-1}$ was observed in acetic acid solvent. Similar rates to methanol are obtainable in other polar solvents, but ethylene glycol is not observed under these conditions except in the presence of carboxylic acids. Studies of this reaction, including infrared measurements under reaction conditions, were carried out to determine the nature of the catalyst and the mechanism of glycol formation. A reaction scheme is proposed in which the function of the carboxylic acid is to assist in converting a coordinated formaldehyde intermediate into a glycol precursor.

## Literature Cited


1. Masters, C. "Advances in Organometallic Chemistry", Vol. 17; Stone, F.G.A.; West, R., eds.; Academic Press: New York, 1979; p. 61.

2. Gresham, W. F. (to DuPont), U. S. Patent 2,636,046 (1953).

3. Pruett, R. L.; Walker, W. E. (to Union Carbide Corp.), U. S. Patent 3,833,634 (1974).

4. Kaplan, L. (to Union Carbide Corp.), U. S. Patent 4,162,261 (1979).

5. Fonseca, R.; Jenner, G.; Kiennemann, A.; Deluzarche, A. "High Pressure Science and Technology", Timmerhaus, K. D.; Barber, M. S., eds.; Plenum Press: New York, 1979; pp. 733-738.

6. Deluzarche, A.; Fonseca, R.; Jenner, G.; Kiennemann, A. Erdöl und Kohle, 1979, 32, 313.

7. Keim, W.; Berger, M.; Schlupp, J. J. Catalysis, 1980, 61, 359.

8. Bradley, J. S. J. Am. Chem. Soc., 1979, 101, 7419.

9. Williamson, R. C.; Kobylinski, T. P. (to Gulf Res. and Dev. Co.), U. S. Patents 4,170,605 (1979) and 4,170,606 (1979).

10. Demitras, G. C.; Muetterties, E. L. J. Am. Chem. Soc., 1977, 99, 2796.

11. Rathke, J. W.; Feder, H. M. J. Am. Chem. Soc., 1978, 100, 3623.

12. Dombek, B. D. J. Am. Chem. Soc., 1980, 102, 0000.

13. Knox, S. A. R.; Koepke, J. W.; Andrews, M. A.; Kaesz, H. D. J. Am. Chem. Soc., 1975, 97, 3942.

14. Crooks, G. R.; Johnson, B. F. G.; Lewis, J.; Williams, I. G.; Gamlen, G. J. Chem. Soc. (A), 1969, 2761.

15. Johnson, B. F. G.; Johnston, R. D.; Lewis, J. J. Chem. Soc. (A), 1968, 2865.

16. Canty, A. J.; Johnson, B. F. G.; Lewis, J.; Norton, J. R. J. Chem. Soc., Chem. Commun., 1972, 1331.

17. Doyle, M. J.; Kouwenhoven, A. P.; Schaap, C. A.; Van Oort, B. J. Organomet. Chem., 1979, 174, C55.

18. Whyman, R. J. Organometal. Chem., 1973, 56, 339.

19. Henrici-Olive, G.; Olive, S. Angew. Chem. Int. Ed. Engl., 1976, 15, 136.

20. Strohmeier, W.; Weigelt, L. J. Organomet. Chem., 1978, 145, 189.

21. Imai, H.; Nishiguchi, T.; Fukuzumi, K. J. Org. Chem., 1976, 41, 665.

22. Kaesz, H. D.; Saillant, R. B. Chem. Rev., 1972, 72, 231.

23. Dobson, A.; Robinson, S. D. Inorg. Chem., 1977, 16, 137.

24. Chaudret, B. N.; Cole-Hamilton, D. J.; Nohr, R. S.; Wilkinson, G. J. Chem. Soc., Dalton Trans., 1977, 1546.

25. Roth, J. A.; Orchin, M. J. Organomet. Chem., 1979, 172, C27.

26. Rolf, A. C.; Hinshelwood, C. N. Trans. Faraday Soc., 1934, 30, 935.

27. Brown, K. L.; Clark, G. R.; Headford, C. E. L.; Marsden, K.; Roper, W. R. J. Am. Chem. Soc., 1979, 101, 503.

28. Dombek, B. D. J. Am. Chem. Soc., 1979, 101, 6466.

RECEIVED December 8, 1980.

15

# Syngas Homologation of Aliphatic Carboxylic Acids

JOHN F. KNIFTON

Texaco Chemical Company, P.O. Box 15730, Austin, TX 78761

In this paper we disclose the syngas homologation of carboxylic acids via ruthenium homogeneous catalysis. This novel homologation reaction involves treatment of lower MW carboxylic acids with synthesis gas ($CO/H_2$) in the presence of soluble ruthenium species. e.g., $RuO_2$, $Ru_3(CO)_{12}$, $H_4Ru_4(CO)_{12}$, coupled with iodide-containing promoters such as HI or an alkyl iodide (1).

$$CH_3COOH \xrightarrow{CO/H_2} C_nH_{2n+1}COOH \qquad (1)$$

Where acetic is the starting acid (eq. 1), homologation selectively yields the corresponding $C_3$+ aliphatic carboxylic acids. Since acetic acid is itself a "syngas" chemical derived from methanol via carbonylation (2,3), this means the higher MW carboxylic acids generated by this technique could also be built exclusively from $CO/H_2$ and would thereby be indepent of any petroleum-derived coreactant.

The scope and mechanism of carboxylic acid homologation is examined here in relation to the structure of the carboxylic acid substrate, the concentrations and composition of the ruthenium catalyst precursor and iodide promoter, synthesis gas ratios, as well as $^{13}C$ labelling studies and the spectral identification of ruthenium iodocarbonyl intermediates.

## Results

A summary of typical preparative data for the acid homologation regime is presented in Tables I and II.

Effect of Catalyst Composition. Where acetic is the typical acid substrate, effective ruthenium catalyst precursors include ruthenium(IV) oxide, hydrate, ruthenium(III) acetylacetonate, triruthenium dodecacarbonyl, as well as ruthenium hydrocarbonyls, in combination with iodide-containing promoters like HI and alkyl iodides. Highest yields of these higher MW acids are achieved with the $RuO_2$-MeI combination,

0097-6156/81/0152-0225$05.00/0
© 1981 American Chemical Society

Table I

Acetic Acid Homologation

| Catalyst composition[a] | | | (mole %)[c] | | | |
|---|---|---|---|---|---|---|
| Ruthenium source | Promoter structure | conversion (%) | Propionic | Butyric | Valeric | Butyric acids n/iso ratio |
| $RuO_2$ | MeI | 52 | 37 | 6.9 | 1.0 | 1.6 |
| $H_4Ru_4(CO)_{12}$ | MeI | 42 | 34 | 5.5 | <1 | |
| $Ru_3(CO)_{12}$ | MeI | 70 | 25 | 7.2 | 3.2 | 5.1 |
| $Ru(acac)_3$ | MeI | 63 | 29 | 8.5 | <1 | 9.2 |
| $RuCl_3$ | MeI | 70 | 20 | 4.3 | <1 | 7.8 |
| $RuCl_2(PPh_3)_3$ | MeI | 10 | 7.6 | | | |
| $RuO_2$ | EtI | 48 | 37 | 3.2 | 0.7 | 1.9 |
| $RuO_2$ | HI | 50 | 37 | 3.7 | 1.0 | 2.0 |
| $RuO_2$ | CsI | 26 | <1 | <1 | | |
| $RuO_2$ | $Bu_4PI$ | 16 | 29 | 7.3 | 2.0 | |
| $RuO_2$ | EtBr | 96[d] | 6.6 | | | |
| $RuO_2$ | - | <5 | - | | | |

[a] Reaction charge: ruthenium, 4.0 mmole; iodide, 40 mmole; acetic acid, 830 mmole.

[b] Typical operating conditions: 220°C; 272 atm initial pressure $CO/H_2$ (1:1).

[c] Carboxylic acid yield basis acetic acid converted.

[d] Two-phase liquid product, low (49%) liquid yield.

B-19

Table II

Carboxylic Acid Homologation

| Acid substrate[a] | Acid[b] conv. (%) | Major acid homologues Composition | (mmole) | n/iso ratio |
|---|---|---|---|---|
| $CH_3-CH_2-COOH$ | 69 | $C_3H_7-COOH$ | 35 | 8.5 |
| n-$C_4H_9-COOH$ | 67 | $C_5H_{11}-COOH$ | 35 | 4.2 |
| $\begin{array}{c}CH_3\diagdown\\CH_3\diagup\end{array}CH-CH_2-COOH$ | 43 | $\begin{array}{c}CH_3\diagdown\\CH_3\diagup\end{array}CH-(CH_2)_2-COOH$ <br><br> $CH_3-CH_2-\underset{\underset{CH_3}{\vert}}{\overset{\overset{CH_3}{\vert}}{C}}-COOH$ | 8<br><br>7 | |
| $\begin{array}{c}CH_3\diagdown\\CH_3\diagup\end{array}CH-COOH$ | 45 | $CH_3-\underset{\underset{CH_3}{\vert}}{\overset{\overset{CH_3}{\vert}}{C}}-COOH$ | 26 | |
| $\begin{array}{c}C_3H_7\diagdown\\CH_3\diagup\end{array}CH-COOH$ | 48 | $\begin{array}{c}C_3H_7\diagdown\\CH_3\diagup\end{array}CH-CH_2-COOH$ <br><br> $C_3H_7-\underset{\underset{CH_3}{\vert}}{\overset{\overset{CH_3}{\vert}}{C}}-COOH$ | 13<br><br>12 | |
| $CH_3-\underset{\underset{CH_3}{\vert}}{\overset{\overset{CH_3}{\vert}}{C}}-COOH$ | 43 | $CH_3-CH_2-\underset{\underset{CH_3}{\vert}}{\overset{\overset{CH_3}{\vert}}{C}}-COOH$ | 26 | |

[a] Reaction charge:  $RuO_2xH_2O$, 2.0 mmole; MeI, 20.0 mmole; RCOOH, 245 mmole.

[b] Operating conditions:  220°C, 272 atm. initial pressure $CO/H_2$ (1:1).

C-19

here total selectivity to $C_3$+ acids is ca. 45% and turnover numbers are typically ca. 110 per g atom Ru. The ruthenium catalyst remains in solution throughout the homologation sequence, and the crude liquid products typically display no metal precipitate. Homologation is not observed in the absence of halogen promoter.

The principal competing reactions to ruthenium-catalyzed acetic acid homologation appear to be water-gas shift to $CO_2$, hydrocarbon formation (primarily ethane and propane in this case) plus smaller amounts of esterification and the formation of ethyl acetate (see Experimental Section). Unreacted methyl iodide is rarely detected in these crude liquid products. The propionic acid plus higher acid product fractions may be isolated from the used ruthenium catalyst and unreacted acetic acid by distillation in vacuo.

Effect of Operating Conditions. Yield data, summarized in Figures 1 and 2, point to acetic acid homologation activity being sensitive to at least four operating variables, viz. ruthenium and methyl iodide concentrations, syngas composition and operating pressure.

Figure 1 illustrates the first order dependence upon initial ruthenium oxide concentrations for [MeI]/[Ru] ratios in the range >10. This first order dependence is observed only up to a ruthenium(IV) oxide charge of 2 mmole (i.e. [Ru] ~ 70 mM) under our selected experimental conditions. Higher initial quantities of ruthenium lead to the presence of significant quantities of yellow precipitates at the end of each run, lower overall yields of $C_3$+ acids, but proportionally higher yields of butyric and valeric acids.

The yields of propionic and higher acids are also definitely improved with increasing initial methyl iodide concentration (Figure 1) but here the relationship is complicated by the fact that at low methyl iodide concentrations (and thereby [MeI]/[Ru] <5) the corresponding esters, particularly ethyl acetate and ethyl propionate, become the principal products, rather than the corresponding free acids, while at much higher iodide concentrations, where [MeI]/[Ru] ratios are 30 or more, there is a phase separation of the product liquids into aqueous-rich and water-poor fractions.

The importance of syngas composition and operating pressure is illustrated in Figure 2. It may be noted that maximum yields of propionic acid are achieved with 1:1 $CO/H_2$ even though the stoichiometry of this synthesis calls for 2 moles of hydrogen per mole of CO (eq. 2). No acetic acid homologation is observed in the absence of the hydrogen component, but while acetic acid conversion generally increases with increasing hydrogen content of the fresh syngas this is also paralleled by an increase in $CO_2$, ethane and propane concentrations in the product off-gas.

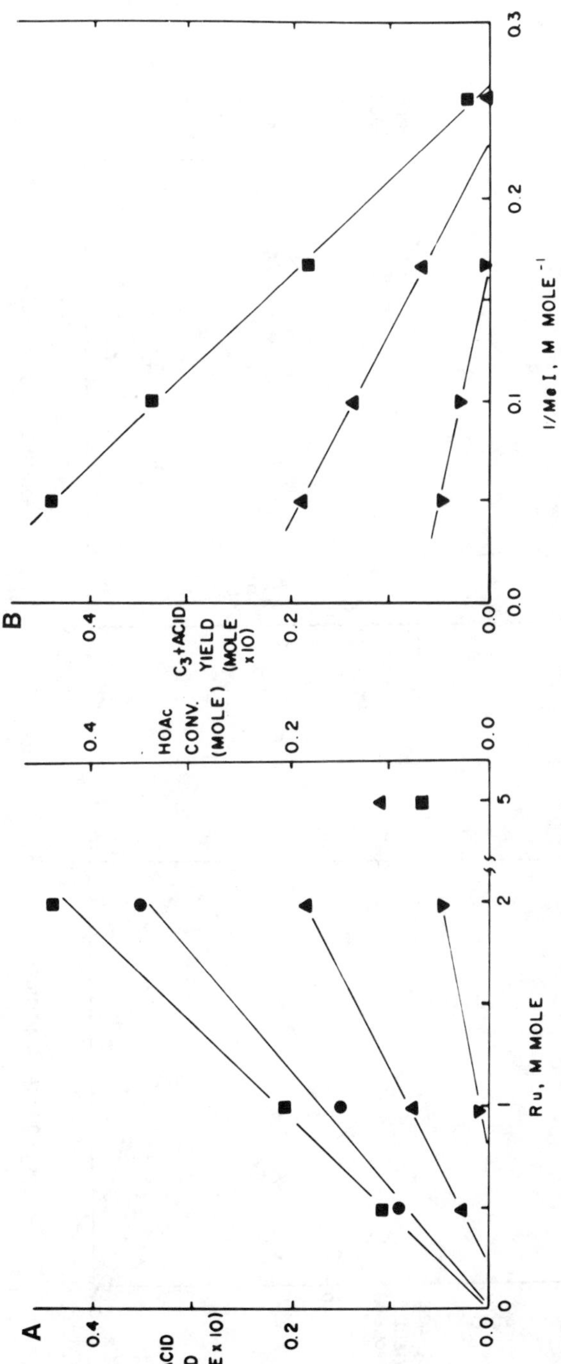


*Figure 1.* Acetic acid homologation: (●) acetic acid, (■) propionic acid, (▲) butyric acids, and (▼) valeric acids; A, effect of Ru (operating conditions: acetic acid, 417 mmol; MeI, 20 mmol; 220°C; 476 atm constant pressure; $CO/H_2 = 1/1$); B, effect of MeI (operating conditions: acidic acid, 417 mmol; Ru(IV) oxide, 2.0 mmol; 220°C; 476 atm constant pressure; $CO/H_2 = 1/1$)

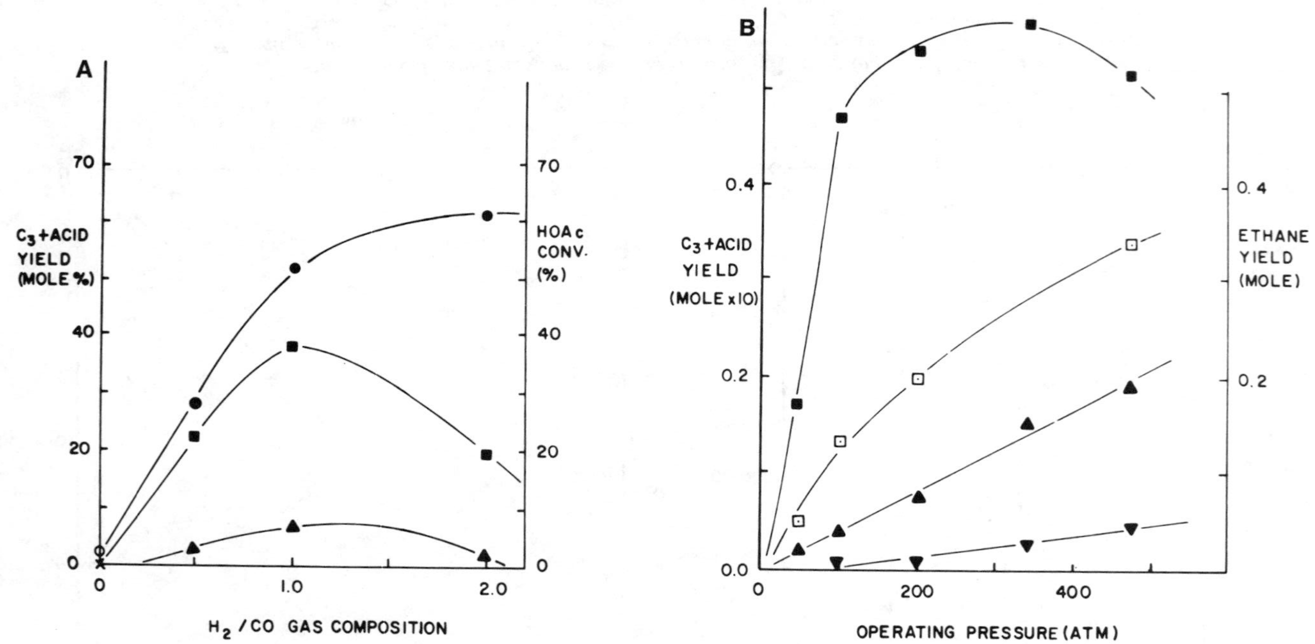


*Figure 2. Acetic acid homologation: (●) acetic acid, (■) propionic acid, (▲) butyric acids, and (▼) valeric acids, (□) ethane; A, effect of syngas composition (operating conditions: acetic acid, 833 mmol; Ru(IV) oxide, 4.0 mmol; MeI, 40 mmol; 220°C; 272 atm initial pressure); B, effect of operating pressure (operating conditions: acetic acid, 417 mmol; Ru(IV) oxide, 2.0 mmol; MeI, 20 mmol; 220°C; constant pressure; $CO/H_2 = 1/1$)*

The effect of operating pressure upon the degree of acetic acid homologation is illustrated also in Figure 2. Although homologation has been observed over a 10-fold change in operating pressure (50 → 500 atm), the yield and selectivity to propionic acid appears to reach a maximum at ca. 100 atm, while the yields of higher acids, butyric and valeric, continue to increase at least to the upper operating limit of our equipment (500 atm).

$$CH_3COOH + CO + 2H_2 \longrightarrow CH_3CH_2COOH + H_2O \qquad (2)$$

Homologation of Higher Acids. The versatility of the homologation technique is illustrated in Table II for a series of typical linear and branched aliphatic carboxylic acids. In the case of straight-chain, $C_nH_{2n+1}COOH$, acid substrates, such as propionic and n-valeric acids, the principal homologated products are the corresponding higher acids containing one additional carbon per molecule, e.g., butyric and hexanoic acids respectively. Here it is the linear chain isomer that generally predominates, but isobutyric and 2-methylvaleric acids are detected, and n/iso ratios are normally in the range 4-8. The principal by-products are water, $CO_2$, and the corresponding hydrocarbon, e.g., propane and n-butane in the case of $C_3$ acid homologation (see Experimental Section) and n-pentane and n-hexane in the case of valeric acid, plus variable quantities of acetic acid. Generally we conclude that increasing the chain length of the alkyl portion of the carboxylic acid substrate, from $C_1$ to $C_4$, does not dramatically change the conversions or yields of higher acids with this class of ruthenium(IV) oxide-methyl iodide catalyst couple.

Homologation of branched-chain aliphatic acids, by contrast, many times is accompanied by substantial rearrangement, and while the principal products are once again those acids containing one additional carbon per molecule, very often there is a tendency to generate "tertiary" acids wherein the α-carbon atom is bonded to three alkyl groups. Typical examples of the rearrangement that may accompany Ru-catalyzed syngas homologation are illustrated in Table II for a series of three classes of branched-chain carboxylic acids, where the carbon α to the carboxylic acid function is methylene, methine or tertiary substituted. Thus typical examples include:

iso-valeric acid ⟶ 2,2-dimethylbutyric acid
2-methylvaleric acid ⟶ 2,2-dimethylvaleric acid
trimethylacetic acid ⟶ 2,2-dimethylbutyric acid

As in the case of the linear carboxylic acids, the principal by-products are water, $CO_2$ and the corresponding hydrocarbons. Substantial quantities of iso-butane are formed for example during iso-butyric acid homologation (see Experimental Section) while 2-methylpentane accompanies the formation of 2,2-dimethylvaleric acid during syngas treatment of 2-methylvaleric acid.

Labelling Studies. Labelling studies employing carbon-13 and deuterium have also been used to gain a better understanding of the mechanism of this novel acid homologation reaction, as well as to confirm the source of carbon for the higher MW acid products. In the more critical set of experiments, starting with an acetic acid charge enriched at the carbonyl carbon to the extent of ca. 580% of natural abundance, and conducting the homologation by the standard procedure (see Experimental Section) with $RuO_2$-10MeI as the catalyst couple and 1/1 ($CO/H_2$) syngas, $^{13}C$ NMR analysis of the propionic acid product fraction (22 wt %), indicated significant $^{13}C$ enrichment only at the methylene carbon (see Table III, eq. 3) and essentially none at the carboxylic function. Likewise, an analysis of the n-butyric acid fraction (3.4 wt %) showed enrichment only at the α- and β-methylene carbons and none at the carbonyl carbon (Table III, eq. 4).

Deuteration studies with acetic acid-$d_4$ (99.5% atom D) as the carboxylic acid building block, ruthenium(IV) oxide plus methyl iodide-$d_3$ as catalyst couple and 1/1 ($CO/H_2$) syngas, were less definitive (see Table III). Typical samples of propionic and butyric acid products, isolated by distillation in vacuo and glc trapping, and analyzed by $^{1}H$ NMR, indicated considerable scrambling had occurred within the time frame of the acid homologation reaction.

Solution Spectra. Additional insight into the ruthenium species involved in acid homologation comes from studies of the solution spectra by FTIR and the metallic complexes isolated from the final product mixtures.

Typical crude product solutions from acetic acid homologation exhibit strong bands at 2112 and 2047 $cm^{-1}$ (Figure 3), characteristic (4,5) of the ruthenium iodocarbonyl ion, $[Ru(CO)_3I_3]^-$. This ruthenium species in acetic acid appears very robust, storage of these solutions for six weeks leads to no significant changes in the spectra, while distillation of the crude product liquids in vacuo leaves a deep-red residual liquid still exhibiting the two characteristic bands at 2107 and 2040 $cm^{-1}$. Treatment of these same solutions with triphenylphosphine in acetic acid yields pure $RuI_2(CO)_2(PPh_3)_2$, as reported by Cleare and Griffith (5) for $[Ru(CO)_3I_3]^-$. No ruthenium carboxylates have been detected.

For catalyst combinations containing initial I/Ru ratios ≤5, the product solutions also show strong new bands at 1999 and 2036 $cm^{-1}$ characteristic (6) of ruthenium pentacarbonyl. Where acetic acid homologation is run at [Ru] > 0.2 M, then another ruthenium iodocarbonyl, $Ru(CO)_3I_2$, may be isolated from the product mix as a yellow crystalline solid. A typical spectrum of this material is illustrated in Figure 3b.

During homologation of higher MW acids, e.g., starting from propionic acid or trimethylacetic acid, the product

## TABLE III

Homologation of $^{13}C$-Enriched Acetic Acid

Catalyst:  $RuO_2-10CH_3I$

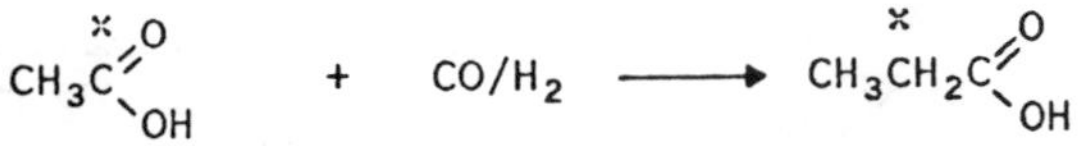

$$CH_3\overset{\times}{C}\!\!\begin{array}{c}O\\OH\end{array} \quad + \quad CO/H_2 \quad \longrightarrow \quad CH_3CH_2\overset{\times}{C}\!\!\begin{array}{c}O\\OH\end{array} \qquad (3)$$

$$CH_3CH_2\overset{\times}{C}\!\!\begin{array}{c}O\\OH\end{array} \quad + \quad CO/H_2 \quad \longrightarrow \quad CH_3CH_2CH_2\overset{\times\ \times}{C}\!\!\begin{array}{c}O\\OH\end{array} \qquad (4)$$

Homologation of Acetic Acid-$d_4$

Catalyst:  $RuO_2-10CH_3I-d_3$

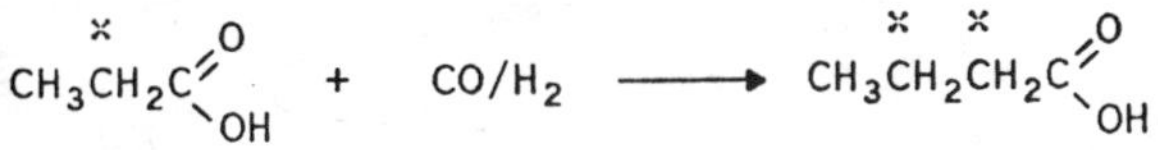

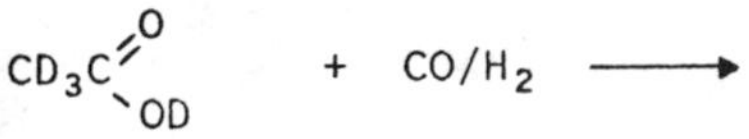

$$CD_3C\!\!\begin{array}{c}O\\OD\end{array} \quad + \quad CO/H_2 \quad \longrightarrow \qquad\qquad\qquad (5)$$

| $CH_3CH_2COOH$ Product Function | : | $CH_3-$ | $-CH_2-C\!\!\begin{smallmatrix}O\\O\end{smallmatrix}$ | $-OH$ |
|---|---|---|---|---|
| Ratio, $^1H$ NMR Response To Theoretical Proton Content | : | 1.0 | 1.2 | 1.5 |

| $CH_3CH_2CH_2COOH$ Product function | : | $CH_3-$ | $-CH_2-$ | $-CH_2C\!\!\begin{smallmatrix}O\\O\end{smallmatrix}$ | $-OH$ |
|---|---|---|---|---|---|
| Ratio, $^1H$ NMR Response To Theoretical Proton Content | : | 1.0 | 1.4 | 1.4 | 1.4 |

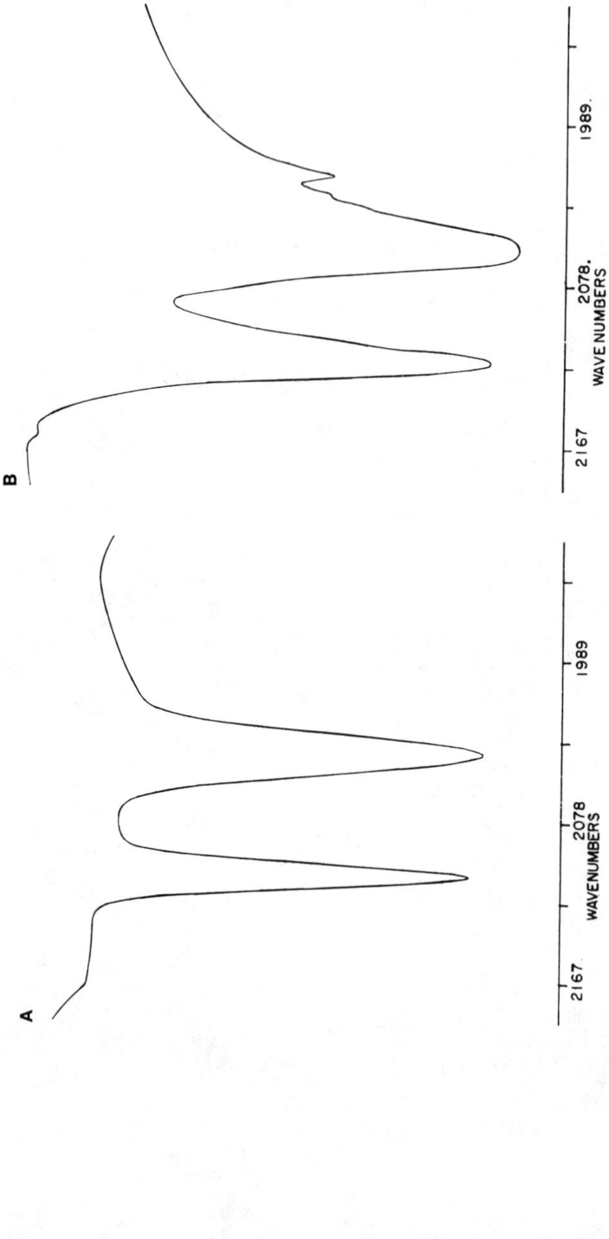


*Figure 3. Acetic acid homologation: A, IR spectrum of typical product solution,* $[Ru(CO_3)I_3]^-$*; B, spectrum of isolated* $Ru(CO_3)I_2$

solutions also exhibit bands similar to those depicted in Figure 3a.

## Discussion

The selective syngas homologation of carboxylic acids via ruthenium homogeneous catalysis (eq. 1) is believed to be a novel reaction. Although higher acids have been reported as by-products of methanol carbonylation reactions (7), prior homologation technology is generally confined to:

a. The cobalt-catalyzed homologation of saturated alkyl and benzyl alcohols, as well as substituted benzyl alcohols (8,9).

b. The homologation of carboxylic acid esters to higher MW esters having at least one additional methylene group (4,10), or to the corresponding acid anhydrides (11).

c. The conversion of noncyclic dialkyl ethers to the corresponding esters and anhydrides, e.g., the conversion of dimethyl ether to methyl and ethyl acetates (11,12).

From product distribution and comparative rate data, Wender and coworkers conclude that cobalt-catalyzed benzyl alcohol homologation involves the intermediate formation of carbonium ions (8). However, since the methyl cation ($CH_3$+) is unstable and difficult to form (9), it is more likely that methanol homologation to ethanol proceeds via nucleophilic attack on a protonated methyl alcohol molecule. Protonated dimethyl ether and methyl acetate forms have been invoked also by Braca (10), along with the subsequent formation of methyl-ruthenium moieties, to describe ruthenium catalyzed homologation to ethyl acetate.

Some of the more important features of our novel syngas homologation of aliphatic carboxylic acids, catalyzed by ruthenium, include:

1) Homologation of a broad range of aliphatic acid structures and carbon numbers, with extensive rearrangement during the homologation of certain branched-chain acids.

2) Hydrocarbons, of the same carbon skeletal structure as the starting acids, plus water and $CO_2$, as the common by-products of these homologation syntheses.

3) Higher acid yield data that are first order in ruthenium but also positively dependent upon methyl iodide concentrations and operating pressure as well as syngas ratios.

4) Similar acetic acid conversions and higher acid yield distributions using ruthenium(IV) oxide in combination with methyl iodide, ethyl iodide and hydrogen iodide as the added iodide promoter under comparable conditions. This is consistent with these different starting materials ultimately forming the same catalytically active species.

5) The presence of ruthenium iodocarbonyl species, such as $[Ru(CO)_3I_3]^-$, in the reaction product solutions.

6) Labelling studies with $^{13}C$-enriched acetic acid, where $^{13}C$ NMR data are consistent with carbon monoxide addition to the carbonyl carbon of the acid substrate.

Several of these features, particularly items 1) and 2), are reminiscent of the observations made by Wender and coworkers when examining benzyl alcohol homologation (8). Braca also reports hydrocarbon formation (item 2) during ester homologation (4) as well as the isolation of $Ru(CO)_4I_2$. By analogy with other iodide-promoted homogeneous metal-catalyzed carbonylation reactions, the first step in acetic acid homologation is likely the rapid formation of acetyl iodide. The observed dependence of higher acid yields upon methyl iodide concentration (Figure 2) and the fact that no homologation is observed in the absence of iodide (Table I) are consistent with this reasoning.

The subsequent steps in the homologation sequence involving successive hydrogenation of the acetyl function, while coordinated to the iodoruthenium carbonyl (e.g., Figure 3), through intermediate α-hydroxyethyl- and ethylruthenium species (B and C, Scheme 1), would allow rationale of our observed product array (vide infra). Somewhat similar ruthenium (10) and cobalt (9) species have been invoked in related homologation reactions, and certainly during acetic acid homologation we consistently observe ethyl acetate (from esterification of by-product ethanol via B) and ethane (from hydrogenolysis of C) as primary by-products. Likewise during the homologation of higher acids, the corresponding hydrocarbons are always in evidence (see Experimental Section). The homogeneous catalytic hydrogenation of saturated monocarboxylic acids, e.g., the production of ethyl acetate from acetic acid, has in fact been reported only recently by Bianchi et al (13), also via ruthenium carbonyl catalysis.

Subsequent insertion of CO into the newly formed alkylruthenium moiety, C, to form Ru-acyl, D, is in agreement with our $^{13}C$ tracer studies (e.g., Table III, eq. 3), while reductive elimination of propionyl iodide from D, accompanied by immediate hydrolysis of the acyl iodide (3,14) to propionic acid product, would complete the catalytic cycle and regenerate the original ruthenium carbonyl complex.

It should be noted at this point that ruthenium-catalyzed acid homologation as practised here (Experimental Section) and outlined in Scheme 1 is likely in competition with at least four alternative reaction pathways leading to the formation of hydrocarbon, alkanols, still higher MW acids and rearranged products. As an alternative, for example, to elimination of propionic acid product from the newly formed acyl species D, the same could undergo successive hydrogenation and the eventual formation of still higher $C_4$ and $C_5$ acids containing two (or more) carbons greater than the original acid substrate (see Table I). Increases in $H_2/CO$ operating pressure would

Scheme 1. *Acetic acid homologation*

$$CH_3-C\overset{I}{\underset{O}{\diagup}} + Ru(CO)_XI_{Y-1} \longrightarrow CH_3\overset{O}{\overset{\|}{C}}Ru(CO)_XI_Y$$

$$\underline{A}$$

$$CH_3CH_2OH \qquad\qquad CH_3CH_3$$

$$\uparrow H_2 \qquad\qquad\qquad \uparrow H_2$$

$$\underline{A} \xrightarrow{H_2} CH_3\overset{O}{\underset{|}{C}}H\,Ru(CO)_XI_Y \xrightarrow[(-H_2O)]{H_2} CH_3CH_2Ru(CO)_XI_Y \xrightarrow{CO} CH_3CH_2\overset{O}{\overset{\|}{C}}Ru(CO)_XI_Y$$

$$\underline{B} \qquad\qquad\qquad \underline{C} \qquad\qquad\qquad \underline{D}$$

$$\underline{D} \xrightarrow[(-HI)]{H_2O} CH_3CH_2C\overset{\diagup O}{\underset{OH}{\diagdown}} + Ru(CO)_XI_{Y-1}$$

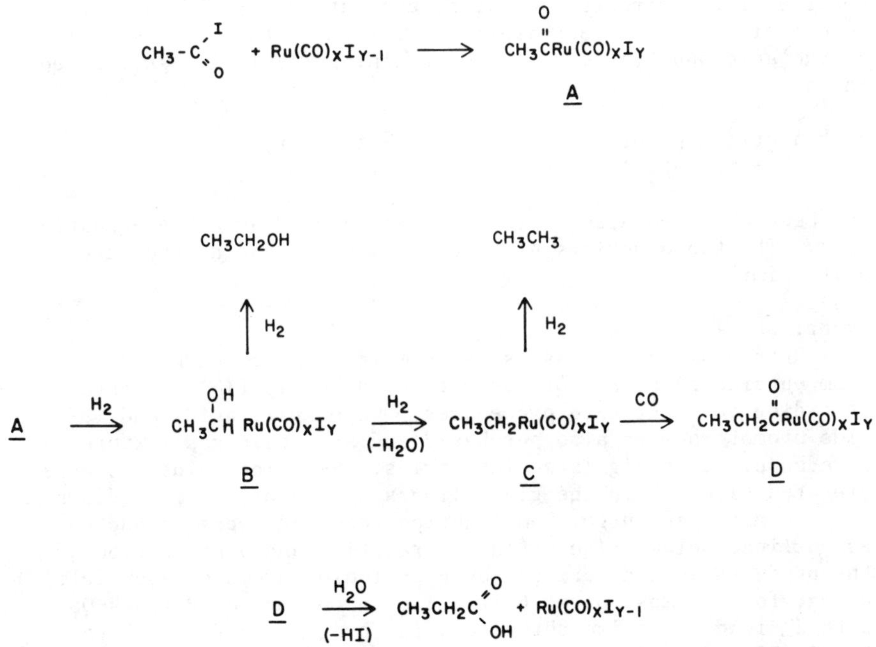

favor this alternative hydrogenation of D and so lead to the improved yields of butyric and valeric acids now documented in Figure 2.

A negative peculiarity of ruthenium syngas chemistry is the high hydrogenation activity (4) that results in hydrocarbon as a side reduction of substrate. During acetic acid homologation, the successive reduction of A and hydrogenolysis of C would yield ethane, and in this work we do observe an increase in hydrogenation-to-carbonylation activity (e.g., ethane/propionic acid product ratio) with increases in hydrogen partial pressure (see Figure 2). Finally, the product distribution observed during homologation of branched-chain carboxylic acids (Table II), as well as the $^{13}C$ distribution in n-butyric acid product from enriched acetic acid (Table III, eq. 4), are suggestive of rapid rearrangement of the ruthenium-alkyl and acyl intermediates. The decrease in reactivity with increase in MW of the acid substrate, and with branching (Table II, summarized in eq. 6),

$$\text{acetic} > \text{propionic} \sim \text{n-valeric} > \text{iso-butyric} \sim \text{trimethylacetic} > \text{iso-valeric acid} \quad (6)$$

has been observed also in ruthenium-catalyzed acid hydrogenation (13). The two orderings of reactivity versus acid structure are quite similar.

## Experimental Section

Ruthenium oxides, salts and complexes were purchased from outside suppliers or, as with $H_4Ru_4(CO)_{12}$ (15), prepared according to literature procedures. Carboxylic acids and iodide promoters were also purchased and synthesis gas mixtures were supplied by Big Three Industries. Reaction solutions were prepared directly in the glass liners of the pressure reactors under a nitrogen purge, homologation reactions were conducted as outlined below. The extent of reaction and distribution of the products were determined by gas-liquid chromatography (glc) using, for the most part, 6 ft x 1/8 in columns of Porapak-QS with 2% loading of iso-phthalic acid, programmed from 120 to 240°C (20 $cm^3$/min He). Water was estimated by Karl Fischer titration. Product acids were isolated by distillation in vacuo or by glc trapping and identified by one or more of the following techniques, glc, FTIR, NMR and elemental analyses. Higher MW acids were analyzed using a 6 ft x 1/8 in column of Porapak-PS with 8% loading of SP-1000 plus 2% iso-phthalic acid.

Syngas Homologation of Acetic Acid. To a $N_2$-flushed liquid mix of acetic acid (50.0 gm) and methyl iodide (5.67 gm, 40 mmole), set in a glass liner is added 0.763 gm of ruthenium(IV) oxide, hydrate (4.0 mmole). The mixture is stirred to partially dissolve the ruthenium and the glass liner plus contents charged to a 450 ml rocking autoclave. The reactor is sealed, flushed with $CO/H_2$, pressured to 272 atm with $CO/H_2$ (1:1) and heated, with rocking, to 220°C. After 18 hr, the gas uptake is 163 atm. Upon cooling, depressuring and sampling the off-gas, the clear deep-red liquid product is recovered from the glass-lined reactor.

Analysis of the liquid fraction by glc shows the presence of:

26.9 wt % propionic acid
2.3 wt % iso-butyric acid
3.6 wt % n-butyric acid
0.4 wt % iso-valeric acid
0.6 wt % n-valeric acid
1.5 wt % ethyl acetate
7.7 wt % water
54.1 wt % unreacted acetic acid

Typical off-gas samples show the presence of:

| | | | |
|---|---|---|---|
| 37% | carbon monoxide | 4.2% | ethane |
| 40% | hydrogen | 1.3% | propane |
| 15% | carbon dioxide | 1.0% | methane |

The propionic and butyric acid product fractions, as well as unreacted acetic acid, may be isolated by fractional distillation in vacuo.

Syngas Homologation of Propionic Acid. To a $N_2$-flushed liquid mix of propionic acid (18.15 gm, 245 mmole) and methyl iodide (2.84 gm, 20.0 mmole) set in a glass liner is added 0.382 gm of ruthenium(IV) oxide, hydrate (2.0 mmole). The mixture is stirred to partially dissolve the ruthenium and the glass liner plus contents charged to a 450 ml rocking autoclave. The reactor is sealed, flushed with $CO/H_2$, pressured to 272 atm with $CO/H_2$ (1:1) and heated with rocking to 220°C. After 18 hr, the gas uptake is 59 atm. Upon cooling, depressuring and sampling the off-gas, the clear amber liquid product (19.4 gm) is recovered from the glass-lined reactor. There is no solid residue.

Analysis of the crude liquid product fraction by glc shows the presence of:

10.4 wt % n-butyric acid
3.3 wt % iso-butyric acid
1.9 wt % n-valeric acid
0.3 wt % iso-valeric acid/2-methylbutyric acid
1.7 wt % acetic acid
33.3 wt % water
25.6 wt % unreacted propionic acid

Typical off-gas samples show the presence of:

| | | | |
|---|---|---|---|
| 29% | carbon monoxide | 6.6% | propane |
| 58% | hydrogen | 1.6% | n-butane |
| 2.6% | carbon dioxide | 0.6% | methane |

The butyric acid product fractions, as well as unreacted propionic acid, may be isolated from the crude liquid product by fractional distillation in vacuo.

Syngas Homologation of iso-Butyric Acid. To a $N_2$-flushed liquid mix of iso-butyric acid (21.6 gm, 245 mmole) and methyl iodide (2.84 gm, 20.0 mmole) set on a glass liner is added 0.382 gm of ruthenium(IV) oxide hydrate (2.0 mmole). The mixture is stirred to partially dissolve the ruthenium and the glass liner plus contents charged to a 450 ml rocking autoclave. The reactor is sealed, flushed with $CO/H_2$, pressured to 272 atm with $CO/H_2$ (1:1) and heated with rocking to 220°C. After 18 hr, the gas uptake is 48 atm. Upon cooling, depressuring and sampling the off-gas, the clear-yellow, two-phase liquid product (19.8 gm, 18 ml) is recovered from the glass-lined reactor. There is no solid residue.

Analysis of the lighter liquid fraction (15 ml) by glc shows the presence of:

| | |
|---|---|
| 14.1% | trimethylacetic acid |
| 0.8% | iso-valeric acid/2-methylbutyric acid |
| 0.8% | acetic acid |
| 2.2% | methyl iodide |
| 3.1% | water |
| 62.3% | unreacted isobutyric acid |

Typical off-gas samples show the presence of:

| | | | |
|---|---|---|---|
| 49% | carbon monoxide | 5.3% | iso-butane |
| 41% | hydrogen | 0.2% | ethane |
| 1.7% | carbon dioxide | 0.6% | methane |

The trimethylacetic acid is isolated from the used ruthenium catalyst and unreacted iso-butyric acid by fractional distillation in vacuo, and identified by NMR and FTIR analyses.

Acknowledgements

The author wishes to thank Texaco Inc. for permission to publish this paper, Messrs M. Swenson, T. D. Ellison, R. Gonzales and D. W. White for experimental assistance, and Messrs C. L. LeBas, R. L. Burke and J. M. Schuster for $^{13}C$ NMR and FTIR data.

## Literature Cited


1. Knifton, J. F., U.S. Patent pending.
2. Spitz, P. H., Chemtech, May 1977, 295.
3. Roth, J. F.; Craddock, J. H.; Hershman, A.; Paulik, F. E., Chemtech, October 1971, 603.
4. Braca, G.; Sbrana, G.; Valantini, G.; Andrich, G.; Gregorio, G.; "Fundamental Research in Homogeneous Catalysis, Vol. 3," Plenum Press, New York, 1979, p. 221.
5. Cleare, M. J.; Griffith, W. P.; J. Chem. Soc. (A), 1969, 372.
6. Doyle, M. J.; Kouwenhoven, A. P.; Schaap, C. A.; Van Ort, B.; J. Organometal. Chem., 1979, 174, C55.
7. Hohenschutz, H.; von Kutepow, N.; Himmele, W., Hydrocarbon Process., November 1966, 45, 141.
8. Orchin, M., "Advances in Catalysis, Vol. V," Academic Press, New York, 1950, p. 385.
9. Wender, I., Catal. Rev. Sci. Eng., 1976, 14, 97.
10. Braca, G.; Sbrana, G.; Valantini, G.; Andrich, G.; Gregorio, G.; J. Amer. Chem. Soc., 1978, 100, 6238.
11. Naglieri, A. N.; Rizkalla, N., U.S. Patent 4,002,677 (1977).
12. Piacenti, F.; Bianchi, M., "Organic Syntheses via Metal Carbonyls, Vol. II," Wiley-Interscience, New York, 1977, p. 1 and references therein.
13. Bianchi, M.; Menchi, G.; Francalanci, F.; Piacenti, F.; J. Organomet. Chem., 1980, 188, 109.
14. Morris, D. E.; Johnson, G. V.; Symp. Rhodium Homogeneous Catal., 1978, 113.
15. Piacenti, F.; Bianchi, M., Frediani, P.; Beneditte, E.; Inorg. Chem., 1971, 10, 2762.


RECEIVED December 8, 1980.

# 16

# Decarbonylation of Aldehydes Using Ruthenium(II) Porphyrin Catalysts

G. DOMAZETIS, B. R. JAMES, B. TARPEY, and D. DOLPHIN

Department of Chemistry, University of British Columbia, Vancouver, British Columbia, Canada V6T 1Y6

During studies on ruthenium(II) porphyrins containing tertiary phosphine ligands (1), we discovered several complexes that readily abstracted carbon monoxide from oxygen-containing organics, particularly aldehydes. A stoichiometric decarbonylation of coordinated NN-dimethylformamide (dmf) within a Ru(TPP)$(dmf)_2$ complex (TPP = dianion of tetraphenylporphyrin) had been observed previously, and this could be made catalytic after photolytic removal of CO from the resulting Ru(TPP)(CO)L complex, L being dmf, or possibly $NHMe_2$ the decarbonylation product (2). Stoichiometric decarbonylation of aldehydes and acid chlorides can be achieved readily using platinum metal complexes (3), but difficulties are encountered in displacing the coordinated CO either thermally (3) or photolytically (4), and few cases of catalytic decarbonylation have been reported (5).

We recently reported briefly on an extremely efficient thermal catalytic decarbonylation of aldehydes using a system based on Ru(TPP)$(PPh_3)_2$ (6), and report here further studies on this system and one based on Ru(TPP)(CO)($^tBu_2$POH).

## Experimental

Materials. The commercially available aldehydes were distilled prior to use and stored at 0°C under argon. The cyclohexene- and cyclopentene- aldehydes, and the indane aldehyde (see Table) were gifts from Professor E. Piers of this department. The Ru(TPP)$(PPh_3)_2$ complex (**1**) was prepared from Ru(TPP)(CO)(EtOH) and $PPh_3$ (1,7), while Ru(TPP)(CO)($^tBu_2$POH), **2**, was prepared from the carbonyl (ethanol) adduct by treatment with $^tBu_2$PCl (1). The phosphines were from Strem Chemicals, and the ruthenium was obtained as $RuCl_3 \cdot 3H_2O$ from Johnson, Matthey Limited.

0097-6156/81/0152-0243$05.00/0
© 1981 American Chemical Society

Decarbonylation Procedure. $Ru(TPP)(PPh_3)_2$ (∿2 mg, ∿1.6 x $10^{-3}$ mmol) was dissolved in 0.5 mL $CH_2Cl_2$ and added to about 30 mL $CH_3CN$. The aldehyde (0.3 - 1.0 mL, ∿2-10 mmol) was added, and CO bubbled through the solution for a few seconds until the uv/vis spectrum showed absorption maxima at 414 and 530 nm that are characteristic of $Ru(TPP)(CO)(PPh_3)$ (1). The solution, as well as that of a blank containing the aldehyde with no Ru complex, was then monitored by g.c. [OV101 and OV17 columns; Hewlett Packard 5830A and Carle 311 instruments]; under these conditions, decarbonylation in the Ru-containing solution was very slow. However, addition of sufficient $^nBu_3P$ (5-10 μL) to cause generation of further Soret bands at 437 nm [$Ru(TPP)(^nBu_3P)_2$] (1) and 420 nm (unknown species X) led to very efficient catalytic decarbonylation of the aldehydes at ambient temperatures; both loss of the aldehyde (compared to the blank) and formation of the product were followed by g.c. Some runs were carried out in toluene or benzonitrile when the $CH_3CN/CH_2Cl_2$ solvent g.c. peaks masked those of the products. The decarbonylation products were generally identified by g.c./m.s., and toluene (from $PhCH_2CHO$) was also identified by n.m.r. In all cases where initial turnover rates of $\geqslant 10^2$ $h^{-1}$ were obtained, the color of the final catalytically inactive solution was green-brown in contrast to the orange-colored active solutions; the intensity in both the Soret and visible regions of the spectrum (Fig. 1) had decreased markedly and the porphyrin skeleton is almost certainly destroyed in these "oxidized" solutions (1).

The same procedure was employed for decarbonylation using $Ru(TPP)(CO)(^tBu_2POH)$, although the CO pretreatment could be omitted.

Cyclic Voltammetry. Voltammograms were obtained using an H-cell with the three compartments separated by sintered glass frits. Potentials were measured at a platinum electrode against a Ag/AgCl reference electrode at 25°C in a 2:1 $CH_2Cl_2/CH_3CN$ solvent mixture with 0.1 M $^nBu_4N^+ClO_4^-$ as supporting electrolyte. The catalyst (complex 1) concentration was ∿$10^{-3}$M, and the aldehyde concentration ∿0.1 M. Voltammograms were recorded for $Ru(TPP)L_2$ and $Ru(TPP)(CO)L$ complexes [L = $PPh_3$, $^nBu_3P$], as well as for $Ru(TPP)(PPh_3)_2$/aldehyde solutions before and after addition of CO and $^nBu_3P$, and at intervals (4,8,20 h) during decarbonylation. Data for the indane aldehyde system are shown in Fig. 2.

Ir. Spectra. These were recorded with a Perkin-Elmer 457, calibrated with polystyrene. Data were obtained during decarbonylation of the indane aldehyde by evaporating a $CH_2Cl_2/CH_3CN$ solution on NaCl plates. Other spectra were measured using a smear of the aldehyde and $Ru(TPP)(PPh_3)_2$ on a NaCl plate, and using concentrated solutions in $CHCl_3$ or $CCl_4$.

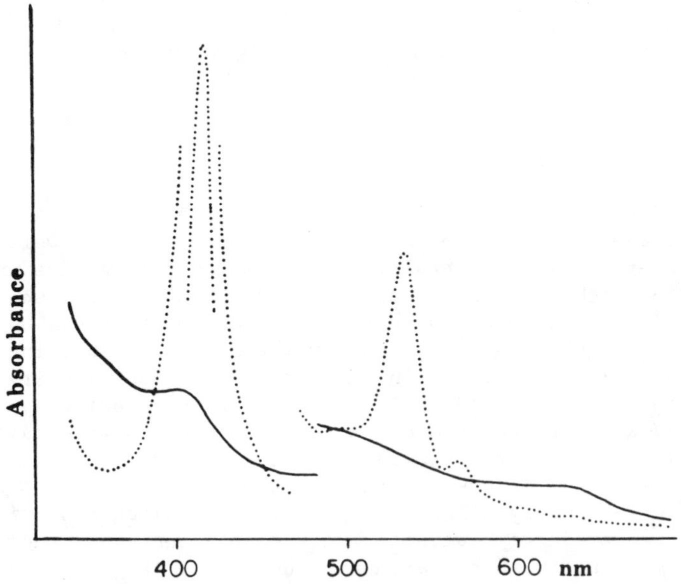


*Figure 1. Visible spectrum typical of solution no longer active for catalytic decarbonylation; that shown here is for solution of $Ru(TPP)(PPh_3)_2/{}^nBu_3P$ after decarbonylation of $PhCH_2CHO$; (– – –) same solution in presence of hydroquinone; inactive for decarbonylation*

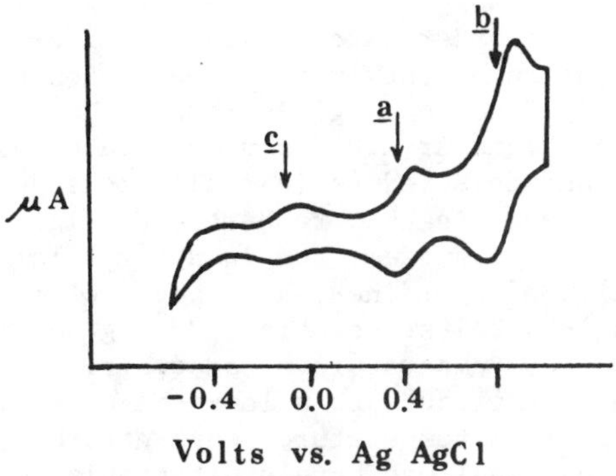


*Figure 2. Cyclic voltammogram (in 2:1 $CH_2Cl_2/CH_3CN$ with 0.1M ${}^nBu_4N^+ClO_4^-$) for $Ru(TPP)(PPh_3)_2$/indane aldehyde system (*see *text)*

E.s.r. Spectra. These were recorded on a Varian E-3 at liquid nitrogen temperature. Fig. 3 shows some data for a number of aldehydes using 1 or 2 as catalysts. Spectra were recorded in 5:1 $CH_2Cl_2/CH_3CN$ media containing ∿$10^{-3}$M catalyst and a 5-fold excess of aldehyde, before and after addition of $CO/^nBu_3P$; no signals were observed in the absence of the Ru complexes.

## Results and Discussion

The Table lists the aldehydes studied, the decarbonylation products, and some turnover numbers. Two to four runs were performed generally with each aldehyde, although twelve runs were carried out with $PhCH_2CHO$. Catalytic decarbonylation also occurred in $CH_2Cl_2$ solution, but the procedure described using essentially $CH_3CN$ gave better turnover numbers. On heating the 2-phenylacetaldehyde system in $CH_3CN$ to ∿60°C, bubbling argon through the solution, and collecting the products using a cold finger (77°K), turnover numbers up to 5 x $10^4 h^{-1}$ were attained, and complete decarbonylation readily achieved.

The decarbonylations, which do not appear to be affected by light, are reasonably selective with aromatic aldehydes, yielding the expected product; however, significant amounts of other products are obtained with non-aromatic substrates (e.g. cyclohexanealdehyde gives methylcyclopentane and small amounts of n-hexane, as well as the expected cyclohexane; and cyclohexen-4-al gives both cyclohexene and cyclohexane). Indeed, the unexpected products perhaps provided a major clue to an understanding of the reaction mechanism(s) involved.

Attempts to study some kinetics were thwarted by poor reproducibility. A very good first-order dependence on aldehyde concentration was observed throughout any one run for several of the aldehydes (monitored by g.c.), but repeat experiments commonly gave pseudo first-order rate constants differing by factors of up to five; variation of catalyst concentration gave very irreproducible data and no meaningful trends. The poor kinetic behavior and some products from C-C bond cleavage indicated a radical mechanism, and this led us initially to some e.s.r. studies. Organic free-radicals were detected (Fig. 3) in a catalytic cyclohexen-4-al system, and in a very slow but catalytic decarbonylation of pyridine-2-aldehyde. Of interest, an active $Ru(TPP)(PPh_3)_2$ catalyst solution, during decarbonylation of $PhCH_2CHO$, gave a low-temperature broad e.s.r. signal centered at g = 2.20, which we attribute to a low-spin $d^5$ Ru(III) species (1, 8); after 6 h at room temperature, when decarbonylation was still occurring, the signal was barely detectable. No e.s.r. signals have been detected in the finally inactive solutions (which are green when diluted - see Decarbonylation Procedure). It should be noted that Ru(III) porphyrin systems generated electrochemically from Ru(II) do not normally give detectable e.s.r. signals (9).

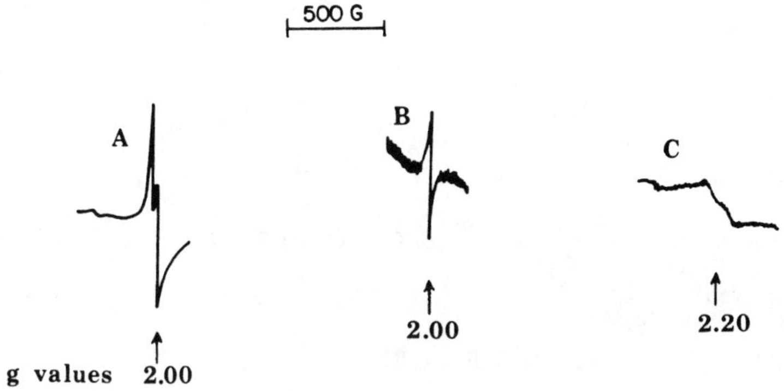


*Figure 3. E.s.r. signals at liquid nitrogen temperature in 5:1 $CH_2Cl_2/CH_3CN$: A, the Ru(TPP)(CO)($^tBu_2POH$)/cyclohexen-4-al system; B, the Ru(TPP)(CO)($^tBu_2$-POH)/pyridine-2-aldehyde system; C, the Ru(TPP)$(PPh_3)_2$/$^nBu_3P$/2-phenylacetaldehyde system*

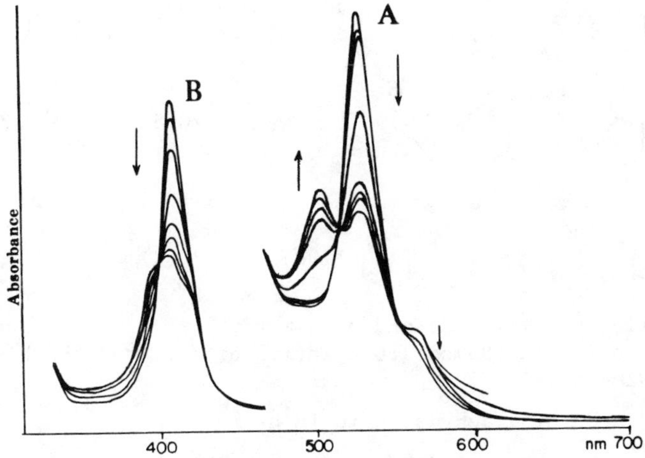


*Figure 4. Visible spectral changes as a function of time (first hour) during decarbonylation of cyclohexen-4-al (~ 0.5M) using Ru(TPP)(CO)($^tBuPOH$) in toluene at room temperature: A, visible region using ~ $10^{-4}$M Ru; B, Soret region using ~ $10^{-5}$M Ru*

Table. Decarbonylation of aldehydes using a $Ru(TPP)(PPh_3)_2/{}^{n}Bu_3P$ catalyst system

| Substrate | Major Product (%)[a] | Conversion (time)[b] | Turn-over[c] |
|---|---|---|---|
| $C_6H_5CHO$ | Benzene (100) | 10(5) | 10 |
| $C_6H_5CH{=}CHCHO$(trans) | Styrene (100) | 20(10) | 20 |
| $C_6H_5CH_2CHO$ | Toluene (95)[d] | 30(1), 90(4) | $10^{3}$[e] |
| p-CN-$C_6H_4CHO$ | Benzonitrile (100) | 15(12) | 20 |
| n-$C_6H_{13}CHO$ | n-$C_6H_{14}$ (65)[f] | 10(1) | $10^2$ |
| 2-Ethylbutanal | n-$C_5H_{12}$ (85)[f] | 30(1) | $10^3$ |
| 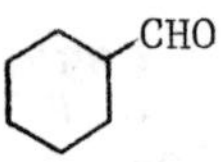 | 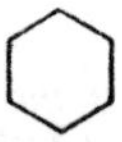 (60); 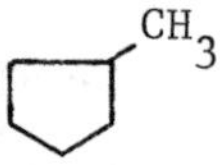 (35) n-$C_6H_{14}$ (5) | 30(1), 50(18), 90(50) | $2 \times 10^2$ |
| 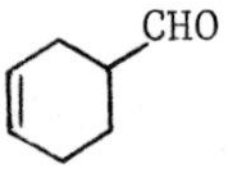 [g] |  (70); 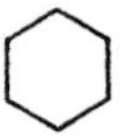 (30) | 10(1), 20(12) | $10^2$ |
| 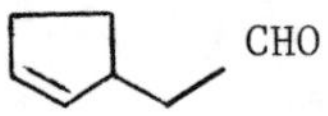 | 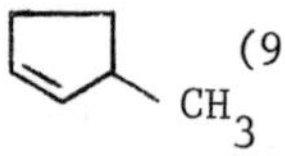 (90)[f] | 10(1), 30(36), 90(150) | $10^2$ |
| 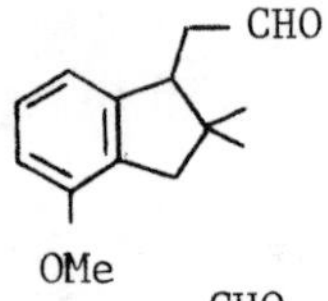 | 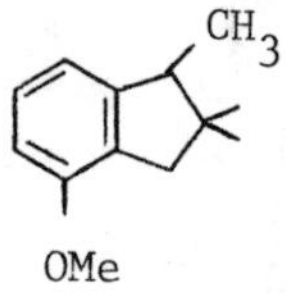 (70)[f] | 20(5) | $10^2$ |
| 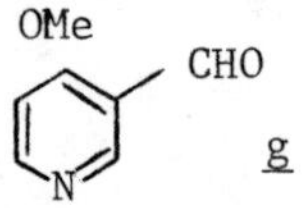 [g] | 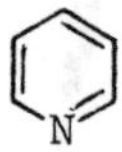 (100) | 20(3) | $4 \times 10^2$ |

[a] Identified by g.c.-m.s. and/or n.m.r.; % refers to amount of major species in the decarbonylation products at the highest conversion noted.

[b] % Conversion of aldehyde; time in h.

[c] For the first hour at ambient temperature, based on loss of aldehyde and/or formation of product as detected by v.p.c.

[d] Small amounts of benzene also detected.

[e] At ~60°C, turnover $5 \times 10^4$ $h^{-1}$.

[f] Other products not yet identified; may be decomposition products.

[g] Using Ru(TPP)(CO)($^{t}Bu_2POH$) in toluene.

Some evidence for the presence of Ru(III) species also comes from cyclic voltammetry (Fig. 2). The $Ru(TPP)(PPh_3)_2$ complex shows a chemically reversible one-electron oxidation at a (0.40 V), which is attributed to a Ru(III)/Ru(II) couple (1, 7, 9). After treatment with CO, a further reversible wave is seen at b (0.83 V); this refers to an oxidation at the porphyrin ring within the carbonyl complex (1, 9, 10), i.e. to the couple,

$$(Ph_3P)Ru^{II}(TPP^{+\cdot})(CO) + e \rightleftharpoons (Ph_3P)Ru^{II}(TPP)(CO) \qquad (1)$$

In the presence of the indane aldehyde, additional somewhat irreversible waves are seen at c (∿-0.1 V), which probably refer to some Ru(III) intermediates (10) involved in the catalysis. In the corresponding phenylacetaldehyde system, additional waves are seen at about -0.08 and +0.08 V. On adding $^nBu_3P$ to the indane aldehyde system, waves are seen at 0.30 V [$Ru^{III}(TPP)(^nBu_3P)_2 + e \rightleftharpoons Ru^{II}(TPP)(^nBu_3P)_2$], and ∿0.90 V [due to the couple shown in eq. 1, with $PPh_3$ likely replaced by $^nBu_3P$], with an extra wave being observed at +0.1 V.

Infrared measurements in the ν(CO) region during the decarbonylation of the indane aldehyde using the $Ru(TPP)(PPh_3)_2$ catalyst system revealed a small peak at 2015 $cm^{-1}$, as well as expected bands in the 1950-1970 $cm^{-1}$ region, characteristic of the Ru(TPP)(CO)L (L = $PPh_3$, $^nBu_3P$) complexes (1). The peak could be that of a trans-dicarbonyl [cf. the 2005 $cm^{-1}$ band of $Ru(TPP)(CO)_2$, (11)]; these species contain an extremely labile CO, which would be germane to our catalytic decarbonylation. Alternatively, the 2015 $cm^{-1}$ peak could be due to a ruthenium(III) hydride species (12); such a $HRu^{III}$ intermediate could provide pathways for formation of hydrogenation products, for example, the cyclohexane observed in the cyclohexen-4-al system. Such hydrides within Ru(III) porphyrins might give rise to the additional reduction potential waves in the range -0.1 to +0.1 V.

Addition of hydroquinone was found to completely inhibit decarbonylation of phenylacetaldehyde using the $Ru(TPP)(PPh_3)_2$ catalyst system. This confirms that a radical-type process is involved, especially as the hydroquinone shows no interaction with the Ru complexes. Indeed, the uv/visible spectrum of the catalyst solution in the presence of the hydroquinone was essentially that of $Ru(TPP)(^nBu_3P)(CO)$ (Fig. 1); there is no indication of the 420 nm Soret band required for catalysis, and no decarbonylation occurs. The spectral changes during a decarbonylation are shown in Fig. 4 for the $Ru(TPP)(CO)(^tBu_2POH)$/cyclohexen-4-al system; there are probably isosbestic points in the Soret and visible region in the earlier stages of the reaction, but eventually the peaks in both regions collapse, and a final build up of absorption above 600 nm (cf. Fig. 1) gives rise to a green tinge. The initial spectral changes are probably due to generation of Ru(III) species, as judged by comparable changes occurring on $Br_2$-oxidation of Ru(II) phosphine complexes (1), but further studies are required to confirm this.

The familiar 'standard' decarbonylation mechanism (3, 5) involving a concerted oxidative-addition of aldehyde, CO migration (with subsequent elimination), and reductive-elimination of product, would seem with metalloporphyrins to require coordination numbers higher than six, and in this case Ru(IV) intermediates. Although this is plausible, the data overall strongly suggest a radical mechanism and Ru(III) intermediates.

A highly tentative mechanism is outlined below:

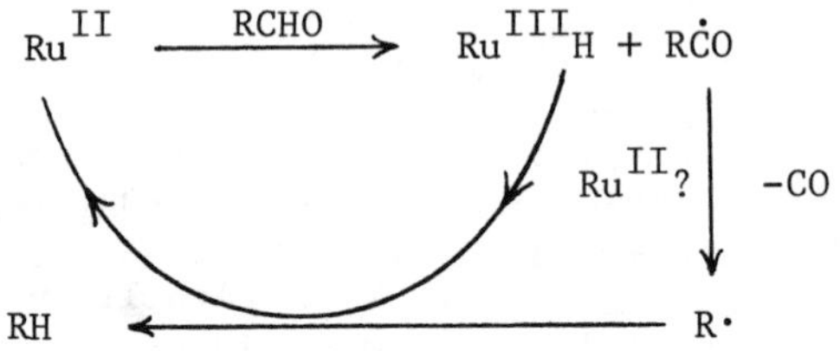

$$Ru^{II} \xrightarrow{RCHO} Ru^{III}H + R\dot{C}O \tag{2}$$

Decarbonylation of the acyl is likely to be metal-assisted ($Ru^{II}$) giving rise to a $Ru^{II}$ carbonyl, which is subsequently decarbonylated by nucleophilic attack by $^{n}Bu_3P$. This phosphine can displace coordinated carbonyl, as exemplified by reaction 3:

$$Ru(OEP)(CO)P + P \underset{}{\overset{K}{\rightleftharpoons}} Ru(OEP)P_2 + CO \tag{3}$$

$$[P = {}^{n}Bu_3P; \; OEP = \text{dianion of octaethylporphyrin}]$$

This reaction occurs thermally in toluene at 30°C with an equilibrium constant (K) equal to 1.5 (13). Both bis(phosphine) and (carbonyl)phosphine Soret bands are present in the active catalyst solutions (see Decarbonylation Procedure), together with the unassigned, and likely critical, band at 420 nm. This could be due to some species giving rise to, or resulting from, a $Ru^{II}$ + $R\dot{C}O$ reaction; this is equivalent, of course, to a ($Ru^{III}$–COR) acyl or a $Ru^{III}(CO)R$ (carbonyl)alkyl species, and the final elimination reaction after loss of CO could be written as:

$$Ru^{III}H + Ru^{III}R \longrightarrow 2Ru^{II} + RH \tag{4}$$

However, formation of an $\dot{R}$ species, either free or within a radical-pair cage with the metal (14), is strongly favored in view of the methylcyclopentane noted in the cyclohexen-4-al decarbonylation, since the rearrangement shown in eq. 5, metal-assisted if necessary, seems plausible (15):

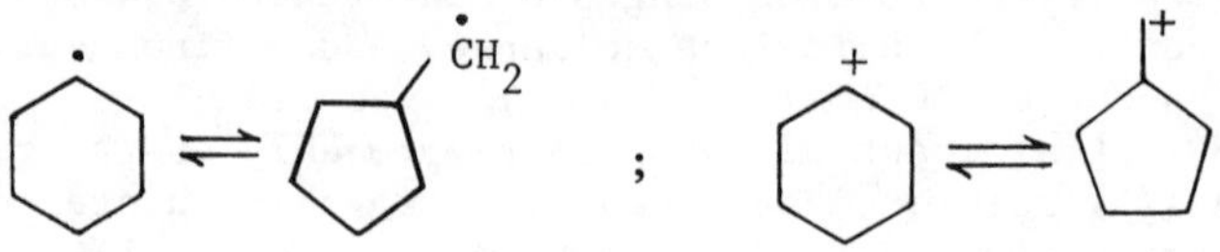

The more well-documented carbonium-ion rearrangement also shown in 5 (16, 17) is also possible but as such this does not require formal Ru(III) intermediates, for example,

$$Ru^{II} + RCHO \longrightarrow Ru^{II}H + R^{+} + CO \qquad (6)$$

The eventual loss of the porphyrin ring, as judged by the final visible spectra (*cf*. Fig. 1), might be due to attack by radicals.

The role of the triphenylphosphine is not clear but is almost certainly related to the fact that Ru(TPP)$(PPh_3)_2$ at the concentrations used, unlike Ru(TPP)$(^nBu_3P)_2$, rapidly dissociates one phosphine (1):

$$Ru(TPP)(PPh_3)_2 \rightleftharpoons Ru(TPP)(PPh_3) + PPh_3 \qquad (7)$$

and the five-coordinate species adds a carbonyl ligand 'instantaneously' using gaseous CO; stoichiometric carbonyl formation from an aldehyde also occurs but less rapidly. Further studies are necessary to see whether equilibria such as 8 are involved;

$$Ru(TPP)(CO)(PPh_3) + {}^nBu_3P \rightleftharpoons \begin{cases} Ru(TPP)(CO)(^nBu_3P) + PPh_3 \\ Ru(TPP)(^nBu_3P)(PPh_3) + CO \end{cases} \qquad (8)$$

both monocarbonyls have Soret bands at ∿415 nm, and are themselves significantly less active than the mixed in situ $PPh_3/^nBu_3P$ system. It is unlikely that the six-coordinate mixed phosphine system would give rise to the 420 nm band, since the bis(tributylphosphine) and bis(triphenylphospine) complexes give bands at 437 and 435 nm, respectively (1). Of interest, the Ru(TPP)$(PPh_3)$ species in toluene shows a Soret absorption at 420 nm, and solvated Ru(TPP)L$(CH_3CN)$, L = tertiary phosphine, species are a possibility. The necessary activation by CO certainly implies that the $Ru^{II}$ species of a scheme such as 2 is already a monocarbonyl or stated in another way, the CO addition prevents formation of inactive Ru(TPP)$(^nBu_3P)_2$ (*cf*. eq. 3).

Although the mechanistic details will be difficult to elucidate, the catalytic system, operating at ambient thermal conditions, appears to have considerable potential in synthesis for removing CO groups from aldehyde moieties of sensitive organic compounds.

Acknowledgment

We thank N.S.E.R.C. for research grants, Johnson,Matthey and Co., Ltd. for a loan of ruthenium, and Professor E. Piers for several of the aldehyde substrates.

Literature Cited


1. Domazetis, G.; Dolphin, D.; James, B.R. To be published.
2. James, B.R.; Addison, A.W.; Cairns, M.; Dolphin, D.; Farrell, N.P.; Paulson, D.R.; Walker, S. in "Fundamental Research in Homogeneous Catalysis"; Vol. 3, ed. Tsutsui, M.; Plenum Press, New York, 1979, p. 751.
3. Tsuji, J. in "Organic Synthesis Via Metal Carbonyls"; Vol. 2, eds. Wender, I. and Pino, P.; Wiley, New York, 1977, p. 595.
4. Geoffroy, G.L.; Denton, D.A.; Keeney, M.E.; Bucks, R.R. Inorg. Chem., 1976, 15, 2382.
5. Doughty, D.H.; McGuiggan, M.F.; Wang, H.; Pignolet, L.H. in "Fundamental Research in Homogeneous Catalysis"; Vol. 3, ed. Tsutsui, M.; Plenum Press, New York, 1979, p. 909.
6. Domazetis, G.; Tarpey, B.; Dolphin, D.; James, B.R. J.C.S. Chem. Comm., in press.
7. Boschi, T.; Bontempelli, G.; Mazzocchin, G-A. Inorg. Chim Acta, 1979, 37, 155.
8. Medhi, O.K. and Agarwala, U. Inorg. Chem., 1980, 19, 1381.
9. Dolphin, D.; James, B.R.; Murray, A.J.; Thornback, J.R. Can. J. Chem., 1980, 58, 1125.
10. Smith, P.D.; Dolphin, D.; James, B.R. J. Organometal. Chem., in press.
11. Eaton, G.R.; Eaton, S.S. J. Am. Chem. Soc., 1975, 97, 235.
12. James, B.R.; Rattray, A.D.; Wang, D.K.W. J.C.S. Chem. Comm., 1976, 792.
13. Walker, S.G.; M.Sc. Dissertation, University of British Columbia, 1980.
14. Walborsky, H.M.; Allen, L.E. J. Am. Chem. Soc., 1971, 93, 5465.
15. Wilt, J.W. in "Free Radicals"; Vol. I, ed. Kochi, J.K.; Wiley, New York, 1973, p. 333.
16. Fry, J.L.; Karabatsos, C.J. in "Carbonium Ions"; Vol. II, eds. Olah, G.A.; Schleyer, P. von R.; Wiley, New York, 1970, p. 521.
17. Olah, G.A.; Olah, J.A. in "Carbonium Ions"; Vol. II, eds. Olah, G.A.; Schleyer, P. von R.; Wiley, New York, 1970, p. 715.


RECEIVED December 8, 1980.

17

# Reactions of ($\eta$-$C_5H_5$)$_2$$NbH_3$ with Metal Carbonyls

## Selective Reduction of Carbon Monoxide to Ethane

J. A. LABINGER and K. S. WONG

Department of Chemistry, University of Notre Dame, Notre Dame, IN 46556

The potential importance of homogeneous catalytic reactions in synthesis gas transformations (i.e., hydrogenation of carbon monoxide) has been widely recognized in recent years. In the first place, such systems could provide structural and mechanistic models for the currently more important, but more difficult to study, heterogeneous catalysts. Secondly, product selectivity is generally more readily achievable with homogeneous catalysts, and this would be an obviously desirable feature in an efficient process converting synthesis gas to useful chemicals and fuels.

We previously presented a rationale for focussing upon organoniobium hydride complexes in attempting to find a homogeneous system capable of activating CO towards reduction (1). To summarize briefly, our approach involves initial attack by a relatively nucleophilic metal hydride on coordinated CO. Such reactivity has been demonstrated repeatedly for main-group metal hydrides; perhaps the most elegantly worked-out system involves $CpRe(CO)_2(NO)^+$ (Cp = $\eta$-$C_5H_5$) which, under varying conditions, can be converted to an entire range of products containing CO at different stages of reduction, including formyl, carbene, hydroxymethyl and methyl species (Scheme 1) (2,3,4,5). Reactions leading to hydrocarbon products are also known; in particular, $AlH_3$ plus a variety of metal carbonyls gives modest yields of hydrocarbon. Notably, with the group VI metal hexacarbonyls, high selectivity for ethylene was observed (6).

Although reactions involving main-group hydrides are not applicable to catalytic reactions of $H_2$, we have previously shown that early transition metal hydrides can exhibit analogous reactivity, with nucleophilic character falling off sharply as one moves to the right of the periodic table (7). Indeed, a number of CO reductions involving Ti and Zr hydrides have been reported in the last few years, as summarized in the following equations:

0097-6156/81/0152-0253$05.00/0
© 1981 American Chemical Society

$$(C_5Me_5)_2ZrH_2 + CO \longrightarrow (C_5Me_5)_2ZrH(OMe) \quad (\underline{inter}\ \underline{alia}) \qquad (\underline{8})$$

$$Cp_2ZrCl_2 + R_2AlH + CO \longrightarrow CH_3(CH_2)_nOH,\ n = 0\text{-}3 \qquad (\underline{9})$$

$$Cp_2Ti(CO)_2 + H_2 \xrightarrow{150^\circ} CH_4 + (CpTi)_6O_8 \qquad (\underline{10})$$

$$Cp_2ZrHCl + CO \longrightarrow (Cp_2ZrCl)_2(CH_2O) \qquad (\underline{11})$$

$$(C_5Me_5)_2ZrH_2 + Cp_2NbH(CO) \xrightarrow{-80^\circ} (C_5Me_5)_2ZrH(OCH)NbHCp_2 \qquad (\underline{12})$$

$$\downarrow CO$$

$$(C_5Me_5)_2ZrH(OCH_2)Nb(CO)Cp_2$$

The last sequence in particular demonstrates the same sort of stepwise transformation observed in the rhenium chemistry described above.

However, these reactions are all stoichiometric, not catalytic; this is a necessary consequence of the fact that all these group IV hydrides are highly sensitive to the hypothetical CO reduction products, alcohols or water (1). In contrast, group V hydrides such as $Cp_2MH_3$ and $Cp_2MH(CO)$ (M = Nb, Ta) are stable to water and alcohol, at least at room temperature, and could conceivably participate in a catalytic cycle. Our initial investigation dealt with trying to hydrogenate $Cp_2NbH(CO)$: a reaction forming methane does occur, but only at 130° C or higher (13). Under such conditions, the necessary stability towards hydroxylic products is lost; hence a much more reactive system is required. Obviously this could be achieved by going back to group IV, since the hydrides are much more active in a nucleophilic sense (note, for example, that the reaction of $Cp'_2ZrH_2$ ($Cp' = C_5Me_5$) with $Cp_2NbH(CO)$ proceeds rapidly even at -80° C! (12)), but this would preclude catalysis. Alternatively, we can try to make the CO more susceptible to nucleophilic attack. Since the CO stretching frequency in $Cp_2NbH(CO)$ is quite low (1900 $cm^{-1}$), it must be a relatively electron rich carbonyl. In order to optimize the desired reactivity, it appears to be necessary to employ a mixed system: an early transition metal hydride, so that the hydride will be nucleophilic, plus a more electrophilic metal carbonyl complex. Accordingly, we have examined the reactions of $Cp_2NbH_3$ with a variety of metal carbonyls.

## Experimental

All manipulations were carried out under inert atmosphere, using standard Schlenk and dry-box techniques. $Cp_2NbH_3$ was prepared from $Cp_2NbCl_2$ and $LiAlH_4$, as previously reported (14). Metal carbonyls and gases were commercial products used without further purification. Solvents were distilled from sodium benzophenone ketyl under argon. NMR spectra were recorded on Varian A-60 and XL-100 spectrometers; mass spectral studies used an AEI MS-9 high-resolution mass spectrometer and a DuPont DP-101 gas chromatograph-mass spectrometer.

Reactions of $Cp_2NbH_3$ with Metal Carbonyls. An equivalent amount of metal carbonyl ($Cr(CO)_6$, $Mo(CO)_6$, $W(CO)_6$, $Mn_2(CO)_{10}$, $Fe(CO)_5$, $Ru_3(CO)_{12}$, $Co_2(CO)_8$) was added to a benzene solution of $Cp_2NbH_3$. The solution was transferred by syringe to a serum-capped NMR tube, and the reaction followed by the disappearance of starting material peaks, as well as by the growth of a new peak or peaks in the Cp region. After complete reaction and evaporation of solvent (and excess metal carbonyl, if present and sufficiently volatile), NMR and IR were used to characterize organometallic products. No organic products could be detected in solution by NMR in any reaction.

The reactions of $M(CO)_6$ (M = Cr, Mo, W) all proceeded to completion in about 2 hr at 50° C. For Cr, the only product detected was $Cp_2NbH(CO)$, formed in about 60-80% yield. With Mo and W, only small amounts of any product giving NMR signals were present under these conditions; however, when the reactions were run under $H_2$ (see below), products were obtained with an NMR singlet at 4.6δ, and IR bands at 2060, 1950, 1880 and 1740 $cm^{-1}$. Recrystallization from toluene-hexane separated the products from unreacted $M(CO)_6$. The Mo product showed a mass spectral peak at m/e = 489, corresponding to $^{98}MoNbC_{16}H_{11}O_6$ (along with other peaks for the less abundant Mo isotopes); fragment peaks included species assigned as $Mo(CO)_n^+$ and $Cp_2NbH(CO)^+$. The W analog showed only fragment peaks in the MS.

The reaction of $Mn_2(CO)_{10}$ was complete in about 30 min at 40°. Again, the only product found by NMR was $Cp_2NbH(CO)$.

The reaction of $Fe(CO)_5$ was complete in about 20 min at room temperature. The product has been previously characterized by X-ray crystallography as $Cp_2(CO)Nb(\mu\text{-}H)Fe(CO)_4$ (15). $Ru_3(CO)_{12}$ reacts faster (about 5 min at room temperature) to give at least two products: one, soluble in benzene, has spectral properties very similar to the above Nb-Fe compound. Additional material is insoluble in benzene, soluble in THF, and exhibits several NMR signals in the Cp region, as well as several CO stretches in the IR. These products have not yet been completely characterized.

The reaction of $Co_2(CO)_8$ was complete immediately on mixing at room temperature. The product, $Cp_2(CO)Nb(\mu\text{-}CO)Co(CO)_3$, has been characterized by X-ray crystallography (16).

Intermediate in the $Cp_2NbH_3$-$Fe(CO)_5$ Reaction. A solution of 0.35 mmol $Cp_2NbH_3$ in 1 ml $C_6D_6$ was transferred to a serum-capped NMR tube and cooled to just above freezing. 0.15 ml $Fe(CO)_5$ was added, the tube was inserted into the XL-100 spectrometer, and a program for kinetic study data acquisition was initiated: a set of transients was accumulated for 30 sec and stored; data acqui-

*Scheme 1.* i, $R_3BH^-$, L = *CO or* $PPh_3$; ii, $NaBH_4/THF/H_2O$ *or* $NaAlEt_2H_2$, L = *CO;* iii, $NaBH_4$; iv, $BH_3$; v, $CF_3COOH$, *−70°;* vi, *warming (disproportionation reaction),* L = $PPh_3$; vii, $Ph_3C^+$

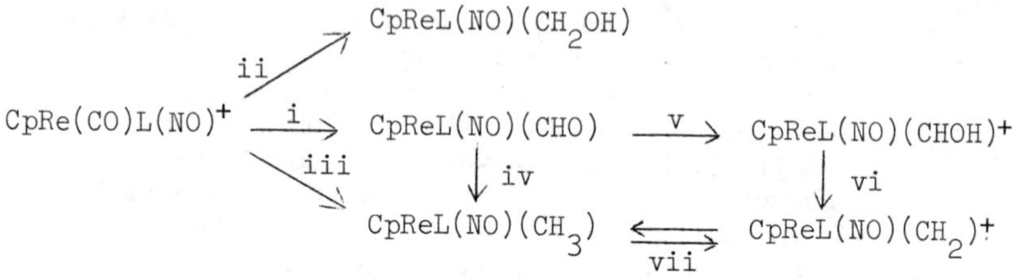


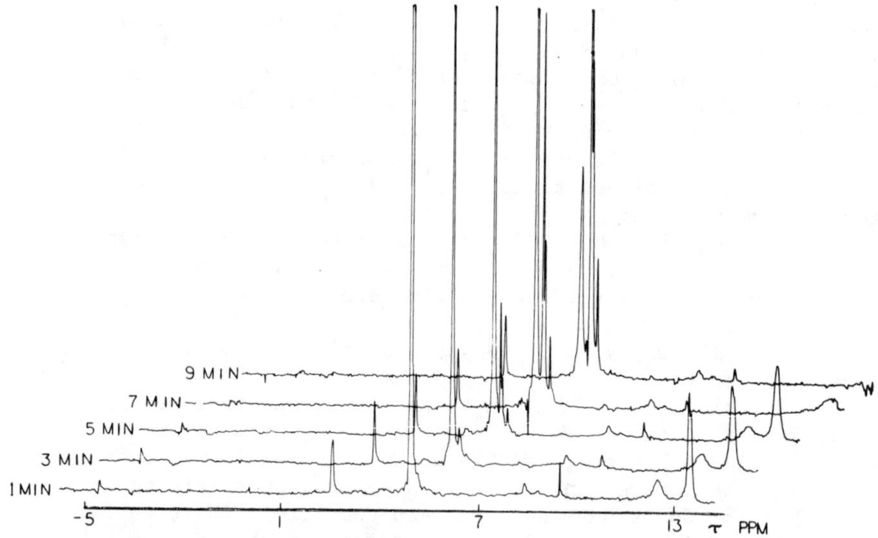


*Figure 1. Evolution of NMR spectrum during reaction of* $Cp_2NbH_3$ *with* $Fe(CO)_5$. *(Peak marked * is* $C_6D_5H$; *weak peaks around 8–10τ are attributable to impurities.)*

sition was halted for 30 sec; and then another 30 sec worth of transients accumulated. This sequence was continued for 20 min, after which the sequential FID's were Fourier transformed, and a selected group of resulting spectra, showing the progress of the reaction, plotted out in stacked form (Figure).

Reactions under $H_2$ for Gaseous Products. Solutions containing about 0.3 mmol each of $Cp_2NbH_3$ and metal carbonyl in several ml benzene were loaded into a Fischer-Porter pressure bottle through a ball valve; this was then pressurized to 1-1/2 - 2 atm $H_2$ and maintained at the reaction temperature for the reaction time appropriate to the particular metal carbonyl (see above). The gaseous atmosphere was then vented through a trap at -78° (to remove most of the benzene vapor) into an evacuated vessel. Samples were removed by gas-tight syringe and injected into a Hewlett-Packard 5790 gas chromatograph, equipped with a 4 ft, 1/8 in Porapak P column and a flame ionization detector. Use of known samples of hydrocarbons (methane and ethane) established that the minimum detectable amounts of product by this procedure were about 0.5-1.0 % (based on starting Nb complex). Several of the reactions ($Mo(CO)_6$, $W(CO)_6$ and $Ru_3(CO)_{12}$) gave small amounts (around 1-2 %) of these alkanes; only with $Cr(CO)_6$ was a substantial yield of hydrocarbon product consistently observed (see below).

Labelling Studies on $Cp_2NbH_3$-$Cr(CO)_6$ Reaction. Reactions were carried out under $D_2$ in one case, and utilizing $^{13}C$-labelled $Cr(CO)_6$ in another. The latter was prepared by the $Bu_3PO$-catalyzed incorporation of $^{13}CO$ into $Cr(CO)_5(pyr)$, as described recently by Darensbourg et al (17). Mass spectral analysis showed that the resulting $Cr(^{13}CO)_x(CO)_{6-x}$ was about 75 % $^{13}C$-labelled. After reaction with $Cp_2NbH_3$ as described above, the atmosphere was vented into a vessel which could be attached directly to the inlet of the MS-9 mass spectrometer.

On-line Sampling of $Cp_2NbH_3$-$Cr(CO)_6$ Reaction. The reaction was carried out in a flask with four stopcocks attached; two of these were used for loading the solution and for evacuating the flask and refilling with $H_2$ or other atmosphere, respectively. The other two were connected in a circuit containing a bellows-type gas circulating pump (Metal Bellows Corp. Model MB-21) to a gas sampling valve mounted in a Carle AGC-311 gas chromatograph. The latter is equipped with two columns, Porapak (80% N, 20% Q, 8 ft, 1/8 in) and 5 A molecular sieve (6 ft, 1/8 in) connected via a switching valve, as well as both thermal conductivity and flame ionization detectors. This arrangement provides monitoring of gases such as $O_2$ and $N_2$, to determine when a satisfactory air-free atmosphere has been established, as well as high sensitivity to hydrocarbon products-- on the same scale as before, yields on the order of 0.01 % can be easily detected. Reactions were carried out by heating the flask to 50°, and gaseous products monitored at any time by simply rotating the sampling valve.

## Results and Discussion

The reactivity of these metal hydride-metal carbonyl reactions can be correlated with the nature of the reactants in a manner consistent with the proposed mechanism: nucleophilic attack by hydride on coordinated CO. Thus reactions involving the highly nucleophilic group IV hydride, $Cp'_2ZrH_2$, are much faster than those of group V metal hydrides. On the other hand, the relatively electrophilic neutral binary metal carbonyls all react with $Cp_2NbH_3$ under mild conditions (20-50° C), whereas more electron-rich complexes such as cyclopentadienylmetal carbonyls $(Cp_2NbH(CO), CpV(CO)_4)$ or anionic carbonyls $(V(CO)_6^-)$ show no reaction under these conditions.

The significance (if any!) of the sequence of reactivities found for the various carbonyls examined-- a steady increase on moving to the right in the periodic table-- is still unclear. There does not appear to be any strong trend in electrophilicity along this series; for example, the IR spectra of the various compounds show quite similar CO stretching frequencies (18). It is also conceivable that not all these reactions follow the same mechanism; for example, a radical pathway might be a reasonable alternative for reactions involving the dimeric carbonyls. More direct evidence for the proposed pathway has been obtained from the $Fe(CO)_5$ reaction. As shown in the Figure, at early stages the NMR shows a peak at $14.3\delta$, in addition to the $Cp_2NbH_3$ signals. As the reaction proceeds, the former grows to a maximum (corresponding to about 20 % yield, assuming it is due to a single proton) and then gradually disappears, while the $Cp_2NbH_3$ peaks also decrease and are replaced by the Cp signals of the product.

The extreme downfield shift of the intermediate signal is characteristic either of a metal formyl, or a hydridocarbene complex, suggesting one of the following structures (or, perhaps, a rapid equilibrium between the two):

$$Cp_2NbH_2^+ \quad HC\overset{O}{\overset{\|}{}}Fe(CO)_4^- \quad \rightleftharpoons \quad Cp_2Nb\overset{H}{\underset{O}{\diagdown}}{-H}$$
$$\underset{H}{\overset{}{\diagup}}C{=}Fe(CO)_4$$

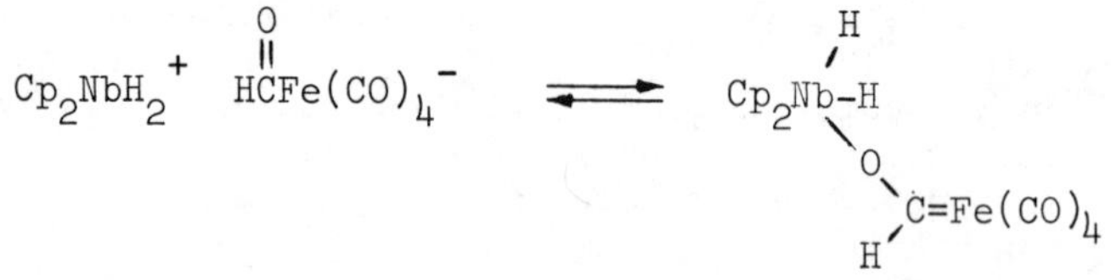

We tentatively prefer the latter, since compounds of this type have been isolated and, in one case, characterized crystallographically (12). The analog of this intermediate has not yet been observed with other metal carbonyls, but since the relative rates of its formation and decomposition may well differ from one system to the next, it is quite possible that an NMR-detectable amount never accumulates, at least at ambient temperatures.

The subsequent chemistry in these systems clearly involves (at least) two competing pathways. In most cases little or no CO reduction product is observed; rather $H_2$ is evolved and hydrido (carbonyl) complexes form. Hence in these cases, loss of $H_2$ from the above intermediate must be more efficient than further transfer of hydride to CO. The resulting coordinatively unsaturated species can undergo i) transfer of CO to Nb; ii) migration ("de-insertion") of the formyl-like hydrogen from CO to metal (a reaction observed for all known metastable formyl complexes); and iii) in some cases, dissociation or loss of further hydrogen to give the appropriate product ($Cp_2NbH(CO)$ or bimetallic compound). Only for $Cr(CO)_6$ were large amounts of hydrocarbon product observed, and hence this system was examined in detail.

When the reaction of $Cp_2NbH_3$ with $Cr(CO)_6$ is carried out under Ar, little or no hydrocarbons are produced; instead $Cp_2NbH(CO)$ is formed in good yield, with a small amount of precipitate depositing. In contrast, when the reaction is run under $H_2$, large amounts of solid precipitate; only a small amount of $Cp_2NbH(CO)$ remains in solution, and gas sampling indicates the formation of ethane in about 10 % yield (based on Nb). With the syringe sampling method, no other hydrocarbon product was detected. The origin of the ethane was probed by labelling methods. Reaction under $D_2$ gave ethane which, by mass spectroscopy, contained significant quantities of $d_0$, $d_1$ and $d_2$-ethane; more highly enriched species could have been present in quantities up to a few percent without being detected.

The method recently published by Darensbourg et al (<u>17</u>) provides an excellent route to highly-enriched $^{13}CO$-labelled compounds, including $Cr(CO)_6$. Starting with 90 % labelled $^{13}CO$, the compound was obtained with about 75 % labelling. When this was subjected to hydrogenation, the ethane produced contained substantial amounts of $^{13}C_2H_6$, as shown by high-resolution mass spectroscopy (in which the $^{13}C_2H_6^+$ peak was well separated from the background peak due to $O_2^+$). A peak with exact mass 32.054 was observed; the calculated value for $^{13}C_2H_6$ is 32.0535. It was not possible to accurately determine the relative amounts of the variously isotopically labelled ethanes because of the complexity in this region: ethane shows M-1, M-2, ... peaks of about the same intensity as the parent ion, and benzene (which could not be entirely removed from the vapor phase) gives fragment peaks at m/e = 30, 29, ... as well. By making approximate corrections for these two factors, we were able to estimate that the amount of $^{13}C$ in the ethane is <u>at least</u> 75 % of that in the starting carbonyl compound, and hence most, if not all, of the ethane arises from CO reduction. Most likely it all does; we have seen evidence previously for formation of ethane from an alternate source, most probably the Cp rings, in related systems, but these reactions require considerably higher temperatures (<u>13</u>).

The solids formed in this reaction have not been completely characterized, but they almost certainly contain some form of niobium oxide species: they are insoluble in all organic solvents but do dissolve in strong mineral acids. (The IR of the solid shows no characteristic or informative bands.) Thus the oxygen atom of the reduced CO appears as metal-bound oxygen, rather than free $H_2O$, a fact which would make catalytic reduction very unlikely for this particular system. When the reaction is run under a mixed CO-$H_2$ atmosphere, the only product observed is $Cp_2NbH(CO)$, in nearly quantitative yield; no hydrocarbons or insoluble products form. This is not simply due to a direct reaction of $Cp_2NbH_3$ with CO, since that reaction is too slow at 50°; rather an intermediate in the reaction of $Cp_2NbH_3$ with $Cr(CO)_6$ must be efficiently trapped by CO. $Cp_2NbH(CO)$ is not sufficiently nucleophilic to attack $Cr(CO)_6$; a mixture of these two compounds shows no reaction until well above 100°, conditions under which the niobium compound begins to react by itself.

Selective formation of ethane in this reaction is of key interest, since one of the major reasons for investigating homogeneous systems was the hope of achieving such selectivity. While a large variety of mechanisms leading to alkane formation might be constructed, few would explain this selectivity; a route involving successive CO insertion into metal alkyls, followed by reduction, for example, would not. The result of Masters cited earlier is most suggestive in this regard: $AlH_3$ plus $M(CO)_6$ is also highly selective for $C_2$ hydrocarbon, although most of the product (95 %) is ethylene, not ethane (6). Masters proposed a mechanism involving a carbene intermediate to account for this. We could account for our result by the same mechanism, if the $Cp_2NbH_3$-derived system also contains (or generates) a species capable of catalyzing the hydrogenation of ethylene to ethane. To check this, we designed a system permitting on-line analysis of the gaseous products during the course of the reaction, and found that considerable amounts of ethylene are indeed present during early stages. Thus the $C_2$ products after one hour consist of 20 % $C_2H_4$ and 80 % $C_2H_6$; after 2 hr, the percentage of $C_2H_4$ has decreased to 8 %; after 5 hr, to 0.6 %. This system also gave us much greater sensitivity, and we were able to determine that small amounts (0.5 - 1 %) of methane and propane are also formed in this reaction. Similarly, if small amounts of ethylene are added to the reaction mixture, they are gradually hydrogenated to ethane during the course of the reaction. With an equal mixture of $H_2$ and $C_2H_4$ as the atmosphere, almost no reduction is observed; rather the $Cp_2NbH_3$ is converted to $Cp_2Nb(C_2H_5)(C_2H_4)$ in high yield. Again, this does not form readily from $Cp_2NbH_3$ at this temperature (19), and is probably due to trapping a reaction intermediate as in the $H_2$-CO reaction; also, this shows that the hydrogenation catalyst is not starting material but a reaction product.

*Scheme 2. Proposed mechanism of reactions*

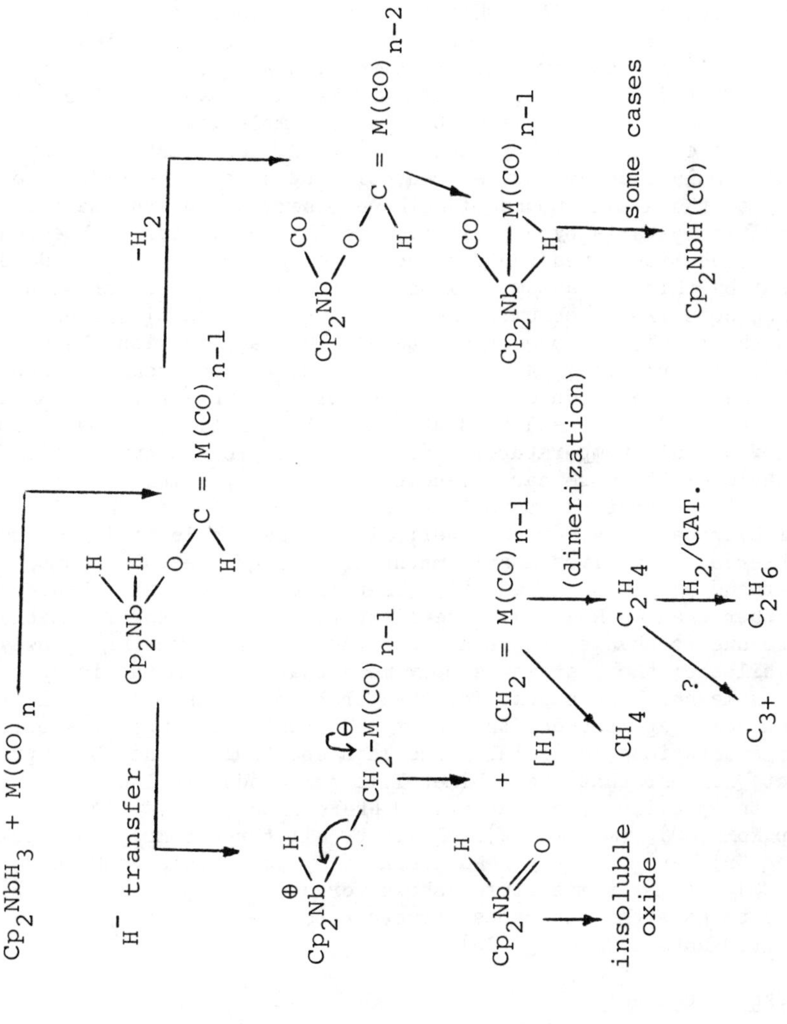

These observations permit us to construct a plausible mechanism for the reaction, shown in Scheme 2. The initial step, involving nucleophilic attack by Nb-H on CO, is probably general to all the metal carbonyls examined here, although as noted above direct evidence is available only for $Fe(CO)_5$. The resulting intermediate can react in one of two ways. Loss of $H_2$ generates an unsaturated species which eventually leads to products containing no reduced CO. For most of the systems studied, this is the dominant pathway. Alternatively, a second hydrogen can be transferred to carbon. Niobium-assisted cleavage of the carbon-oxygen bond generates a niobium(oxo) complex, which leads to the insoluble metal oxide products observed, and a carbene complex. This could undergo further reduction by niobium hydrides, leading to the small amount of methane observed, but the main decomposition path is by dimerization to form ethylene. Subsequent hydrogenation gives the observed major product, ethane. Note that by this mechanism, all the hydrogens added to carbon up to $C_2H_4$ come from niobium; atmospheric hydrogen participates only in the final, hydrogenation stage. This agrees with the $D_2$ labelling results: a maximum of two deuterium atoms are found in the ethane produced. (Exchange of $D_2$ with starting $Cp_2NbH_3$, which could also lead to deuterium incorporation, is much too slow at this temperature.) Formation of propane by reaction of ethylene with more carbene complex is also possible.

The reactions of $Mo(CO)_6$ and $W(CO)_6$ differ in that i) yields of hydrocarbon are much lower; ii) the same selectivity is not observed: Mo gives mostly methane, W roughly equal amounts of methane and ethane; and iii) bimetallic products are obtained, rather than $Cp_2NbH(CO)$. The first two differences presumably are due to changes in relative rates of the competing pathways available; the last may simply mean that the bimetallic species is somewhat less stable for the first-row transition metal Cr, and comes apart under the reaction conditions. Spectroscopic considerations, especially the mass spectrum of the Nb-Mo product, suggest that these bimetallic compounds are isoelectronic to the previously characterized Nb-Fe product; that is, $Cp_2NbMH(CO)_6$ (M = Mo, W). In fact, all three complexes (including Cr) have recently been prepared by photolysis of $Cp_2NbH_3$ with $M(CO)_6$ at low temperature, shown (crystallographically for M = Cr) to have the analogous structure to the Nb-Fe compound, $Cp_2(CO)Nb(\mu\text{-}H)Cr(CO)_5$ (<u>20</u>).

Acknowledgement

We thank the National Science Foundation for support through grant CHE 77-01585.

Literature Cited


1. Labinger, J. A., Adv. Chem. Series, 1978, 167, 149.

2. Casey, C. P., Andrews, M. A., McAlister, D. R., Rinz, J. F., J. Am. Chem. Soc., 1980, 102, 1927.

3. Tam, W., Wong, W.-K., Gladysz, J. A., J. Am. Chem. Soc., 1979, 101, 1589.

4. Wong, W.-K., Tam, W., Gladysz, J. A., J. Am. Chem. Soc., 1979, 101, 5440.

5. Sweet, J. R., Graham, W. A. G., J. Organometal. Chem., 1979, 173, C9.

6. Masters, C., van der Woude, C., van Doorn, J. A., J. Am. Chem. Soc., 1979, 101, 1633.

7. Labinger, J. A., Komadina, K. H., J. Organometal. Chem., 1978, 155, C25.

8. Manriquez, J. M., McAlister, D. R., Sanner, R. D., Bercaw, J. E., J. Am. Chem. Soc., 1978, 100, 2716.

9. Shoer, L. I., Schwartz, J., J. Am. Chem. Soc., 1977, 99, 5831.

10. Huffman, J. C., Stone, J. G., Krusell, W. C., Caulton, K. G., J. Am. Chem. Soc., 1977, 99, 5829.

11. Fachinetti, G., Floriani, C., Roselli, A., Pucci, S., J. Chem. Soc., Chem. Commun., 1978, 269.

12. Wolczanski, P. T., Threlkel, R. S., Bercaw, J. E., J. Am. Chem. Soc., 1979, 101, 218.

13. Labinger, J. A., Wong, K. S., Scheidt, W. R., J. Am. Chem. Soc., 1978, 100, 3254.

14. Labinger, J. A., Wong, K. S., J. Organometal. Chem., 1979, 170, 373.

15. Wong, K. S., Scheidt, W. R., Labinger, J. A., Inorg. Chem., 1979, 18, 136.

16. Wong, K. S., Scheidt, W. R., Labinger, J. A., Inorg. Chem., 1979, 18, 1709.

17. Darensbourg, D. J., Walker, N., Darensbourg, M. Y., J. Am. Chem. Soc., 1980, 102, 1213.

18. Braterman, P. S., "Metal Carbonyl Spectra", Academic Press: New York, 1975; pp. 179-185.

19. Guggenberger, L. J., Meakin, P., Tebbe, F. N., J. Am. Chem. Soc., 1974, 96, 5420.

20. W. A. Herrmann, personal communication.

RECEIVED December 8, 1980.

# 18

# Formation of Hydrocarbons by Hydridic Reduction of Carbon Monoxide on $Cp_2Fe_2(CO)_4$

ANDREW WONG and JIM D. ATWOOD

Chemistry Department, State University of New York at Buffalo, Buffalo, NY 14214

The search for homogeneous analogues of Fischer-Tropsch catalysts has been vigorous in recent years. Recent reviews cover much of the pertinent literature.[1,2] Homogeneous systems are amenable to mechanistic studies and often have advantages of selectivity and mild reaction conditions which would be important economically. There are a number of complexes which undergo homogeneous CO reduction. Unlike the typical Fischer-Tropsch reactions, most homogeneous reductions yield methane or methanol as primary products; only in a few systems were higher hydrocarbon chains observed.[1,2] Demitras and Muetterties observed formation of methane, ethane, propane and isobutane in the ratio 1:4:trace:trace upon treatment of $Ir_4(CO)_{12}$ with synthesis gas in molten $NaCl{\cdot}2AlCl_3$ at 180°.[3] Schwartz and co-workers observed chain build-up in a zirconium based system ($Cp_2ZrCl_2$, $(i\text{-}Bu)_2AlH$ and CO at a few atmospheres of pressure) which produced alcohols (methanol, ethanol, 1-propanol and 1-butanol in a molar ratio of 1.0:0.12:0.15:0.03) upon hydrolysis at room temperature.[4-6] Masters and co-workers observed the formation of methane, ethane, propane and butane upon treatment of $Ru_3(CO)_{12}$ with $AlH_3$.[7] Ethane or ethylene ($C_2$ hydrocarbons) have been observed in a few cases.[7-9]

Fischer-Tropsch reactions, although they have been known for more than 50 years, are not well understood mechanistically.[10-12]

0097-6156/81/0152-0265$05.00/0
© 1981 American Chemical Society

In the current view the formation of methane and chain growth in the production of higher hydrocarbons occurs via oxygen-free $CH_x$ species, generated on the surface of the catalyst by dissociative chemisorption of CO followed by partial hydrogenation.[12] Although reactive metal carbide complexes have been observed in metal cluster compounds[13,14] and have been postulated as intermediates in the formation of methane,[15] it is unlikely that the initial step in a homogeneous CO reduction cycle will be C≡O bond cleavage. In most homogeneous systems it is believed that formation of a C-H bond is the initial step in the reduction of CO, followed by the successive formation of alkylidene or hydroxymethyl, and methyl complexes.

$$M-C\equiv O \longrightarrow M-\overset{\overset{\displaystyle O}{\|}}{C}-H \longrightarrow M=CH_2$$

$$\downarrow \qquad\qquad\qquad \downarrow$$

$$M-CH_2OH \longrightarrow M-CH_3$$

$$\downarrow \qquad\qquad\qquad \downarrow$$

$$CH_3OH \qquad\qquad CH_4$$

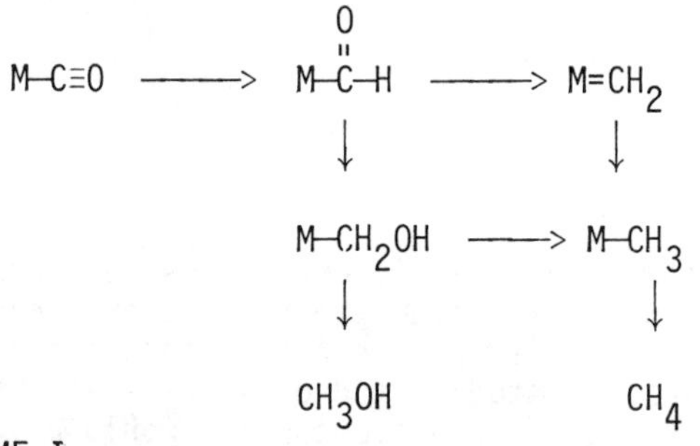

SCHEME I

An excellent example of this sequence was illustrated in the reaction of $CpRe(CO)_2(NO)^+$ with $NaBH_4$ which under carefully controlled conditions produced the formyl, the hydroxymethyl and the methyl complexes, successively.[16,17]

$$[CpRe(CO)_2NO]^+ + NaBH_4 \xrightarrow[\text{0°C, 15 min}]{\text{THF/H}_2\text{O}} CpRe(CO)(NO)CHO$$

$$0°, 30 \text{ min} \downarrow NaBH_4, \text{THF} \qquad\qquad\qquad 0°C, 15 \text{ min} \downarrow +NaBH_4$$
$$\downarrow\text{THF/H}_2\text{O}$$

$$CpRe(CO)(NO)CH_3 \xleftarrow[\text{25°, 5 hr}]{NaBH_4} CpRe(CO)(NO)CH_2OH$$

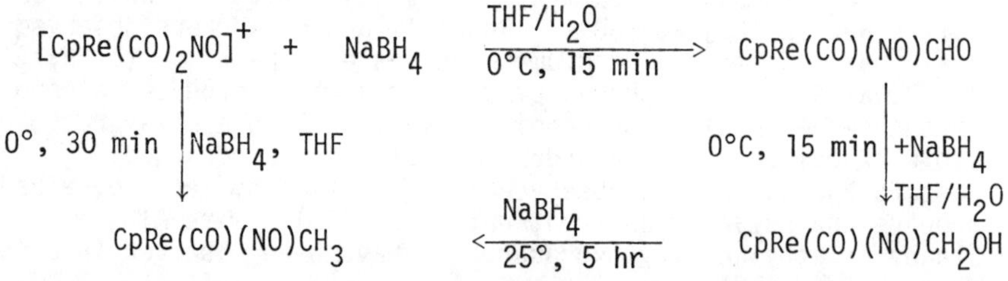

This was the first example in which models for presumed Fischer-Tropsch intermediates have been isolated and their sequential reduction demonstrated. Neither methane nor methanol was observed from further reduction of the methyl and the hydroxymethyl complexes. The use of $THF/H_2O$ as solvent was crucial in this sytem; in THF alone $CpRe(CO)(NO)CH_3$ was the only species observed, probably because the initial formyl complex was further reduced by $BH_3$.[18] When multihydridic reagents are reacted with metal carbonyl complexes, formyl species are usually not observed. The rapid hydrolysis of $BH_3$ by aqueous THF allowed $NaBH_4$ to act as a monohydridic reagent in its reaction with $CpRe(CO)_2NO^+$.

When Group VI carbonyl complexes were reacted with alane or $Cp_2NbH_3$ reduction of CO to ethylene was noted.[7,8] Ethylene was the primary product of the $Cp_2NbH_3$ reduction although it subsequently was hydrogenated to ethane. Masters and co-workers suggested that ethylene was formed through an alkylidene dimerization as shown below.

$$2M=CH_2 \longrightarrow M \overset{CH_2}{\underset{CH_2}{\diamond}} M \longrightarrow C_2H_4$$

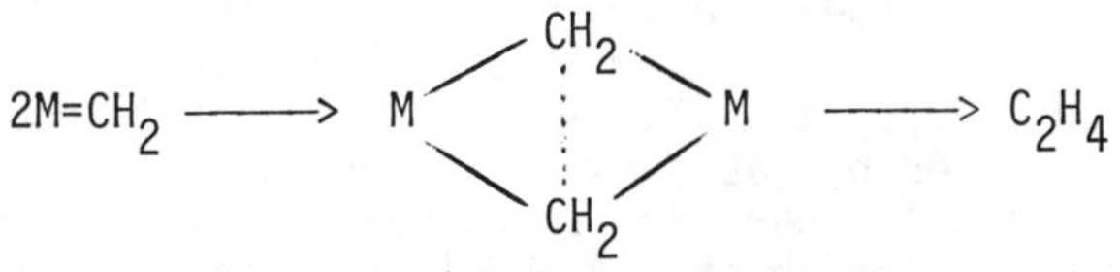

To further explore possible mechanisms of carbon-carbon bond formation, we have examined the reaction of $Cp_2Fe_2(CO)_4$ with $LiAlH_4$ (LAH). This reaction yields hydrocarbon products up to butane.[20] The results of this study are reported herein.

Experimental

In a typical reaction, 0.5 mmole $Cp_2Fe_2(CO)_4$, excess $LiAlH_4$ and a Teflon stir bar were transferred into a glass vessel (120 cm x 2.2 cm) equipped with a Kontes Teflon high vacuum stopcock connected to a ball joint and with a small septum covered stopcock for sample removal. The apparatus was evacuated on a vacuum-line and ∿ 5 ml of solvent, previously dried by extended refluxing over sodium metal and benzophenone, was vacuum distilled into the glass vessel at -196°C. When required, carbon monoxide may be transferred into the glass vessel subsequently on the vacuum-line. The reaction mixture was allowed to warm and react at room temperature. Gas samples were withdrawn through the septum-covered stopcock directly with a gas-tight syringe (Hamilton Co.) and analyzed by a Varian 2440 gas chromatograph with flame ionization detectors. Chromatographic separations were obtained with a 6 ft x 1/8 in. stainless-steel column of Spherocarb (Analab, 80/100 mesh), for separations of permanent gases and light hydrocarbons C-1 to C-3; and a 6 ft x 1/8 in. stainless-steel column of Porapak Q (Analab, 80/100 mesh), for separation of C-1 to C-5 hydrocarbons. Helium flow rate was ∿ 20 ml/min and column temperature ranged from ambient to 150°C. Peaks were identified by comparing retention times with known samples and peak areas were converted into mole quantities by comparison with standard samples.

Products in the reaction mixture were analyzed by infrared and nmr spectroscopy. IR spectra were obtained on a Perkin-Elmer 521 infrared spectrophotometer. The reaction mixture was first suction filtered through a fine porosity glass frit inside a glove box (Vacuum Atmosphere HE-43-2) under an argon atmosphere, and was placed in NaCl cells (International Crystals, Inc.) of path length 0.1 or 0.5 cm with Teflon stoppers. $^1H$ nmr spectra were obtained on a Varian T-60 spectrometer with TMS (0.00 ppm) or benzene (7.23 ppm) as standards.

## Results and Discussion

Reaction of $Cp_2Fe_2(CO)_4$ and $LiAlH_4$ or $LiAlD_4$. The reaction between $Cp_2Fe_2(CO)_4$ and a 10-30 fold excess of LAH in THF at room temperature resulted in the rapid formation of $CH_4$. $C_2H_4$, $C_2H_6$, $C_3H_6$, $C_3H_8$, $C_4H_8$ and $C_4H_{10}$. The reaction was essentially complete in one hour, as determined by gas chromatography, when no more growth in hydrocarbons was observed. In toluene a similar reaction was observed, although at a much slower rate with completion in about two weeks, presumably because of the relative insolubility of LAH in toluene. We decided to study this system in toluene, despite the slow rate, because a side reaction between LAH and THF was observed leading to $CH_4$, $C_2H_4$ and $C_2H_6$ in small amounts (< 5% of those observed with $Cp_2Fe_2(CO)_4$ and LAH).

The exact ratio of the hydrocarbons formed from the reaction of $Cp_2Fe_2(CO)_4$ and LAH in toluene varied with time as shown in Table 1. Ethylene was the predominant product initially (up to 36 hr.) and then decreased dramatically with time. Propylene also decreased with time, although not as dramatically. Butene did not show this change with time but the amounts were small and experimental error could be a factor.

Use of $LiAlD_4$ produced perdeutero-hydrocarbons, confirming reduction of CO with $LiAlD_4$ as the only hydrogen source. Small amounts of partially or non-deuterated hydrocarbons were also observed (< 10% of the total) probably as a result of impurity in the LAD (98% D).

Table I. Gaseous products from the reduction of CO on $Cp_2Fe_2(CO)_4$ by 30 fold excess LAH in toluene.[a]

| | $CH_4$ | $C_2H_4$ | $C_2H_6$ | $C_3H_6$ | $C_3H_8$ | $C_4H_8$ | $C_4H_{10}$ |
|---|---|---|---|---|---|---|---|
| 36 hour | 0.028 | 0.044 | 0.024 | $5x10^{-3}$ | $4x10^{-3}$ | $9x10^{-5}$ | $7x10^{-5}$ |
| 14 days | 0.178 | $6x10^{-4}$ | 0.14 | $9x10^{-4}$ | 0.023 | $5x10^{-5}$ | $5x10^{-4}$ |

[a]The yields reported are mmole of the hydrocarbon product per mmole of iron complex.

The total yield of hydrocarbon for the reaction between $Cp_2Fe_2(CO)_4$ and LAH was ∿ 40% (0.4 mmole of hydrocarbon product per mmole of iron starting material). $LiCpFe(CO)_2$ and $Cp_2Fe_2(CO)_4$ were both in solution after the reaction as shown by infrared spectra even when a 30-fold excess of LAH was used. We believe that $HCpFe(CO)_2$ was formed initially, but because of its thermal instability, decomposed to $H_2$ and $Cp_2Fe_2(CO)_4$. Support for this came from the observations that (1) hydrogen was observed and (2) when the reaction was carried out with ∿ 0.01 mmole of $PBu_3$ and a 10-fold excess of LAH, $LiCpFe(CO)_2$ and the thermally stable $HCpFe(CO)PBu_3$ were the only products in solution as determined by IR and NMR spectra. The presence of $PBu_3$ in the reaction mixture also changed the ratio of hydrocarbon products with a significant enhancement in the production of ethylene as shown by the distribution of products in Table II.

Table II. Hydrocarbons from the Reaction of LAH with $Cp_2Fe_2(CO)_4$ in the presence of a Ligand.[a]

| L | $CH_4$ | $C_2H_4$ | $C_2H_6$ | $C_3H_6$ | $C_3H_8$ | $C_4H_8$ | $C_4H_{10}$ |
|---|---|---|---|---|---|---|---|
| $PBu_3$ | 0.108 | 0.376 | 0.224 | 0.022 | 0.016 | $7x10^{-4}$ | $9x10^{-4}$ |
| CO | 0.144 | 0.511 | 0.045 | 0.011 | $6x10^{-3}$ | $3x10^{-4}$ | $2x10^{-4}$ |

[a]The yields reported are mmole of the hydrocarbon product per mmole of iron complex. $PBu_3$ = 1 equivalent, CO = 1 atmosphere.

The formation of higher hydrocarbons ($C_2$ to $C_4$) requires forming C-C bonds. Several possibilities have been suggested for C-C bond formation in Fischer-Tropsch systems.[2] Methyl migration (CO insertion) has been suggested as one possibility.[21] To assess the importance of CO insertion in this system we have investigated the reaction of $Cp_2Fe_2(CO)_4$ with LAH under a CO atmosphere. The results are reported in Table II. A net increase in CO reduction product is observed under CO, but the most dramatic difference is the greatly enhanced production of ethylene. This is similar though greater in magnitude than the production of $C_2H_4$ in the presence of $PBu_3$ as shown in Table II. Reaction of LAH with $Cp_2Fe_2(CO)_4$ was also investigated under a $^{13}CO$ atmosphere with substantial incorporation of $^{13}C$ into the hydrocarbon product as shown in Table III.

Table III. Products of the reduction of $Cp_2Fe_2(CO)_4$ with LAH in the presence of $^{13}CO$.

| | |
|---|---|
| methane | $CH_4 > {}^{13}CH_4$ (< 30%) |
| ethylene | ${}^{13}CCH_4 \simeq {}^{13}C_2H_4 > C_2H_4$ (50%) |
| ethane | ${}^{13}CCH_6 \simeq {}^{13}C_2H_6 > C_2H_6$ (50%) |
| propene | ${}^{13}C_2CH_6 > {}^{13}CC_2H_6 \simeq C_3H_6$ (70%) $> {}^{13}C_3H_6$ (50%) |
| propane | ${}^{13}C_2CH_8 \simeq {}^{13}CC_2H_8 > C_3H_8$ (40%) $\simeq {}^{13}C_3H_8$ |

The number in parenthesis is the percentage of that species compared to the predominant species. The uncertainty in the propene and propane is much larger because of the much smaller amounts and because fragmentation is much more extensive. The wide distribution of label in the hydrocarbon product indicates a complicated reaction scheme with considerable mixing of the label. The amount of label in methane cannot be determined precisely because $^{13}CO$ contains $^{13}CH_4$. For each hydrocarbon, the species with only one C-12 is among the predominant products, which suggests that a CO insertion mechanism is operative.

Reaction of $Cp_2Fe_2(CO)_4$ with $NaBH_4$ led to similar reduction products though the yield was less than 5% of that seen with LAH. Reaction of $LiBEt_3H$ effected reduction in less than 0.3% of that observed with LAH.

Reaction of $CH_3FeCp(CO)_2$, $C_2H_5FeCp(CO)_2$, and $CH_3C(O)FeCp(CO)_2$ with LAD. To further investigate the importance of CO insertion in the build-up of hydrocarbon chains upon reduction of CO on $Cp_2Fe_2(CO)_4$ with LAH we have prepared the possible intermediates $CH_3FeCp(CO)_2$, $C_2H_5FeCp(CO)_2$ and $CH_3C(O)FeCp(CO)_2$ and investigated their reactivity towards LAD. The results are presented in Table IV.

Table IV. Products from the reaction of $RFeCp(CO)_2$ with LAD.

| $RFeCp(CO)_2$ | Primary Hydrocarbons Formed | (Ratio) |
|---|---|---|
| R = $CH_3$ | $CH_3D$, $C_2H_3D_3$ | (1:2) |
| R = $C_2H_5$ | $C_2H_5D$, $C_3H_5D_3$ | (1:2) |
| R = $CH_3C(O)$ | $C_2H_3D_3$, $C_3H_3D_3$, $C_3H_3D_5$ | (3:0.8:1) |

These reactions were carried out in toluene with a 10-fold excess of LAD with total yields of hydrocarbons at 50-60% (mmole hydrocarbon per mmole $RFeCp(CO)_2$). These reactions were stopped after fairly short times so that the reduction would occur primarily at the R group with only a small contribution from CO reduction. The observation of primarily propane and propene upon reduction of $C_2H_5FeCp(CO)_2$ and $CH_3C(O)FeCp(CO)_2$ suggests that CO insertion is responsible for chain propagation. Formation of an ethylidene complex and dimerization would lead to butane or butene. In addition to the primary products listed in Table IV, other hydrocarbons were also observed ranging from $C_1$-$C_4$ in considerably smaller amounts. Many of these had quite high deuterium enrichments and we believe that these are the result of competing CO reduction. The reaction of LAH with the acetyl complex was different in several respects from the reaction with the methyl and ethyl complexes. (1) The predominant gaseous products in the reduction of the methyl and ethyl complexes were $C_{n+1}$ hydrocarbons while for the acetyl complex the $C_n$ and $C_{n+1}$ hydrocarbons were formed in comparable yields. (2) In the reduction of the acetyl complex an olefin (propene) was observed; in the methyl and ethyl reductions only traces of olefins were observed. (3) $Cp_2Fe_2(CO)_4$ and $LiCpFe(CO)_2$ were observed as products in solution after reduction of the acetyl complex, but only $LiCpFe(CO)_2$ was observed for the methyl and ethyl complexes. (4) The reduction of the acetyl complex proceeded much more rapidly (completion in 1 hr) than reduction of the methyl and ethyl complexes (completion in 2 days).

The differences in products (both gaseous and in solution) between the reduction of the acetyl complex, $CH_3C(O)FeCp(CO)_2$, and the methyl and ethyl complexes suggest that different intermediates are involved. The rapidity of the reduction of the acetyl complex to hydrocarbons rules out the ethyl complex as an intermediate in the reduction of $CH_3C(O)FeCp(CO)_2$ with LAH. Reduction of the acetyl to the ethyl complex has been observed.[19]

$$CH_3C(O)FeCp(CO)_2 \xrightarrow[THF]{BH_3} CH_3CH_2FeCp(CO)_2$$

A possible intermediate was prepared by Gladysz and Selover by reaction of $CH_3C(O)FeCp(CO)_2$ with $LiEt_3BH$.[22]

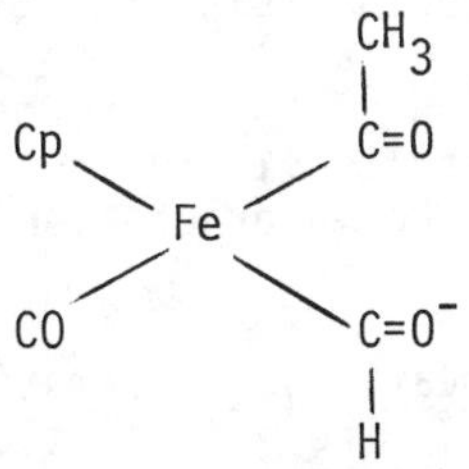


While this species certainly may be in solution, the relative importance of this intermediate or direct transformations upon the acetyl group in the production of hydrocarbons cannot be assessed at this point.

For the reductions of the methyl and ethyl complexes with LAH the $C_{n+1}$ alkane was the major product with the $C_n$ product present to a significant extent. These two products could arise by the following sequence.

$$
\begin{array}{ccccccc}
\overset{\displaystyle CO}{\underset{\displaystyle CO}{\overset{\displaystyle |}{\underset{\displaystyle |}{CpFe\text{-}R}}}} & \xrightarrow[-AlH_3]{LiAlH_4} & \overset{\displaystyle CO}{\underset{\displaystyle \overset{\displaystyle |}{\underset{\displaystyle H}{C=OLi}}}{\overset{\displaystyle |}{CpFe\text{-}R}}} & \xrightarrow{-CO} & \overset{\displaystyle H}{\underset{\displaystyle \overset{\displaystyle |}{\underset{\displaystyle Li^+}{CO^-}}}{\overset{\displaystyle |}{CpFe\text{-}R}}} & \longrightarrow & RH
\end{array}
$$

$\Big\downarrow AlH_3$

$$
\begin{array}{c}
H\!\!-\!\!AlH_2 \\
| \\
O \\
\parallel \\
CpFe - C - R \longrightarrow \longrightarrow RCH_3 \\
| \\
CO \\
| \\
H
\end{array}
$$

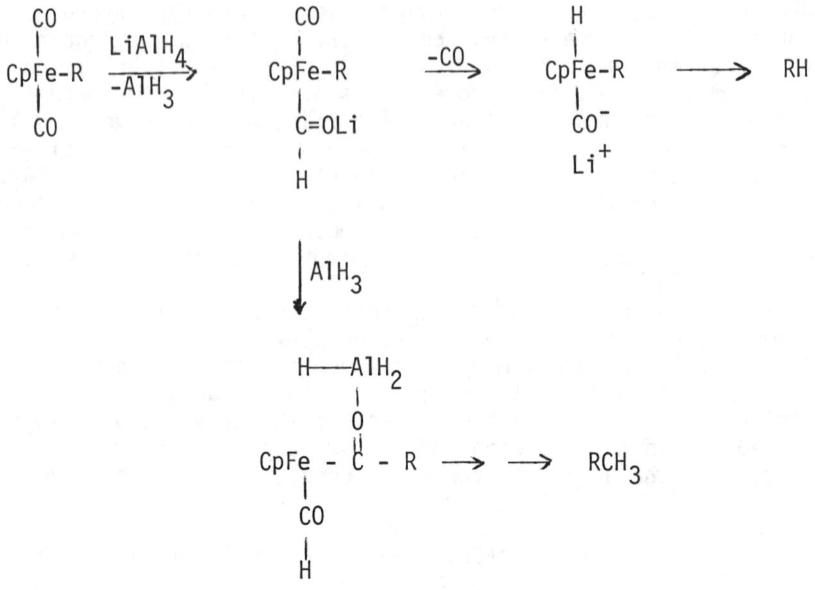

Scheme 2.   Suggested Scheme for the Reduction of $RFeCp(CO)_2$
            with LAH (R = Me, Et).

The presence of $AlH_3$ would assist the alkyl migration and the
intermediate suggested is analogous to those Shriver has isolated.[23,24] The reaction of $CH_3C(O)FeCp(CO)_2$ and $CH_3FeCp(CO)_2$
with LAH under $^{13}CO$ resulted in no incorporation of $^{13}C$ into the
$C_n$ and $C_{n+1}$ products.  No incorporation would be expected for
Scheme 2.

Mechanistic observations on formation of hydrocarbons in
$Cp_2Fe_2(CO)_4$ with LAH.     There is little doubt that the initial
step in the reaction of $Cp_2Fe_2(CO)_4$ with LAH involves formation
of a formyl complex by addition of a hydride to coordinated CO.
We believe the initial site of attack is on a terminal CO which
should be more susceptible than the more electron rich bridging
CO's.[25] The formyl complex will not be "free" but will almost
certainly have aluminum coordinated to the oxygen.  Further
reduction to a methyl could occur as was observed in $NaBH_4$
reduction of $CpRe(CO)_2NO$.  We would concur with the statement
that the intermediates will all have coordination of the aluminum
to the oxygen during the reduction.[17] We have demonstrated in a
separate experiment that methane is formed when $CH_3FeCp(CO)_2$ is
reacted with LAH.

$$CH_3FeCp(CO)_2 + LAD \rightarrow CH_3D + LiCpFe(CO)_2 + AlD_3$$

More interesting is the formation of higher hydrocarbons which involves forming carbon-carbon bonds. We believe from the data discussed in an earlier section that the higher hydrocarbons are formed as a result of a CO insertion (alkyl migration) process. Two distinct mechanisms have been proposed for chain extension in Fischer-Tropsch sytems - CO insertion and alkylidene (carbene) oligomerization. While our system has the characteristics of a CO insertion, we do not think that the insertion occurs by alkyl migration as has usually been postulated. Comparison of the results of the LAH reduction of $CH_3FeCp(CO)_2$, $C_2H_5FeCp(CO)_2$ and $CH_3C(O)FeCp(CO)_2$ with LAH reduction of $Cp_2Fe_2(CO)_4$ show that propagation by alkyl migration does not occur. We believe that the CO insertion occurs at a partially hydrogenated stage, probably via an alkylidene migration reaction. An alternate scheme for chain extension in this iron system is presented in Scheme 3 (Figure 1). Intermediate A is an iron-formyl with aluminum coordinated to the oxygen of the formyl and is very similar to the suggested intermediate in $CH_3FeCp(CO)_2$ reduction. Elimination of the aluminum and oxygen and transfer of the hydride to carbon allows formation of an iron alkylidene (B) which has precedence in Brookhart's work.[26,27] The CO insertion occurs to an iron-ketene complex (C) as Stevens and Beauchamp have previously observed.[28] This type of insertion has also been observed by Herrmann and Plank with conversion of diphenylcarbene and CO into a diphenylketene manganese complex.[29] Further action of LAH on the ketene complex, C, must lead to an ethylene complex (D) and also to a species which can undergo insertion to increase the chain length. We suggest that this species is the ethylidene (E). The ethylene complex could react with a nucleophile (CO, $PBu_3$) to eliminate ethylene (in agreement with the enhanced ethylene production in the presence of CO and $PBu_3$) or be further reduced to ethane by an iron hydride or by LAH.[30] The ethylidene could insert a new CO forming a ketene with three carbons and extend the chain. The importance of coordination of $AlH_3$ to the oxygen of the CO undergoing reduction is shown by the relatively poor yield of hydrocarbons upon reduction of $Cp_2Fe_2(CO)_4$ with $NaBH_4$ or $LiEt_3BH$. While each of these hydrides is sufficiently hydridic tc reduce a CO to a formyl, the successive reduction steps are not as efficient as with LAH.

The predominant formation of ethylene in the early stages of the reduction of $Cp_2Fe_2(CO)_4$ with LAH raises the possibility that ethylene is formed by an alternate mechanism. An especially attractive mechanism for ethylene formation in this system is shown in Scheme 4.

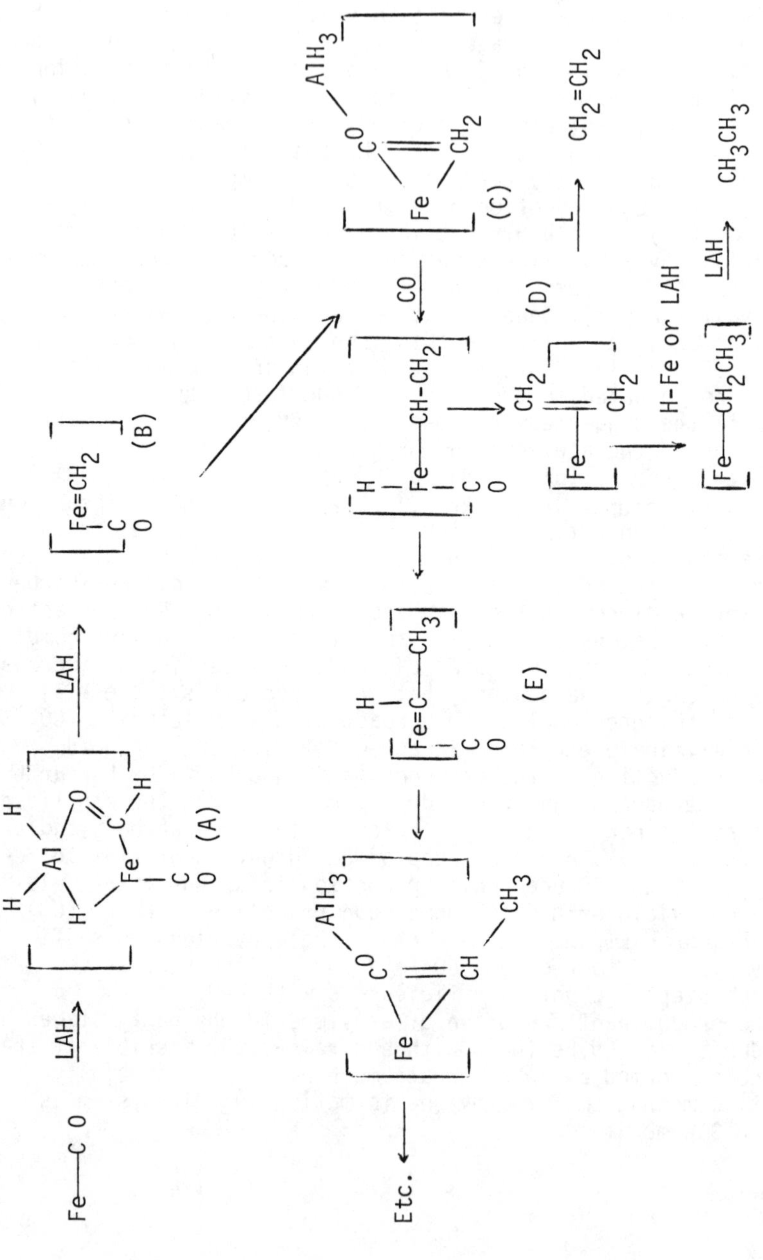


*Figure 1. Suggested scheme for hydrocarbon formation*

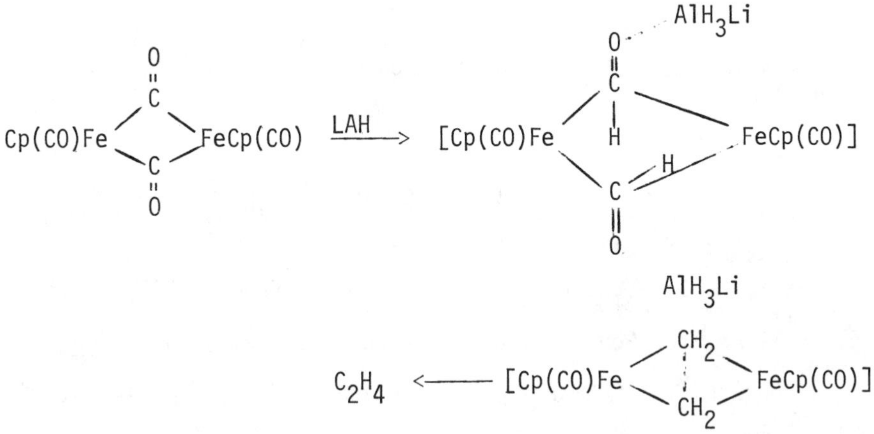


Scheme 4. Possible scheme for selective formation of ethylene.

Adducts of $AlEt_3$ with the bridging CO's on $Cp_2Fe_2(CO)_4$ are known.[31] The dramatic decrease in the amount of olefins during the reaction would arise by recoordination and further reaction.

The seemingly plausible Scheme shown in 4 is inconsistent with the results of the $^{13}CO$ labeling study as are most schemes which do not involve CO insertion for the chain propagation. We believe that ethylene arises from the same sequence of steps as the other hydrocarbon products. The role of the second metal center in the reduction cannot be described. We believe that the iron-iron bond is cleaved early in the reaction since the reduction in the presence of $PBu_3$ produced the unsubstituted species, $LiCpFe(CO)_2$. While there is too little information currently available to assess the importance of Scheme 3, our results on reduction in this iron system are not consistent with the normal CO insertion mechanism or with carbene oligomerization. We suggest Scheme 3 until further research can be accomplished.

In each of the homogeneous systems where C-C chains of three or more are built up the possibility of Lewis-acid assisted insertions exists.[23,24] In Muetterties' system the solvent $NaCl{\cdot}2AlCl_3$ provides $AlCl_3$ as a Lewis-acid which could assist insertions.[3] Schwartz's system uses $i\text{-}Bu_2AlH$ as the reductant and certainly has the potential for Lewis-acid coordination.[4-6] Olivé and Olivé performed alkylation of benzene with $H_2$ and CO with $W(CO)_6$ and $AlCl_3$ building hydrocarbon chains of up to $C_5$.[32] Scheme 3 suggests the importance of Lewis-acid coordination in our system. Whether or not this Lewis-acid coordination (specifically aluminum) is essential to build up the carbon chain will be an important question to resolve in homogeneous CO reductions.

We have presented a new system for formation of hydrocarbons from coordinated carbon monoxide. By preparing and reducing possible intermediates we have shown that an insertion step is important in the chain formation and suggest a scheme involving CO insertion into a carbene.

## Acknowledgement

We acknowledge the financial support of the National Science Foundation.

## Literature Cited

1. Masters, C. Adv. Organomet. Chem., 1979, 17, 61.
2. Muetterties, E. L.; Stein, J. Chem. Rev. 1979, 79, 479.
3. Demitras, G. C.; Muetterties, E. L. J. Am. Chem. Soc., 1977, 99, 2796.
4. Gell, K. I.; Schwartz, J. J. Organomet. Chem., 1978, 162, C11.
5. Gell, K. I.; Schwartz, J. J. Organomet. Chem., 1978, 100, 3246.
6. Shoer, L. I.: Schwartz, J. J. Organomet. Chem., 1977, 99, 5831.
7. van der Woude, C.; van Doorn, J. A.; Masters, C. J. Am. Chem. 1979, 101, 1633.
8. Wong, K. S.; Labinger, J. A. J. Am. Chem. Soc., 1980, 102, 3652.
9. Wong, A.; Harris, M.; Atwood, J. D. J. Am. Chem. Soc., 1980, 102, 4529.
10. Storch, H. H.; Golumbic, N.; Anderson, R. B. "The Fischer-Tropsch and Related Synthesis", John Wiley and Sons, Inc., 1951, and references therein.
11. Vannice, M. A. J. Catal., 1975, 37, 462.
12. Biloen, P.; Helle, J. N.; Sachtler, W. M. H. J. Catal. 1979, 102, 4541.
13. Tachikawa, M.; Muetterties, E. L. J. Am. Chem. Soc., 1980, 102, 4541.
14. Beno, M. A.; Williams, J. M.; Tachikawa, M. Muetterties, E. L.; J. Am. Chem. Soc., 1980, 102, 4542.
15. Whitmire, K.; Shriver, D. F. J. Am. Chem. Soc., 1980, 102, 1456.
16. Stewart, R. P.; Okamoto, N.; Graham, W. A. G. J. Organomet. Chem. 1972, 42, C32.
17. Sweet, J. R.; Graham, W. A. G. J. Organomet. Chem., 1979, 173, C9.
18. Reduction of transition metal acetyl complexes to ethyl complexes with $BH_3$ has been previously observed.[19]
19. van Doorn, J.A.; Masters, C.; Volger, H. C. J. Organomet. Chem., 1976, 105, 245.
20. Wong A.; Atwood, J. D. J. Organometal. Chem., in press.
21. Henrici-Olivé, G.; Olivé, S. Angew. Chem. Int. Ed. Eng. 1976, 15, 136.
22. Gladysz, J. A.; Selover, J. C. Tett. Lett, 1978, 319.
23. Butts, S. B.; Holt. E. M.; Strauss, S. H.; Alcock, N. W.; Stimson, R. E.; Shriver, D. F. J. Am. Chem. Soc., 1979, 101, 5864.

24a. Stimson, R. E.; Shriver, D. F. Inorg. Chem. 1980, 19, 1141.
b. Butts, S. B.; Strauss, S. H.; Holt, E. M.; Stimson, R. E.; Alcock, N. W.; Shriver, D. F. J. Am. Chem. Soc., 1980, 102, 5093.
25. The presence of a bridging CO is not a requirement for formation of hydrocarbons as treatment of $Fe(CO)_5$ with LAH gives a range of alkanes.
26. Brookhart, M.; Nelson, G. O. J. Am. Chem. Soc., 1977, 99, 6099.
27. Brookhart, M; Tucker, J. R.; Flood, T. C.; Jensen, J. J. Am. Chem. Soc., 1980, 102, 1203.
28. Stevens, A. E.; Beauchamp, J. L.; J. Am. Chem. Soc., 1978, 100, 2584.
29. Herrmann, W. A.; Plank, J. Angew. Chem. Int. Ed. Engl. 1978, 17, 525.
30. Bodnar, T.; laCroce, S. J.; Cutler, A. R. J. Am. Chem. Soc. 1980, 102, 3294.
31. Nelson, N. J.; Kime, N. E.; Shriver, D. F. J. Am. Chem. Soc. 1969, 91, 5173.
32. Henrici-Olivé, G.; Olivé, S. Angew. Chem. Int. Ed. Engl. 1979, 18, 77.

RECEIVED December 8, 1980.

19

# Aspects of Homogeneous Carbon Monoxide Fixation

## Selective Conversion of Carbonyl Ligands on $(\eta_5\text{-}C_5H_5)Fe(CO)_3^+$ to $C_2$ Organic Compounds

ALAN CUTLER, THOMAS BODNAR, GENE COMAN, STEPHEN LACROCE, CAROL LAMBERT, and KEVIN MENARD

Department of Chemistry, Wesleyan University, Middletown, CN 06457

Procurement of transition metal catalysts that convert carbon monoxide and dihydrogen to organic molecules represents an important research objective of modern inorganic chemistry. Impetus for this research derives from the use of coal, a source of $CO/H_2$ synthesis gas mixtures, as a future source of petrochemicals (1). Although heterogeneous catalysis in the Fischer-Tropsch synthesis (2,3) of complex hydrocarbon mixtures from synthesis gas has long been established, homogeneous catalysis of these reactions would have the added advantage of a potentially high and manipulative product selectivity. A selective homogeneous catalytic synthesis of $C_2$ organics — especially $C_2$-oxygenates such as ethylene glycol (4), acetaldehyde, and acetic acid — would provide a much desired coal-based source of organic feedstocks (1).

One approach favored by us for the rational design of such catalysts first requires an understanding of reaction pathways by which ligated CO undergoes hydrogenation and subsequent synthesis reactions (ie. chain growth of the $C_1$ ligand), before eliminating the desired organic molecule. Fully characterized organometallic complexes are used as model systems to collect this mechanistic information. Then more labile (and hopefully catalytic) systems are designed and tested. Limited mechanistic data is now available for stoichiometric conversion of CO ligands to the $C_2$ organic compounds ethane (5,6,7), ethylene (6,8), acetaldehyde (9), methylacetate (10), and a coordinated enediol of glycolaldehyde (11). Clearly a need exists for more extensive information on CO fixation and synthesis reaction pathways.

We have delineated viable coordinated ligand reactions and their attendant intermediates for the stoichiometric conversion of CO ligands selectively to the $C_2$ organics ethane, ethylene, methyl (or ethyl) acetate, and acetaldehyde. We now outline results from three lines of research: (1) $\eta^1$-Alkoxymethyl iron complexes $CpFe(CO)_2CH_2OR$ (**2**) are available by reducing coordinated CO on $CpFe(CO)_3^+$ (**1**) [$Cp = \eta^5\text{-}C_5H_5$]. Compounds **2** then form $\eta^1$-alkoxyacetyl complexes via migratory-insertion (i.e. CO

0097-6156/81/0152-0279$07.00/0
© 1981 American Chemical Society

insertion) reactions. (2) Iron $\eta^1$-ethyl, $\eta^2$-ethylene, $\eta^1$-carboalkoxymethyl, $\eta^1$-formylmethyl, or $\eta^2$-vinyl ether complexes are available by selectively carrying out coordinated ligand reactions (Scheme 1) on the $\eta^1$-alkoxyacetyl compounds. The free $C_2$ organic molecules mentioned above are then liberated. (3) Transition organometallic Lewis acids and hydride complexes are available for replacing the carbocation and borohydride reagents, respectively, used in activating and reducing organometallic acyl complexes. We can now demonstrate reaction pathways for transforming two carbonyls on a discrete organometallic complex selectively to $C_2$ organic molecules.

The choice of CpFe carbonyl complexes for studying "Fischer-Tropsch chemistry" warrants comment. Iron $\eta^1$-alkyl complexes Cp(CO)LFe-R [L = CO, $PPh_3$, $P(OMe)_3$], representing 18-electron and coordinatively saturated Fe(II) compounds, are endowed with both kinetic inertness towards ligand dissociation and thermodynamic stability under ambient conditions (12). Their synthesis poses no problems. General preparative procedures for $CpFe(CO)_2R$ include metalation of organic halides with $CpFe(CO)_2^{-1}$ (13) and modification of other coordinated ligands, usually $\eta^2$-olefin cations (14-18). The corresponding $\eta^1$-alkyl and cationic $\eta^2$-olefin complexes employing the phosphine [L = $PPh_3$] (19,20,21,22) and phosphite [L = $P(OPh)_3$, $P(OMe)_3$] (23,24) substituted systems are also conveniently available from the appropriate $CpFe(CO)_2R$ precursors. Moreover, the usefulness of CpFe(CO)L(R) comes from the Fe-C σ-bond serving as a good representation for transition metal-C σ-bonds, at least for those systems where the highest occupied molecular orbital (HOMO) resides on the metal (25,26). Reactions of these complexes are prototypal of organometallic σ-alkyl complexes.

A large body of mechanistic information concerning the reaction chemistry of the Fe-C bond in CpFe(CO)L(R) complexes also exists. Reaction mechanisms involving the coordinated alkyl ligand in conversion to $\eta^1$-acyl complexes by alkyl migration (27,28), generation of $\eta^2$-olefin salts (15,17,18,29), formation of cationic alkoxycarbene (30) and alkylidene (31) compounds, addition of electrophiles to the ligand (14, 17,18), and the electrophilic cleavage of the Fe-C bond (21,22,26,32,33,34) have been scrutinized. Absence of metal-oxygen bond formation as a driving force distinguishes this reaction chemistry from that of earlier transition metals (35).

## Reduction of Coordinated CO on $CpFe(CO)_3^+$; Conversion of Two CO Ligands to an Alkoxyacetyl Ligand

An equimolar quantity of sodium cyanoborohydride ($Na^+BH_3CN^-$) in methanol reduces $CpFe(CO)_3^+BF_4^-$ (1) over one hour to the known (36,37) methoxymethyl complex $CpFe(CO)_2(CH_2OMe)$ (2). After solvent and varying amounts of $CpFe(CO)_2H$ were removed under vacuum from the reaction mixture, 2 (45%) was separated

Scheme 1. Stoichiometric CO Fixation and Synthesis Reactions of $CpFe(CO)_3^+$

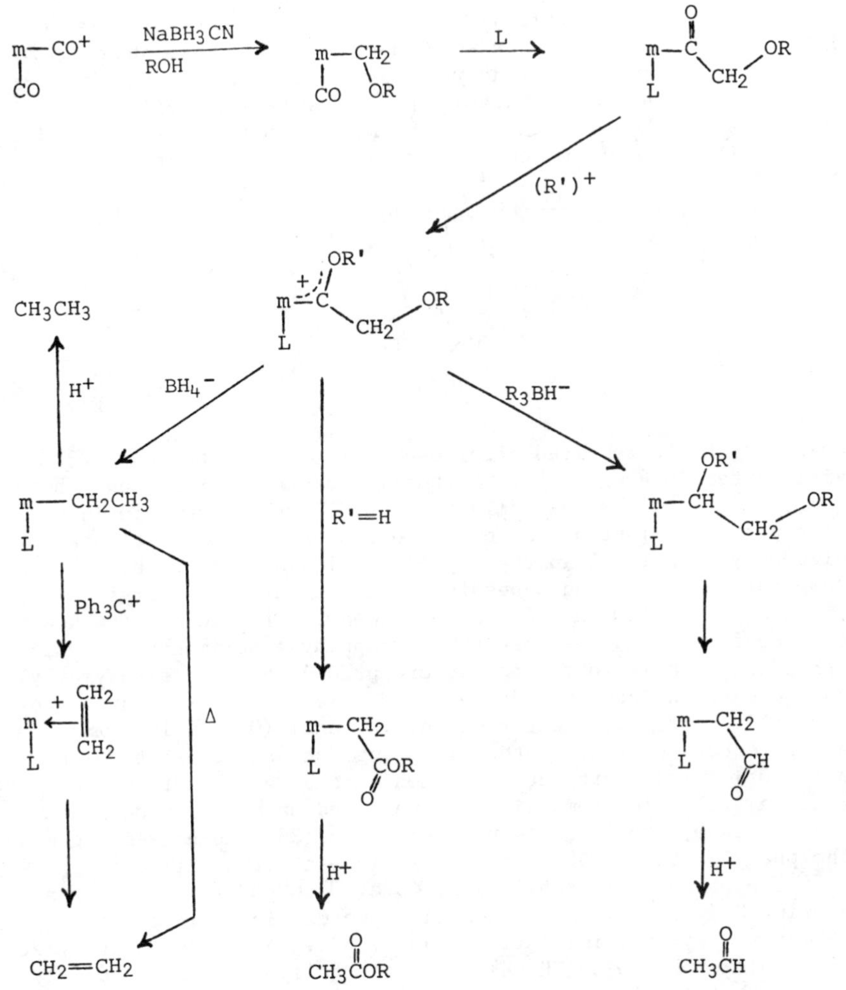


m = Cp(CO)Fe

L = CO, $PPh_3$, $P(OMe)_3$

R = Me, Et; R' = H, Me, Et

from the dimeric byproduct $[CpFe(CO)_2]_2$ (3) by column chromatography (activity-3 neutral alumina, 3 : 1 pentane-ether). Both

$$CpFe(CO)_3^+ \ + \ NaBH_3CN \ \xrightarrow{MeOH} \ CpFe(CO)_2CH_2OMe$$
$$\underset{\underset{\sim}{1}}{} \qquad\qquad\qquad\qquad\qquad\qquad \underset{\underset{\sim}{2}}{}$$

the methyl complex $CpFe(CO)_2CH_3$, and the product of hydride addition to the Cp ligand, $(\eta^4-C_5H_6)Fe(CO)_3$ (46), were absent.

Sodium cyanoborohydride reaction with $\underset{\sim}{1}$ entails the intermediacy of a thermally unstable $\eta^1$-hydroxymethyl complex $CpFe(CO)_2(CH_2OH)$ (4). When the reduction sequence was carried out in ethanol at room temperature, only trace amounts of

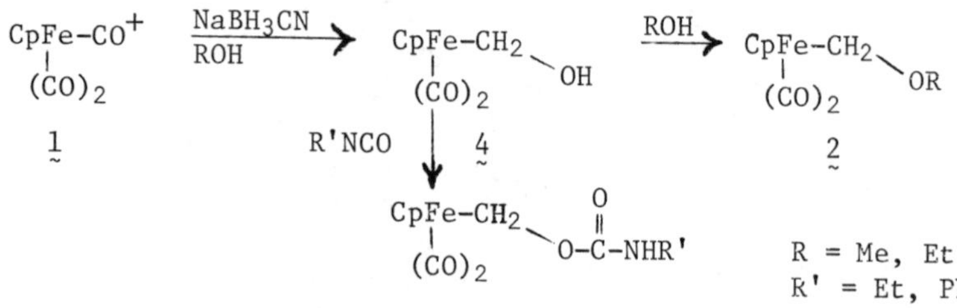

$$\underset{\substack{\text{(CO)}_2 \\ \underset{\sim}{1}}}{CpFe-CO^+} \ \xrightarrow[ROH]{NaBH_3CN} \ \underset{\text{(CO)}_2}{CpFe-CH_2}\diagdown_{OH} \ \xrightarrow{ROH} \ \underset{\substack{\text{(CO)}_2 \\ \underset{\sim}{2}}}{CpFe-CH_2}\diagdown_{OR}$$

$$R'NCO \Bigg\downarrow \ 4$$

$$\underset{\text{(CO)}_2}{CpFe-CH_2}\diagdown_{O-\overset{\overset{\textstyle O}{\|}}{C}-NHR'} \qquad\qquad R = Me, Et$$
$$\qquad\qquad\qquad\qquad\qquad\qquad\qquad\qquad\qquad R' = Et, Ph$$

$CpFe(CO)_2CH_2OEt$ resulted after one hour of reaction and a 23% yield after 10 hours. Both methanol and ethanol reactions, however, dissolved the initially insoluble $\underset{\sim}{1}$ as yellow solutions within thirty minutes. Attempts to isolate or to detect unambiguously putative $\underset{\sim}{4}$ in these yellow solutions, at or below room temperature, proved unsuccessful. Only dimer $\underset{\sim}{3}$ and varying amounts of $CpFe(CO)_2H$ and $\underset{\sim}{2}$ were evident. Transience of $\underset{\sim}{4}$ however conforms with the results of trapping experiments, in which ethyl and phenyl isocyanates intercepted $\underset{\sim}{4}$ (but not alkoxymethyl complexes $\underset{\sim}{2}$) as urethane derivatives. We prepared the urethane derivatives of $\underset{\sim}{4}$ by first removing methanol (0° and $10^{-1}$ mm) after reduction of $\underset{\sim}{1}$ and then reacting excess isocyanate with cold (0°) toluene extracts. Column chromatography of the reaction residue, following removal of solvent and unchanged isocyanate under vacuum, provided the ethylurethane (28%) as a yellow gum and the phenylurethane (38%) as a yellowish-tan crystalline solid.

The observed instability of $CpFe(CO)_2CH_2OH$ (4) augments a growing body of evidence concerning thermal instability of α-hydroxyalkyl complexes (38,39,40,41,42). A similar hydroxymethyl complex $CpRe(CO)NO(CH_2OH)$, the only fully characterized α-hydroxymethyl complex to date (38,39), likewise converts to its methoxymethyl complex in methanol.

The selection of an alcoholic $NaBH_3CN$ reducing medium for CO fixation on $\underset{\sim}{1}$ is critical. Sodium cyanoborohydride serves as an excellent reducing agent for Lewis acids (43), as well as reacting as a milder and more selective reducing agent than $BH_4^-$ or $Et_3BH^-$ towards coordinated ligands (44). Previous workers demonstrated that $NaBH_4$ in tetrahydrofuran (THF) transfers hydride to the iron center on 1 (giving $CpFe(CO)_2H$ and ultimately 3) (45), but $NaBH_3CN$ in THF transfers hydride to the Cp ring (producing $(\eta^4\text{-}C_5H_6)Fe(CO)_3$) (46). After confirming these results in THF, we then established that $NaBH_4$-methanol (0°) and $Ph_3PMe^+BH_4^-$-$CH_2Cl_2$ also transform 1 into $CpFe(CO)_2H$, whereas $Ph_3PMe^+BH_3CN^-$-$CH_2Cl_2$ gives $(\eta^4\text{-}C_5H_6)Fe(CO)_3$. We find it revealing that an organometallic carbonyl complex $CpFe(CO)_3^+$ (1) possessing an established resistance towards reduction of a CO ligand does finally undergo facile CO fixation with the appropriate reducing medium.

Examples of isolable complexes resulting from hydride reduction of coordinated CO on cationic organometallic complexes are rare, since most cationic $L_xM(CO)_y$ systems suffer either reduction at the metal or hydride addition to a coordinated ligand (L) other than CO (47). Indeed, the facile reduction of a CO ligand on $CpRe(CO)_2NO^+$ (38,39,48) or on $CpMo(CO)_3PPh_3^+$ (48,49) with borohydride reagents (giving for example isolable $\eta^1$-formyl, hydroxymethyl, and methyl complexes of CpRe(CO)NO) appear anomolous. Perhaps this paucity of examples for stoichiometric CO fixation reflects in part the limited number of reducing media investigated. We are accordingly extending our studies with 1 and alcoholic $NaBH_3CN$ to see if the results will be prototypal for other $L_xM(CO)_y^+$ systems.

We utilized the alkoxyacetyl complexes $CpFe(CO)L(COCH_2OR)$ [L=CO (5), $PPh_3$ (6), $P(OMe)_3$ (7); R=Me (a), Et (b)], which are derived from $CpFe(CO)_2CH_2OR$ (2) and hence 1, as our common intermediate in the selective synthesis of other $C_2$ ligands and organic products. Note that both carbon atoms of the alkoxyacetyl and successive ligands (Scheme 1) derive from CO: the inserted carbonyl of $CpFe(CO)L(COCH_2OR)$ (5-7) comes from alkyl-acyl migratory-insertion on 2. Subsequent synthetic reactions on 5-7 then provide the desired $C_2$ organics.

Alkoxyacetyl compounds 6 and 7a,b were procured by refluxing acetonitrile solutions of $CpFe(CO)_2CH_2OR$ (2) and excess phosphine or phosphite [L=$PPh_3$, $P(OMe)_3$] for 4 and 10 days respectively.

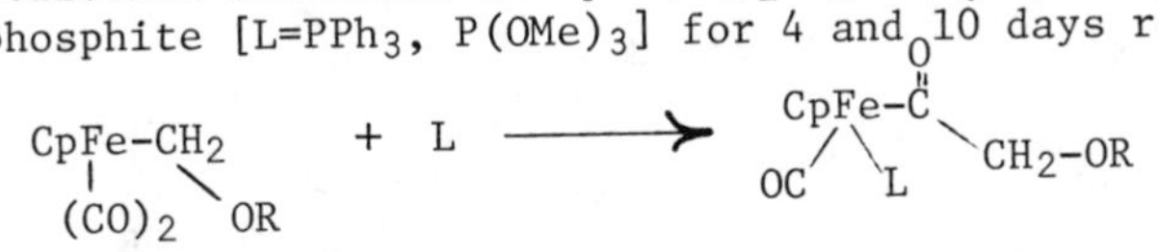

2

5, L=CO  a, R=Me
6, L=$PPh_3$  b, R=Et
7, L=$P(OMe)_3$

Products 6 and 7a,b, air-stable yellow solids, were obtained in 30-50% yields after recrystallization from $CH_2Cl_2$-heptane. The relatively vigorous reaction conditions required for generating 6,7 (vs. less than one day refluxing for greater than 80% yields of $CpFe(CO)L(COCH_3)$ from $CpFe(CO)_2CH_3$) further demonstrates the reluctance of alkoxymethyl ligands to undergo alkyl-acyl migratory-insertion (50). $CpFe(CO)_2(COCH_2OR)$ (5) is in principal available from $CpFe(CO)_2(CH_2OR)$ (2) by either high pressure carbonylation (51) or by Lewis acid promoted alkyl-acyl migratory insertion under lower CO pressures (52). We however secured the previously prepared $CpFe(CO)_2(COCH_2OMe)$ (5a) by metalation of methoxyacetyl chloride (53). An analogous Mn complex $(CO)_5Mn-COCH_2OMe$ has recently been prepared by the same procedure (50).

Other examples of alkoxyacetyl compounds or their derivatives occur in homogeneous catalysis. Compounds bearing hydroxyacetyl ligands thus have been formulated as catalysts in ethylene glycol syntheses from synthesis gas (4,50) and in hydroformylation of formaldehyde to glycolaldehyde (54). Although only one β-hydroxyacyl complex $(CO)_5MnCOCH(OH)Ph$ has been isolated (41), an analogous acyloxyacetyl complex $(CO)_5MnCOCH_2OCOCMe_3$ serves as an intermediate during hydrogenation of $(CO)_5MnCH_2OCOCMe_3$ to an ethyleneglycol monoester (55). The only $C_2$ organic molecules previously derived from alkoxy - or hydroxyacetyl organometallic complexes, until now, are ethylene glycol (free and as an ester) and its glycolaldehyde precursor.

## Conversion of Alkoxyacetyl Complexes $CpFe(CO)PPh_3(COCH_2OR)$ to Ethylene or to Ethane

The alkoxyacetyl complexes $CpFe(CO)PPh_3(COCH_2OR)$ (6a,b) were transformed to the known (56,57) ethyl complex $CpFe(CO)PPh_3(CH_2CH_3)$ (8), and then to free ethane or ethylene via the reactions depicted in Scheme 2. Previous workers established that ethane results from protonation of 8 (21), but ethylene eliminates from 8 above 60° (56,57). Also, hydride abstraction from 8 with $Ph_3C^+PF_6^-$ affords $CpFe(CO)PPh_3(\eta^2-CH_2{=}CH_2)^+$ (58). This ethylene complex then reacts slowly with $PPh_3Me^+I^-$ in $CH_2Cl_2$ to deliver $CpFe(CO)PPh_3(I)$ and ethylene gas. The first step in Scheme 2, however, entails activation of the alkoxyacetyl ligand on 6a,b before a $BH_4^-$ reagent can reduce it to the ethyl complex 8.

Scheme 2

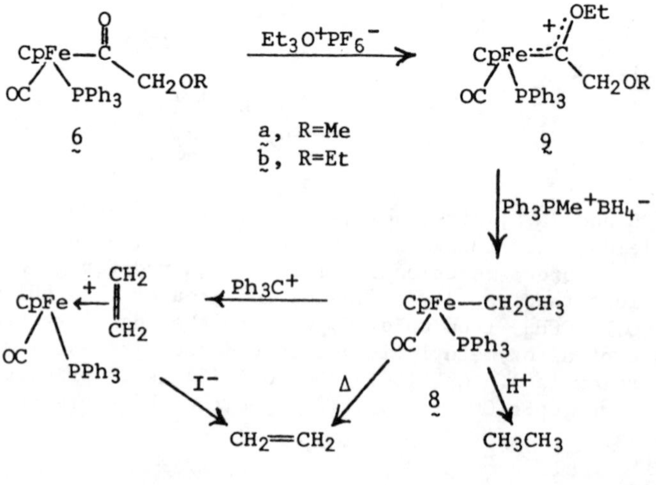

Until properly activated, transition metal acyl complexes generally do not add nucleophiles (59), including hydride from nucleophilic hydride-donating reagents (60,61,62), to the acyl functionality. We established, for example, that both acetyl complexes $CpFe(CO)L(COCH_3)$ ($L=CO,PPh_3$) remain unchanged by $PPh_3Me^+BH_4^-$ in $CH_2Cl_2$ at room temperature. But when the acetyl ligand is alkylated, giving a cationic α-alkoxyethylidene compound, the α-carbon is rendered sufficiently electrophilic (i.e., activated) for hydride addition. Davison and Reger, accordingly, demonstrated that $NaBH_4$ in ethanol reduces $CpFe(CO)PPh_3(C(OEt)CH_3)^+$ (10) to the ethyl complex 8 plus the α-ethoxyethyl compound $CpFe(CO)PPh_3(CH(OEt)CH_3)$ (63). By using $PPh_3Me^+BH_4^-$ in $CH_2Cl_2$, we obtained the desired selectivity; the same α-alkoxyethylidene compound exclusively gave 8 in 78% yield.

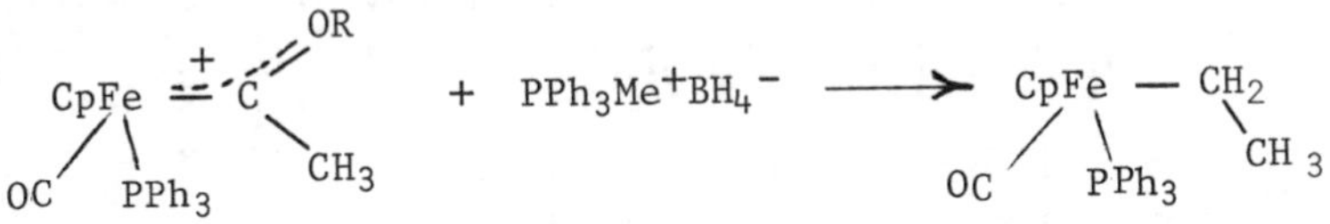

10 R=Me,Et 8

Absence of the α-alkoxyethyl compound under these conditions depends on the susceptibility of α-alkoxyalkyl ligands to further reduction with the $BH_3$ byproduct (30).

We converted the alkoxyacetyl complexes 9 to the ethyl complex 8, by employing the aforementioned acyl ligand activation and $BH_4^-$ reduction procedures, as outlined in Scheme 2. The α,β-dialkoxyethylidene salts 9a,b resulted from alkylation of 6a,b with $Et_3O^+PF_6^-$ in $CH_2Cl_2$; recrystallization from $CH_2Cl_2$-ether provided 9 in 72% yields as air-stable yellow $PF_6^-$ salts. NMR and IR spectra (Table 1), as well as the results of reactions with iodide (reversion to 6), are in accord with α-alkoxyethylidene structures (30,63). We found no evidence for isomerization at ambient conditions to the ethylated carboalkoxymethyl complexes $CpFe(CO)PPh_3(CH_2C(OEt)(OR)^+)$, vide infra. Borohydride reduction of 9a,b using an equimolar quantity of $PPh_3Me^+BH_4^-$ in $CH_2Cl_2$, followed by recrystallization from $CH_2Cl_2$-heptane, affords $CpFe(CO)PPh_3(CH_2CH_3)$ (8) in 69% yield. Absence of β-alkoxyethyl complexes $CpFe(CO)PPh_3(CH_2CH_2OR)$ after reduction of 9 follows from the susceptibility of the β-alkoxyethyl ligand towards Lewis acid (e.g. $BH_3$) induced reduction to the ethyl ligand (16,64). Overall, we can now account for the selective conversion of two carbonyls on $CpFe(CO)_3^+$ (1) to the ethyl ligand on $CpFe(CO)PPh_3(CH_2CH_3)$ (8), and then to the $C_2$-hydrocarbons ethane or ethylene.

## Conversion of Alkoxyacetyl Complexes to Alkyl Acetates

In order to produce free alkyl acetates, the alkoxyacetyl complexes 5-7 must first isomerize to carboalkoxymethyl compounds

Table 1 Spectral Data

| Compound | IR($CH_2Cl_2$) $\nu_{c\equiv o}$ ($cm^{-1}$) | Solvent | NMR $\delta^{TMS}$ |
|---|---|---|---|
| $CpFe(CO)_2CH_2OMe$ (2a) | 2005, 1943 | ($CDCl_3$) | 4.83 (s, 2, $CH_2$)<br>4.77 (s, 5, Cp)<br>3.21 (s, 3, $OCH_3$) |
| $CpFe(CO)_2CH_2OEt$ (2b) | 2000, 1940 | ($CDCl_3$) | 4.84 (s, 2, Fe-$CH_2$)<br>4.73 (s, 5, Cp)<br>3.34 (q, J=7.0 Hz, 2, $OCH_2$)<br>1.14 (t, J=7.0 Hz, 3, $CH_2\underline{CH_3}$) |
| $CpFe(CO)_2CH_2OCONHEt$ | 2016, 1955<br>$\nu_{c=o}$ 1693 | ($CDCl_3$) | 5.24 (s, 2, Fe-$CH_2$)<br>4.81 (s, 5, Cp)<br>4.57 (m, 1, NH)<br>3.15 (m, 2, $N\underline{CH_2}CH_3$)<br>1.10 (t, J=7.0 Hz, 3, $NCH_2\underline{CH_3}$) |
| $CpFe(CO)_2CH_2OCONHPh$ | 2022, 1961<br>$\nu_{c=o}$ 1722 | ($CDCl_3$) | 7.4 (m, 5, Ph)<br>5.43 (s, 2, Fe-$CH_2$)<br>4.84 (s, 5, Cp) |
| $CpFe(CO)PPh_3(COCH_3)$ (27) | 1909<br>$\nu_{c=o}$ 1599 | (acetone-$d_6$) | 7.38 (s, 15, $PPh_3$)<br>4.42 (d, $J_{P-H}$=1.5 Hz, 5, Cp)<br>2.26 (s, 3, $CH_3$) |
| $CpFe(CO)PPh_3(COCH_2OCH_3)$ (6a) | 1916<br>$\nu_{c=o}$ 1614 | ($CDCl_3$) | 7.40 (br s, 15, $PPh_3$)<br>4.44 (d, $J_{P-H}$=1.5 Hz, 5, Cp)<br>4.20, 3.55 (AB mult, J=17 Hz, 2, $COCH_2$)<br>2.91 (s, 3, $CH_3$) |

Table 1 (continued)

| Compound | IR($CH_2Cl_2$) | Solvent | NMR |
| --- | --- | --- | --- |
| $CpFe(CO)PPh_3(COCH_2OEt)$ (6b) | 1917<br>$\nu_{c=o}$ 1615 | ($CDCl_3$) | 7.40 (br s, 15, $PPh_3$)<br>4.45 (d, $J_{P-H}$=1.5 Hz, 5, Cp<br>4.24, 3.60 (AB mult, J=17 Hz, 2, $COCH_2$)<br>2.90 (m, 2, $O\underline{CH_2}CH_3$)<br>0.96 (t, J=7 Hz, 3, $OCH_2\underline{CH_3}$) |
| $CpFe(CO)P(OMe)_3(COCH_2OCH_3)$ (7a) | 1935<br>$\nu_{c=o}$ 1618 | ($CDCl_3$) | 4.61 (s, 5, Cp)<br>4.07 (s, 2, $COCH_2$)<br>3.67 (d, $J_{P-H}$=11 Hz, 9, POMe)<br>3.29 (s, 3, $OCH_3$) |
| $CpFe(CO)P(OMe)_3(COCH_2OEt)$ (7b) | 1933<br>$\nu_{c=o}$ 1619 | ($CDCl_3$) | 4.62 (s, 5, Cp)<br>4.13 (s, 2, $COCH_2$)<br>3.64 (d, $J_{P-H}$=11 Hz, 9, POMe)<br>3.5 (m, 2, $O\underline{CH_2}CH_3$)<br>1.14 (t, J=7 Hz, 3, $OCH_2\underline{CH_3}$) |
| $CpFe(CO)PPh_3[C(OCH_3)CH_2OCH_3]^+$ $PF_6^-$ | 1984 | (acetone-$d_6$) | 7.55 (br s, 15, $PPh_3$)<br>5.06 (s, 5, Cp)<br>4.17 (s, 3, FeC-$OCH_3$)<br>3.91 (br s, 2, $CH_2$)<br>3.48 (s, 3, $FeCCH_2O\underline{CH_3}$) |

Table 1 (continued)

| | IR($CH_2Cl_2$) | NMR |
|---|---|---|
| $CpFe(CO)PPh_3[C(OEt)CH_2OCH_3]^+$ $PF_6^-$ (9a) | 1985 | (acetone-$d_6$) 7.55 (br s, 15, $PPh_3$) <br> 5.03 (s, 5, Cp) <br> 4.37 (m, 2, $FeCOCH_2CH_3$) <br> 3.96 (br s, 2, $\underline{CH_2}OCH_3$) <br> 3.47 (s, 3, $OCH_3$) <br> 1.43 (t, J=7 Hz, 3, $FeCOCH_2\underline{CH_3}$) |
| $CpFe(CO)PPh_3[C(OEt)CH_2OEt]^+$ $PF_6^-$ (9b) | 1985 | (acetone-$d_6$) 7.57 (br s, 15, $PPh_3$) <br> 5.04 (s, 5, Cp) <br> 4.36 (q, J=7 Hz, 2, $FeCO\underline{CH_2}CH_3$) <br> 4.06 (br s, 2, $\underline{CH_2}OEt$) <br> 3.68 (q, J=7 Hz, 2, $CH_2-O\underline{CH_2}CH_3$) <br> 1.40 (t, J=7 Hz, 3, $FeCOCH_2\underline{CH_3}$) <br> 1.32 (t, J=7 Hz, 3, $CH_2OCH_2\underline{CH_3}$) |
| $CpFe(CO)_2[C(OEt)CH_3]^+$ $PF_6^-$ (22b) | 2070, 2022 | ($CF_3CO_2H$) 5.32 (s, 5, Cp) <br> 4.82 (q, J=7 Hz, 2, $O\underline{CH_2}CH_3$) <br> 3.11 (s, 3, $CH_3$) <br> 1.66 (t, J=7 Hz, 3, $OCH_2\underline{CH_3}$) |
| $CpFe(CO)_2[C(OEt)CH_2OMe]^+$ $PF_6^-$ (14a) | 2070, 2030 | ($CD_3NO_2$) 5.40 (s, 5, Cp) <br> 4.81 (q, J=7 Hz, 2, $O\underline{CH_2}CH_3$) <br> 4.31 (s, 2, $FeC\underline{CH_2}$ + solvent) <br> 3.55 (s, 3, OMe) <br> 1.64 (t, J=7 Hz, 3, $OCH_2\underline{CH_3}$) |
| $CpFe(CO)_2\{CO[Fe(CO)_2Cp]CH_3\}^+$ $PF_6^-$ (28) | 2065, 2043 <br> 2020, 1986 | (acetone-$d_6$) 5.61 (s, 5, CpFeO) <br> 5.19 (s, 5, CpFeC) <br> 2.73 (s, 3, $CH_3$) |

Table 1 (continued)

| | IR($CH_2Cl_2$) $\nu_{c\equiv o}$($cm^{-1}$) | NMR $\delta^{TMS}$ | |
|---|---|---|---|
| $CpFe(CO)_2\{CO[Mo(CO)_3Cp]CH_3\}^+$ $PF_6^-$ (29) | 2068, 2045<br>1990 (br) | ($CD_3NO_2$) | 6.21 (s, 5, CpMo)<br>5.17 (s, 5, CpFe)<br>2.69 (s, 3, $CH_3$) |
| $CpFe(CO)(PPh_3)\{CO[Fe(CO)_2Cp]CH_3\}^+$ $PF_6^-$ (28) | 2065, 2008<br>1941 | (acedone-$d_6$) | 7.55 (m, 15, $PPh_3$)<br>5.36 (s, 5, CpFeO)<br>4.65 (d, $J_{P-H}$=1.5, 5, CpFeC)<br>2.68 (s, 3, $CH_3$) |
| $CpFe(CO)(PPh_3)\{CO[Mo(CO)_3Cp]CH_3\}^+$ $PF_6^-$ (29) | 2058, 1973 (br) | (acetone-$d_6$) | 7.55 (m, 15, $PPh_3$)<br>6.11 (s, 5, CpMo)<br>4.76 (s, 5, CpFe)<br>2.42 (s, 3, $CH_3$) |
| $CpFe(CO)_2\{CO[Mo(CO)_3Cp]CH_2OMe\}^+$ $PF_6^-$ | 2065, 2048<br>1995 (br) | (acetone-$d_6$) | 6.10 (s, 5, CpMo)<br>5.37 (s, 5, CpFe)<br>4.95 (s, 2, $CH_2$)<br>3.73 (s, 3, $OCH_3$) |
| $CpFe(CO)_2[OC(CH_3)_2]^+$ $PF_6^-$ | 2075, 2033 | (acetone-$d_6$)<br><br>($CD_3NO_2$) | 5.56 (s, 5, Cp)<br>2.34 (s, 6, $CH_3$)<br>5.43 (s, 5, Cp)<br>2.32 (s, 6, $CH_3$) |
| $CpFe(CO)_2CH_2COOEt$ (11b) | 2022, 1970<br>$\nu_{c=o}$ 1678 | ($CDCl_3$) | 4.84 (s, 5, Cp)<br>4.03 (q, J=7 Hz, 2, O$\underline{CH_2}CH_3$)<br>1.50 (s, 2, Fe-$\underline{CH_2}$)<br>1.24 (t, J=7 Hz, 3, $OCH_2\underline{CH_3}$) |

Table 1 (continued)

| | IR($CH_2Cl_2$) | NMR | |
|---|---|---|---|
| $CpFe(CO)(PPh_3)CH_2COOMe$<br>(12a) | 1928<br>$\nu_{c=o}$ 1666 | ($CDCl_3$) | 7.30 (br s, 15, $PPh_3$)<br>4.28 (s, 5, Cp)<br>3.50 (s, 3, $OCH_3$)<br>1.50 (m, 1, Fe-$\underline{CH_2}$)<br>0.77 (m, 1, Fe-$\underline{CH_2}$) |
| $CpFe(CO)P(OMe)_3CH_2COOMe$<br>(13a) | 1947<br>$\nu_{c=o}$ 1669 | ($CDCl_3$) | 4.53 (s, 5, Cp)<br>3.57 (d, J=11 Hz, 9, POMe)<br>3.53 (s, 3, $OCH_3$)<br>1.45 (m, 1, Fe-$\underline{CH_2}$)<br>0.88 (m, 1, Fe-$\underline{CH_2}$) |
| $CpFe(CO)(PPh_3)CH_2CHO$<br>(21) | 1927<br>$\nu_{c=o}$ 1634 | ($CDCl_3$) | 9.15 (m, 1, CHO)<br>7.33 (br s, 15, $PPH_3$)<br>4.28 (s, 5, Cp)<br>1.74 (m, 1, Fe-$\underline{CH_2}$)<br>1.03 (m, 1, Fe-$\underline{CH_2}$) |
| $CpFe(CO)_2CH(OEt)CH_3$<br>(23b) | 1998, 1938 | ($CS_2$) | 4.83 (q, J=6 Hz, 1, FeCH)<br>4.60 (s, 5, Cp)<br>3.25 (q, J=7 Hz, 2, $O\underline{CH_2}CH_3$)<br>1.62 (d, J=6 Hz, 3, $FeCH\underline{CH_3}$)<br>1.08 (t, J=7 Hz, 3, $OCH_2\underline{CH_3}$) |
| $CpFe(CO)_2CH(OEt)CH_2OMe$<br>(17) | 2003, 1944 | ($CDCl_3$) | 4.9 (m, 1, Fe-CH)<br>4.77 (s, 5, Cp)<br>4.0-3.2 (m, 4, $CH(O\underline{CH_2}CH_3)\underline{CH_2}OMe$)<br>3.33 (s, 3, $OCH_3$)<br>1.8-0.73 (m, 3, $OCH_2\underline{CH_3}$) |

CpFe(CO)L($CH_2CO_2R$) [L=CO(11), $PPh_3$(12), $P(OMe)_3$(13); R=Me(a), Et(b)]. Acid promotes this isomerization; and subsequent protolytic cleavage, as outlined in Scheme 3, generates free methyl or ethyl acetate.

Scheme 3

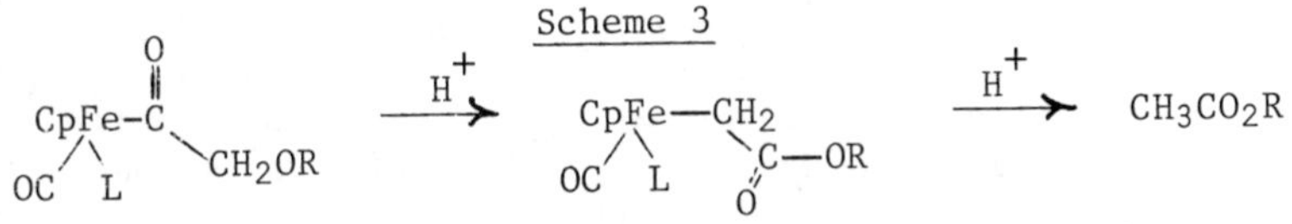

5, L=CO
6, L=$PPh_3$
7, L=$P(OMe)_3$

11, L=CO
12, L=$PPh_3$
13, L=$P(OMe)_3$

a, R=Me
b, R=Et

Reactions of acid with the alkoxyacetyl complexes 5-7 can be run so as to give either their corresponding carboalkoxymethyl compounds 11-13 or, starting with 6 or 7a,b only, free methyl and ethyl acetates. The carboalkoxymethyl compounds 11-13 were intercepted and isolated in 60-77% yields after treatment of 5-7 with one equivalent of trifluoromethanesulfonic acid in $CH_2Cl_2$, neutralization with triethylamine, and column chromatography. Previous preparations of these carboalkoxymethyl compounds include metalation of methyl or ethyl chloroacetate (53,65), giving 11a,b, and photolytic CO replacement on 11a by $PPh_3$(20), giving 12a. The direct conversion of the carboalkoxy compounds 12 and 13 to methyl or ethyl acetate, however, was driven by using excess acid over 24 hours. NMR and IR monitoring identified the alkyl acetate; and quantitative IR analysis, using the acetate carbonyl absorption at 1735 $cm^{-1}$, established at least 55% conversion. The same results obtain whether carboalkoxy compounds 12,13 or alkoxyacetyl complexes 6,7 are treated with excess acid. Alkyl acetate liberated in these experiments, of course, derives from two CO ligands on $CpFe(CO)_3^+$ (1).

Intermediates occurring in the acid-promoted isomerization of an alkoxyacetyl complex, $CpFe(CO)_2COCH_2OMe$ (5a) to its carboalkoxymethyl compound $CpFe(CO)_2CH_2CO_2Me$ (11a) have been observed by IR. Upon addition of acid to 5a, its IR terminal $\nu_{c\equiv o}$ (2025, 1965 $cm^{-1}$) and acyl $\nu_{c=o}$ (1658 $cm^{-1}$) stretching frequencies

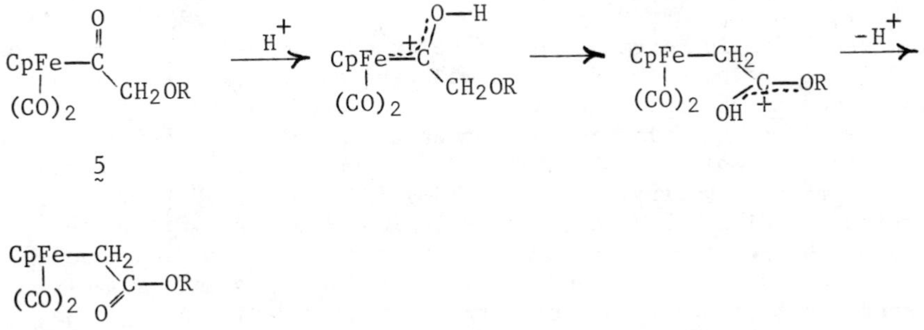

5

11

converted to $\nu_{C\equiv O}$ (2071, 2035 $cm^{-1}$). These latter absorptions are consistent with $CpFe(CO)_2C(OH)CH_2OMe^+$, cf. the data for $CpFe(CO)_2C(OEt)CH_2OMe^+$ (14a) in Table 1. Over a one hour period the IR absorptions of this yellow-brown solution transformed to $\nu_{C\equiv O}$ (2038, 1993 $cm^{-1}$), which correspond to the previously observed $CpFe(CO)_2CH_2C(OH)OMe^+$ during protonation of 11a (53). Deprotonation then left the carbomethoxymethyl compound 11a, $\nu_{C\equiv O}$(2027, 1968 $cm^{-1}$), $\nu_{C=O}$ (1695 $cm^{-1}$). Since the mechanistic details of the overall rearrangement step (α-hydroxy-β-alkoxyethylidene complex to protonated carboalkoxymethyl compound) are unknown, solution stability of the analogous α,β-dialkoxyethylidene complexes is of obvious interest.

The α,β-dialkoxyethylidene complexes $CpFe(CO)L[C(OEt)CH_2OR]^+$ [L=CO(14), $PPh_3$(9), $P(OMe)_3$(15); R=Me(a), Et(b)] were synthesized by alkylation of the alkoxyacetyl complexes 5-7 with $Et_3O^+PF_6^-$ in $CH_2Cl_2$. Reaction conditions are critical; otherwise the salts $CpFe(CO)_2L^+$ [L=$PPh_3$, $P(OMe)_3$] accumulate during alkylation of 6 and 7. Nevertheless, facile alkylation of 6 occurred, and the

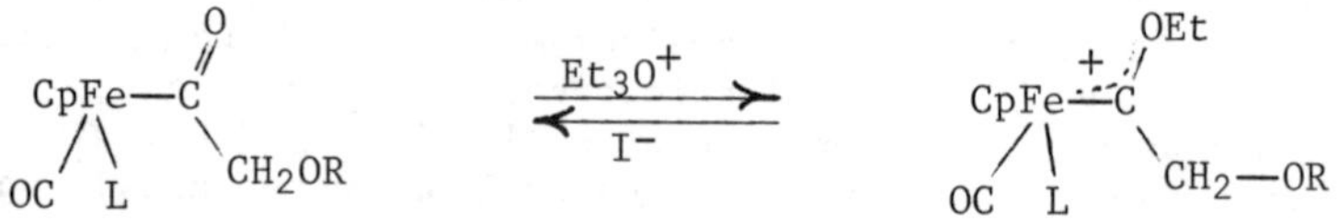

5, L=CO
6, L=$PPh_3$
7, L=$P(OMe)_3$

a, R=Me
b, R=Et

14, L=CO
9, L=$PPh_3$
15, L=$P(OMe)_3$

$PF_6^-$ salts $CpFe(CO)PPh_3[C(OEt)CH_2OR]^+$ (9a,b) and $CpFe(CO)PPh_3[C(OMe)CH_2OEt]$ (using $Me_3O^+PF_6^-$)formed within one hour in $CH_2Cl_2$. Precipitation with ether and recrystallization from $CH_2Cl_2$-ether gave 70-80% yields of the yellow, air-stable salts. Although 5 and 7 likewise underwent facile alkylation with $Et_3O^+PF_6^-$, the products 14a and 15a,b did not crystallize: continued handling of these gums regenerated 5a from 14a, and $CpFe(CO)_2P(OMe)_3^+$ from 15a,b. The presence of extraneous acid as a complicating factor during alkylation of 5 and 7 remains unlikely, since the $Et_3O^+PF_6^-$ had been rigorously freed of acid by recrystallization from nitrobenzene-ether. IR and NMR data of 14a and 15a,b are, however, in accord with the α,β-dialkoxyethylidene complexes as the only organometallic compounds present.

We found no evidence for isomerization of the α,β-dialkoxyalkylidene complexes $CpFe(CO)L[C(OEt)CH_2OR]$ (9,14,15) to alkylated carboalkoxymethyl salts $CpFe(CO)L(CH_2C(OR)(OEt)^+)$ under ambient conditions. Indeed the solution chemistry of α,β-dialkoxyethylidene complexes parallels that of α-alkoxyethylidene compounds $CpFe(CO)L[C(OR)CH_3]^+$, in that 9,14,15 quantitatively revert to 5-7 upon treatment with excess iodide in $CH_2Cl_2$. We did, however, prepare samples of the ethylated carboalkoxymethyl salts $CpFe(CO)P(OMe)_3[CH_2C(OEt)OMe]^+$ and the known $CpFe(CO)_2[CH_2C(OEt)OMe]^+$ (53) from reactions of $Et_3O^+PF_6^-$ with carbomethoxymethyl compounds 13a and 11a respectively. Both ethylated carbomethoxymethyl salts were dealkylated with iodide to their carboalkoxymethyl compounds 13b and 11b. Whereas α-hydroxy-β-alkoxyethylidene intermediates (resulting from protonation of alkoxyacetyl complexes 5-7) isomerize to carboalkoxymethyl compounds 11-13 and selectively liberate methyl or ethyl acetate, the corresponding α,β-dialkoxyethylidene complexes 9,14,15 evidently remain inert under similar conditions. α,β-Dialkoxyethylidene salts 9,14 do however provide acetaldehyde.

## Conversion of Alkoxyacetyl Complexes to Acetaldehyde

The alkoxyacetyl complexes 5 and 6 also serve as precursors, via reactions of the coordinated $C_2$ ligands depicted in Scheme 4, for acetaldehyde. Activation of 5 and 6 as α,β-dialkoxyethylidene

Scheme 4

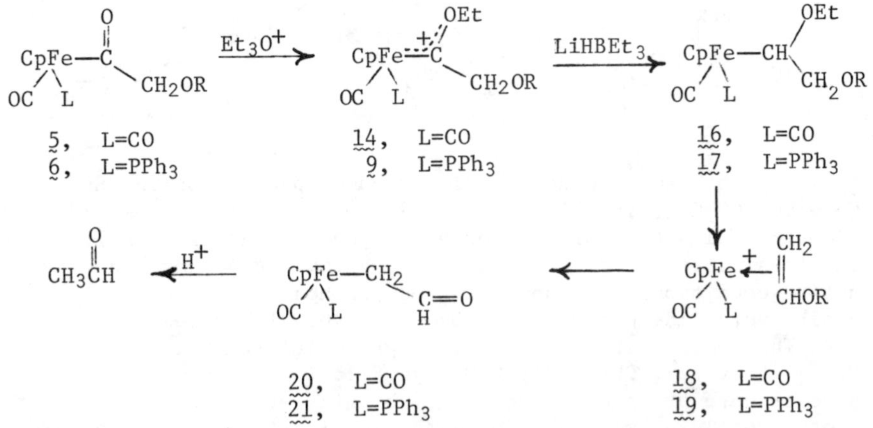


compounds 14 and 9, followed by monohydridic reduction, generates α,β-dialkoxyethyl complexes $CpFe(CO)L[CH(OEt)CH_2OR]$ [L=CO(16), $PPh_3$(17)]. These α,β-dialkoxyethyl complexes then eliminate alkoxide and give $\eta^2$-vinyl ether salts 18 and 19, before undergoing solvolysis to formylmethyl compounds $CpFe(CO)L(CH_2CHO)$ [L=CO(20), $PPh_3$(21)]. A similar sequence of solvolytic reactions has been reported for hydrolysis of a β,β-dialkoxyethyl compound $CpFe(CO)_2CH_2CH(OMe)_2$ (53). We therefore believed that the critical stage of Scheme 4 corresponds to the accessibility of the novel α,β-dialkoxyethyl iron complexes 16,17.

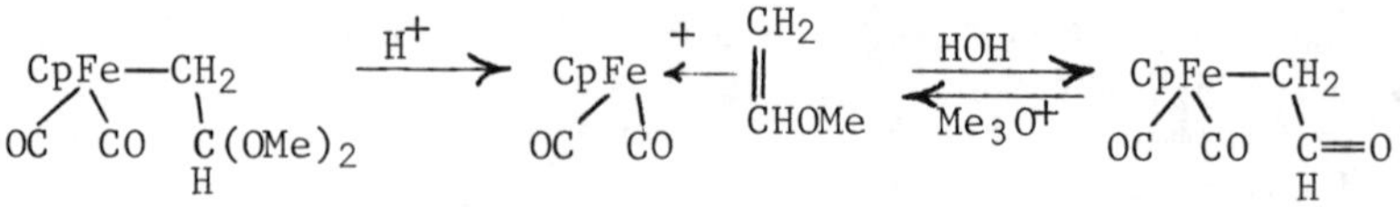


18 20

We were concerned that synthesis of α,β-dialkoxyethyl complexes 16 and 17 would be limited by overreduction of 13 and 14. Both appended α-alkoxyethyl (30,63) and β-alkoxyethyl (16,64) groups separately undergo facile reductive cleavage to ethyl ligands. Model studies were accordingly carried out in order to first establish conditions favoring generation of β-alkoxyethyl and α-alkoxyethyl compounds via monohydridic reduction of suitable substrates.

We found that $LiHBEt_3$ cleanly reduces the ethylvinyl ether salt 18b in THF to a β-ethoxyethyl iron complex in 80% yield. In contrast, $Ph_3PMe^+BH_4^-$ in $CH_2Cl_2$ transforms 18b into 1:1 mix-

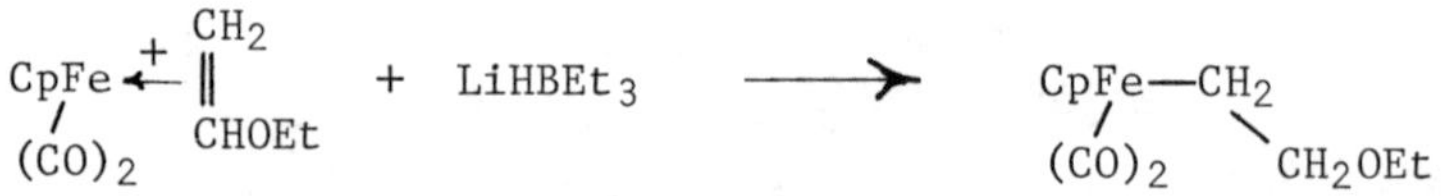


18b

tures of the same β-ethoxyethyl compound $CpFe(CO)_2CH_2CH_2OEt$ and $CpFe(CO)_2CH_2CH_3$, in which the ethyl complex derives from reductive cleavage (with $BH_3$) of the former compound. Having established the feasibility of generating β-alkoxyethyl compounds under reductive conditions, we then examined monohydridic reduction of α-alkoxyethylidene compounds to α-alkoxyethyl complexes.

The α-alkoxyethylidene salts $CpFe(CO)L[C(OR)CH_3]^+$ $PF_6^-$ [L=CO(22), $PPh_3$(10); R=Me(a), Et(b)] used in our model studies were prepared conveniently by alkylation of the requisite acetyl complexes with dialkoxycarbenium ions (generated in situ from $Ph_3C^+PF_6^-$ and trialkyl orthoacetate or orthoformate) in $CH_2Cl_2$. Yellow salts 22 and 10 resulted in 70-80% yields after reprecipitation from $CH_2Cl_2$-$Et_2O$. The salts 10a,b have been prepared previously with other alkylating agents (30), but 22a,b are new. Other alkoxymethylidene (30) and α-alkoxyalkylidene (66,67) salts of $CpFe(CO)_2^+$ have been reported however.

Treatment of 22,10 with one equivalent of $LiHBEt_3$ in THF at -80°, followed by removal of solvent and pentane extraction of the residue, afforded the α-alkoxyethyl complexes 23,24 in 75-95% yields. The corresponding iron ethyl complexes were not present as judged by NMR. Compound 24b, previously reported as the analogous $NaBH_4$-ethanol reduction product mixed with the corresponding ethyl complex 8 (63), represents the only other known Group 8 α-alkoxyethyl complex. Results of our model studies established that both α-alkoxyethyl and β-alkoxyethyl complexes can be synthesized under reductive conditions. Hence transfer of one hydride to α,β-dialkoxethylidene complexes 14 and 9 should afford the corresponding α,β-dialkoxyethyl complexes 16 and 17.

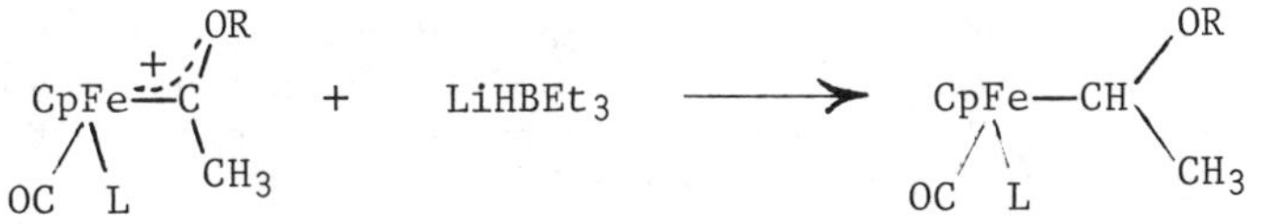

$$\underset{\underset{OC\quad L}{/\backslash}}{CpFe}{\overset{+}{=\!\!=}}\underset{CH_3}{\overset{OR}{\underset{\backslash}{C}}} \quad + \quad LiHBEt_3 \quad \longrightarrow \quad CpFe{\underset{\underset{OC\quad L}{/\backslash}}{-}}CH{\overset{OR}{\underset{CH_3}{\backslash}}}$$

|   |         |    |        |    |           |
|---|---------|----|--------|----|-----------|
| 22, | L=CO    | a, | R=Me   | 23, | L=CO      |
| 10, | L=PPh$_3$ | b, | R=Et   | 24, | L=PPh$_3$ |

One equivalent of LiHBEt$_3$ or LiHB(sec-butyl)$_3$ in THF at -80° consumed the phosphine substituted α,β-dialkoxyethylidene salts 9a,b and delivered 73% yields of the formylmethyl complex CpFe(CO)PPh$_3$(CH$_2$CHO) (21) as the only isolable organometallic compound. The product 21 can be accounted for by an electrophile

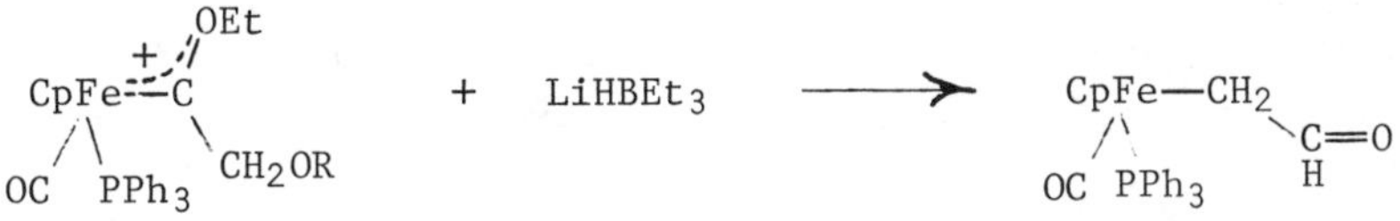

$$\underset{\underset{OC\quad PPh_3}{/\backslash}}{CpFe}{\overset{+}{=\!\!=}}\underset{CH_2OR}{\overset{OEt}{\underset{\backslash}{C}}} \quad + \quad LiHBEt_3 \quad \longrightarrow \quad CpFe{\underset{\underset{OC\quad PPh_3}{/\backslash}}{-}}CH_2{\underset{H}{\overset{}{\diagdown C=O}}}$$

$$9 \qquad\qquad\qquad\qquad\qquad\qquad\qquad\qquad 21$$

(e.g. BEt$_3$) induced ionization of the α,β-dialkoxyethyl intermediate 17, resulting from monohydridic reduction of 9, to an unstable η$^2$-vinyl ether salt 19. This vinyl ether compound 19 then undergoes solvolysis to the observed 21. We would not expect a stable η$^2$-vinyl ether salt 19 under ambient conditions, due to adverse steric interactions involving the bulky PPh$_3$ group. As an example of this adverse steric interaction, solutions of the η$^2$-propene complex CpFe(CO)PPh$_3$(η$^2$-CH$_2$=CHMe)$^+$ also eliminate propene rapidly at room temperature (68). The formylmethyl complex CpFe(CO)PPh$_3$(CH$_2$CHO) (21) was independently synthesized in overall 56% yield by photolytic replacement of CO by PPh$_3$ on CpFe(CO)$_2$CH$_2$CH(OMe)$_2$ and chromatography on alumina.

$$\underset{(CO)_2}{\overset{}{CpFe}}{-}CH_2{\underset{(OMe)_2}{\overset{}{\diagdown CH}}} \quad \xrightarrow[PPh_3]{h\nu} \quad CpFe{\underset{\underset{OC\quad PPh_3}{/\backslash}}{-}}CH_2{\underset{(OMe)_2}{\overset{}{\diagdown CH}}} \quad \xrightarrow{alumina}$$

$$CpFe{\underset{\underset{OC\quad PPh_3}{/\backslash}}{-}}CH_2{\underset{H}{\overset{}{\diagdown C=O}}}$$

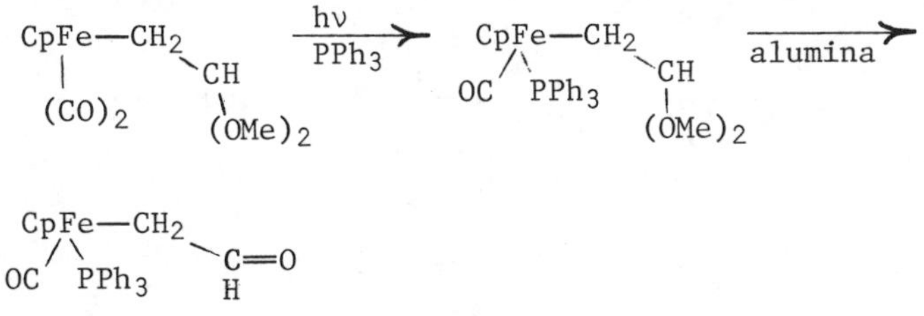

$$21$$

The formylmethyl complex 21 also serves as a source of free acetaldehyde, and one equivalent of trifluoromethanesulfonic acid in $CH_2Cl_2$ releases it from 21 within one hour at room temperature. Acetaldehyde was identified by its 2,4-dinitrophenylhydrazone (isolated in 42% yield), and was determined directly (48%) by quantitative analysis of its IR $\nu_{co}$ 1716 $cm^{-1}$ absorption. The protonation of 21 presumably generates a $\eta^2$-vinyl alcohol compound 19 (R=H) [IR observable $\nu_{co}$ 1983 $cm^{-1}$], which then dissociates acetaldehyde. We have overall converted selectively two carbonyls on $CpFe(CO)_3^+$ (1) to acetaldehyde.

The results of studies employing $CpFe(CO)_2$ complexes within Scheme 4 offer more mechanistic insight into the reaction chemistry of α,β-dialkoxyethyl complexes. One equivalent of $LiHBEt_3$ thus reduced the α-ethoxy-β-methoxyethylidene salt 14a in THF (-80°), but the product corresponded to addition of one hydride—an α-ethoxy-β-methoxyethyl $CpFe(CO)_2$ complex (16a). After pre-

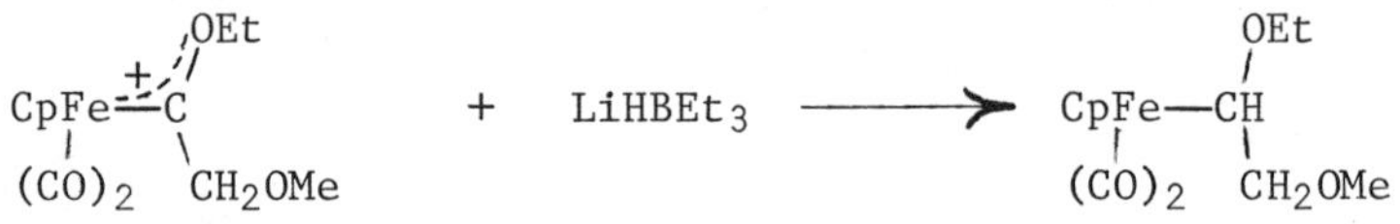

13a 16a

cipitation twice from cold ether-pentane, 16a was obtained in about 40% yield as an impure brown oil. It has not been amenable to further purification. (Studies are in progress with $\eta^5$-$C_5Me_5$ analogues in Scheme 4, L=CO, in an attempt to obtain crystalline samples.) Spectroscopic data (Table 1), however, are in accord with a α,β-dialkoxyethyl structure 16a, but not with $CpFe(CO)_2CH_2CH(OMe)(OEt)$. Thus the NMR of 16a evidences no absorptions in the δ 4.0-4.5 region, where $CpFe(CO)_2CH_2CH(OMe)_2$ absorbs as a triplet for the methine group. Of particular interest is the subsequent conversion of the α,β-dialkoxyethyl complex 16a into $\eta^2$-vinyl ether compounds.

The impure α,β-dialkoxyethyl complex 16a reacted with $HPF_6 \cdot OMe_2$ in $CH_2Cl_2$ and produced a 2:1 mixture of the ethyl and methyl vinyl ether compounds 18a,b in 45% yield. These products

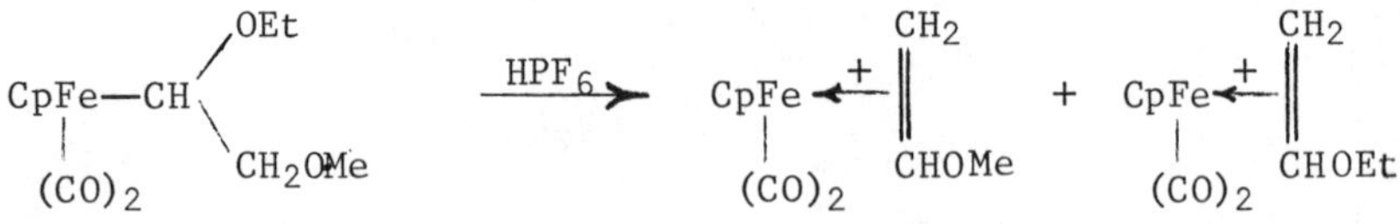

16a 18a 18b

correspond to protonation of 16a and alcohol elimination from both α-and β-positions respectively. Acid lability of the β-alkoxide (in terms of the former pathway: 18b) certainly has precedent, with the facile conversion of $CpFe(CO)_2CH_2CH_2OMe$ to its $\eta^2$-ethylene compound (16). In the latter pathway we postulate the

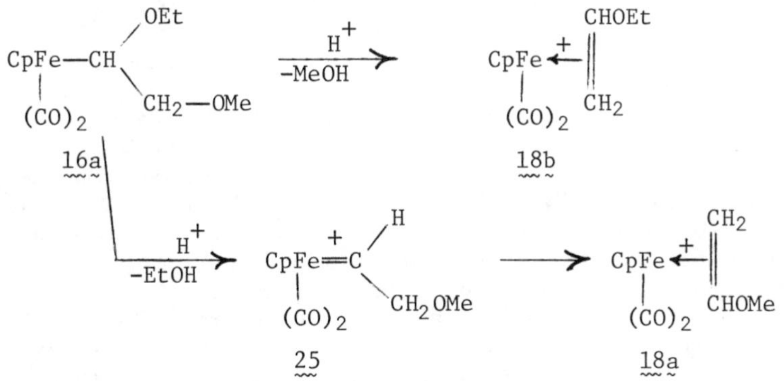


intermediacy of a β-methoxyethylidene salt 25 that rearranges to the observed methyl vinyl ether compound 18a. A similar $\eta^1$-alkylidene to $\eta^2$-alkene ligand rearrangement has been documented during protonation of the ($\eta^1$-cyclopropyl)$Fe(CO)_2Cp$ (69). We nevertheless independently verified the generation of a $\eta^1$-ethyldene complex, via α-alkoxide abstraction with acid, and its rearrangement to a $\eta^2$-ethylene compound.

The results thus far of a model study confirm the acid lability of an α-alkoxyethyl complex 24, followed by generation of a very reactive ethylidene compound 26 (70). It was only by use of $CpFe(CO)PPh_3$ complexes that the reactivity of the ethylidene compound was sufficiently tempered for chemical trapping: the analogous $CpFe(CO)_2$ compounds undergo rapid degradative reactions due to the higher electrophilicity of the ethylidene salt. Compounds 24a,b underwent protonation in $CH_2Cl_2$ below -60° and gave yellow solutions, which produced yellow crystals with excess ether at -80°. Either the crystals or the solutions slowly

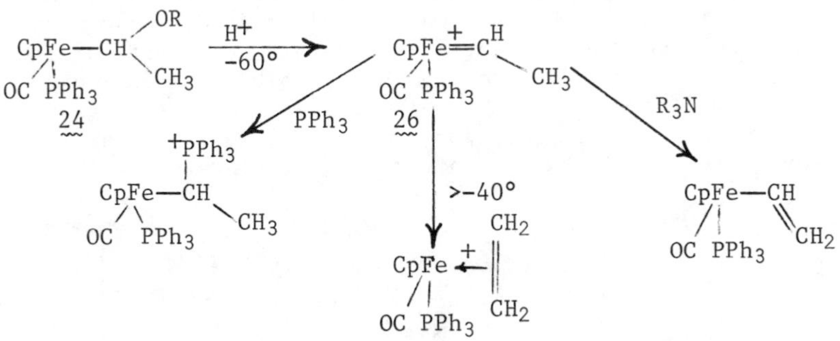

turned reddish-orange above ca.-40° and supplied $CpFe(CO)PPh_3(\eta^2\text{-}CH_2{=}CH_2)^+$ in 40% yield. This rearrangement of 26 is analogous to the postulated rearrangement of the β-methoxyethylidene compound 25 to the $\eta^2$-vinyl ether salt 18a. The ethylidene compound 26 was also trapped at low temperatures by $PPh_3$ as $CpFe(CO)PPh_3(CHMe^+PPh_3)$ in 68% yield and by deprotonation to the vinyl complex (40-50%) with amines. We therefore assume that the intermediary of a β-methoxyethylidene compound 25 in the conversion of 16a to 18a is chemically feasible.

Several examples of alkylidene complexes relevant to the chemistry of 25 and 26 should be mentioned. Gladysz and coworkers recently isolated a stable ethylidene compound $CpRe(NO)PPh_3(CHMe)^+$ (71), and also demonstrated $PPh_3$ addition to the corresponding methylidene salt as an example of derivatizing cationic alkylidene complexes (72). Stable benzylidene compounds $CpFe(CO)L(CHPh)^+$ [L=CO,$PPh_3$], obtained through protonation of the requisite α-alkoxybenzyl iron complexes, have been reported by Brookhart and Nelson (73). Anionic α-alkoxyalkyl complexes in acidic media likewise eliminate alcohol and form neutral alkylidene complexes. For example, Casey et.al. detailed the conversion of $(CO)_5WCPhMe(OMe)^{-1}$ to its unstable α-phenylethylidene complex, which then decomposed to styrene (74).

## Towards Catalytic Relevance: Bimetallic Activation of Acyl Ligands and Transition Organometallic Hydrides as Reducing Agents

Having established viable reaction pathways for selective conversion of CO ligands to $C_2$ organics, we then investigated metal complexes that would replace the carbocation and borohydride reagents used in Scheme 1. Such complexes, especially if they were regenerated under mild conditions, might prove catalytically relevant. Thus cationic metal complexes that activate acyl complexes via formation of $\mu_2$-acyl complexes were studied as replacements for carbocations, and transition metal hydride complexes were examined as replacements for borohydride reagents.

Bimetallic activation of acetyl and alkoxyacetyl ligands -- through formation of cationic $\mu_2$-acyl complexes -- to reaction with nucleophilic hydride donors was established. Cationic transition metal compounds possessing an accessible coordination site bind a neutral $\eta^1$-acyl ligand on another complex as a cationic $\mu_2$-acyl system. These $\mu_2$-acyl systems activate the acyl ligand to reduction analogous to carbocation activation. Several examples of $\mu_2$-acyl complexation have been reported previously.

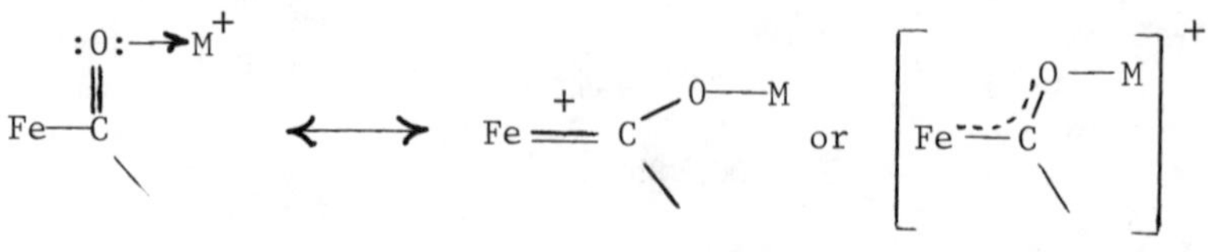

Complexes bearing $\mu_2$-acyl ligands are rare (75), but they are recognized now as a means of stabilizing and activating formyl ligands to reduction (35,76,77). Lewis acid activation of other acyl ligands has been established with $\eta^2$-acyl complexation to a metal (78,79), and with $BH_3$ reduction of $\eta^1$-acetyl (60) and $\eta^1$-formyl (38,48) complexes. Other Lewis acids such as Al(III) (52) or Li(I) (80) also facilitate alkyl-acyl migratory-insertion reactions through complexation of the incipient acyl ligand (75).

We synthesized cationic $\mu_2$-acetyl compounds 28,29 by combining iron acetyl complexes $CpFe(CO)L(COCH_3)$ (27) [L=CO,$PPh_3$] with a coordinatively unsaturated (16-electron) metal carbonyl salt $CpM(CO)_n^+$[M=Fe,n=2;M=Mo,n=3], as indicated in Scheme 5. Thus

Scheme 5

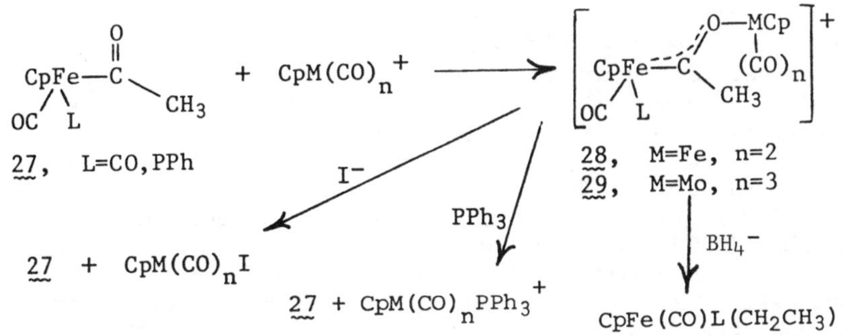


refluxing $CH_2Cl_2$ solutions of the labile isobutylene (18) or THF (81)(L') salts $CpFe(CO)_2(L')^+$ and the requisite acetyl complex 27 provide the $\mu_2$-acetyl $Fe_2$ compounds 28 [L=CO,$PPh_3$]. The coordinatively unsaturated $CpMo(CO)_3^+$, which is generated at -40° by $Ph_3C^+$ hydride abstraction from $CpMo(CO)_3H$ (82), coordinates 27 at -20° and gives $\mu_2$-acetyl FeMo compounds 29 [L=CO,$PPh_3$]. Both series of compounds 28 and 29 were obtained as air-stable red solids in 40-70% yields after recrystallization from $CH_2Cl_2$-ether.

These $\mu_2$-acetyl complexes 28,29 entail σ metal-oxygen bonding (30) analogous to that established for bonding of organic ketones, esters, and amides to $CpFe(CO)_2^+$ (83). Sterically less

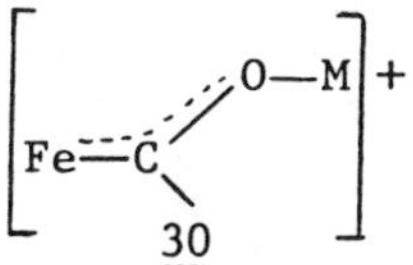

30

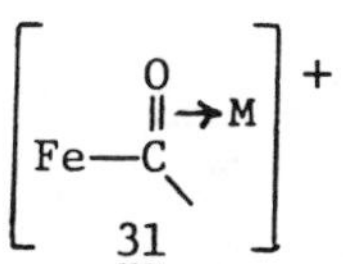

31

demanding E-stereoisomers for 30 have been arbitrarily favored. [$CpFe(CO)_2COPh$ evidently forms compounds similar to 28 and 29, but as mixtures of the NMR distinctive E- and Z-stereoisomers (84).] NMR data of 28,29 (Table 1) are interpreted as favoring σ-structure 30 rather than the π-bonding structure 31 exhibited by other

$\mu_2$-acyl complexes (77). If 28,29 engaged in π-bonding, then we would expect mixtures of diastereomers for L=$PPh_3$; this derives from the prochiral Fe-$COCH_3$ acetyl groups, which are π-complexed, and the chiral CpFe(CO)$PPh_3$ center existing within the same molecule. For example, π-complexation of prochiral propene or 1-butene to CpFe(CO)$PPh_3^+$ gave diastereomeric mixtures that were easily discerned by NMR (29,68). No evidence, however, has been found for diastereomeric mixtures either within the crude reaction products or after recrystallization of 28 and 29, L=$PPh_3$.

Further analysis of the spectral data for the $\mu_2$-acetyl compounds 28 and 29 (Table 1) also indicates that the positive charge is distributed over both metal centers. The electronic environment of the C-bonded iron of 28 and 29 clearly lies intermediate between that of the starting acetyl complex and of the α-alkoxyalkylidene compounds. Similarly the acetyl complexes provide more electron density to $CpFe(CO)_2^+$ within 28 than does acetone or THF within $CpFe(CO)_2L'^+$ [L'=acetone, THF].

The reaction chemistry of the $\mu_2$-acetyl complexes 28,29 evinces both the desired activation of the acetyl ligand to hydride donors, as well as lability of the activating groups $CpM(CO)_n^+$ towards nucleophiles. This latter mode of reactivity resembles that of $CpFe(CO)_2(acetone)^+$ (85,86): iodide or $PPh_3$ quantitatively displace $CpM(CO)_n^+$. Borohydride (equimolar $PPh_3Me^+BH_4^-$ in $CH_2Cl_2$) does however reduce 28,29 (L=CO,$PPh_3$) and eliminate the requisite ethyl complexes in 20-35% yields. The balance of the reaction products, the corresponding acetyl complexes and $[CpFe(CO)_2]_2$ (3) or $[CpMo(CO)_3]_2$, conform with displacement of $CpM(CO)_n^+$ by the hydride donor.

Alkoxyacetyl complex 5a also forms bimetallic $\mu_2$-alkoxyacetyl compounds $Cp(CO)_2Fe(CO[M(CO)_nCp]CH_2OMe)^+$ 32,M=Fe,n=2 and 33,M=Mo,n=3, but these adducts are less stable than the corresponding $\mu_2$-acetyl adducts 28,29. Only 33 was isolable, although

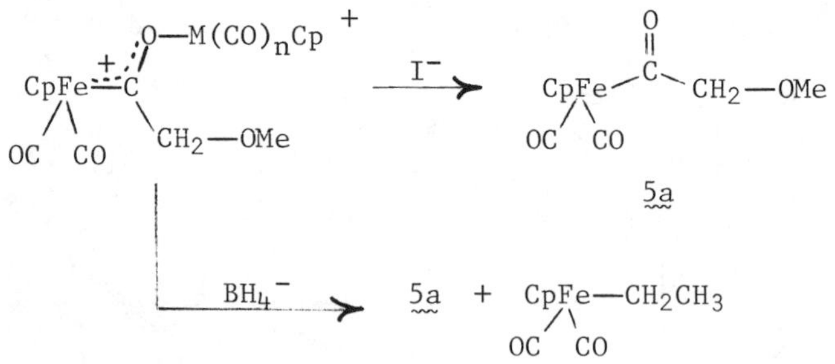


32, M=Fe, n=2
33, M=Mo, n=3

substantial formation of 32 was apparent in its crude reaction product. Treatment of crude 32 with iodide thus reverted the reaction mixture to 5a and $CpFe(CO)_2I$. $PPh_3Me^+BH_4^-$ in $CH_2Cl_2$

reduces the FeMo $\mu_2$-alkoxyacetyl $\underline{33}$ and gives $CpFe(CO)_2CH_2CH_3$ in
21% yield after column chromatography. We have thus used coordinatively unsaturated transition metal salts, of the type generated
by protolytic cleavage of metal-carbon bonds in $CpM(CO)_n$-R $(\underline{32})$,
as activating groups on metal acyl ligands. The resulting
$\mu_2$-acyl complexes then undergo facile reduction of the acyl
ligand. Clearly the next step is to use transition metal hydride
complexes as the hydride donors.

We have demonstrated that a series of first row, Group 8
organometallic hydride complexes effect intermolecular hydride addition to coordinated $\eta^2$-alkene, $\eta^2$-vinyl ether, and $\alpha$-alkoxyethy-
lidene compounds $(\underline{64})$. For example, one equivalent of
$CpFe(CO)PPh_3(H)$ quantitatively reduces $CpFe(CO)_2(\eta^2-CH_2=CH_2)^+$ to
$CpFe(CO)_2CH_2CH_3$ within one-half hour and leaves
$CpFe(CO)PPh_3(CH_3CN)^+$. A mechanism in which the nucleophilic Fe-H

$$CpFe \overset{+}{\underset{(CO)_2}{\longleftarrow}} \overset{CH_2}{\underset{CH_2}{\|}} \quad \xrightarrow{\quad CpFe(CO)PPh_3(H) \quad} \quad CpFe \underset{(CO)_2}{-}CH_2 \diagdown CH_3$$

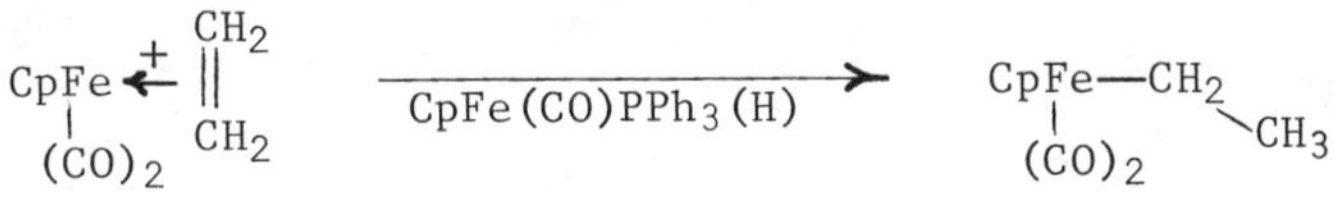

bond attacks the coordinated alkene and eliminates the coordinatively unsaturated $CpFe(CO)PPh_3^+$ agrees with both the substituent
effects studied and the established facility of nucleophilic addition to coordinated alkenes.

Attempts thus far at using the samp Group 8 organometallic
hydride complexes as hydride donors to the coordinated CO on
$CpFe(CO)_3^+$ $(\underline{1})$ have been unsuccessful. Reaction of $\underline{1}$ with

$$CpFe\underset{(CO)_2}{-}L^+ + CpFe\overset{H}{\underset{OC\quad PPh_3}{\diagup\diagdown}} \longrightarrow CpFe\text{-}H\text{-}FeCp^+ + L$$
$$\underset{(CO)_2\ OC\quad PPh_3}{}$$

$\underline{1},\quad L=CO$ \qquad $[CpFe(CO)_2]_2 \quad + \quad CpFe(CO)_2PPh_3^+$
$\quad\quad L=THF$
$\qquad\qquad\qquad\qquad\qquad \underset{\sim}{3}$

$CpFe(CO)PPh_3(H)$ in $CH_2Cl_2$ at room temperature realizes only replacement of CO by the Fe-H bond $(\underline{87})$. The cationic bridging hydride product, which independently forms from the reaction of
$CpFe(CO)_2(THF)^+$ and $CpFe(CO)PPh_3(H)$, subsequently decomposes to
$CpFe(CO)_2H$ (then dimeric $\underline{3}$) and $CpFe(CO)_2PPh_3^+$ (through disproportionation of $CpFe(CO)PPh_3^+$). Essentially the same results
were observed when the solvent was changed to methanol. Two
equivalents each of $NEt_4^+HFe(CO)_4^-$, $CpFe(CO)PPh_3(H)$, or
$CpFe(Ph_2PCH_2CH_2PPh_2)H$ in methanol transform $\underline{1}$ into varying amounts
of the dimer $\underline{3}$ as the only isolable organometallic species $(\underline{88})$.
In contrast precedent for using the hydride complex
$(\eta^5-C_5Me_5)_2ZrH_2$ to transfer hydride to CO ligated on another metal

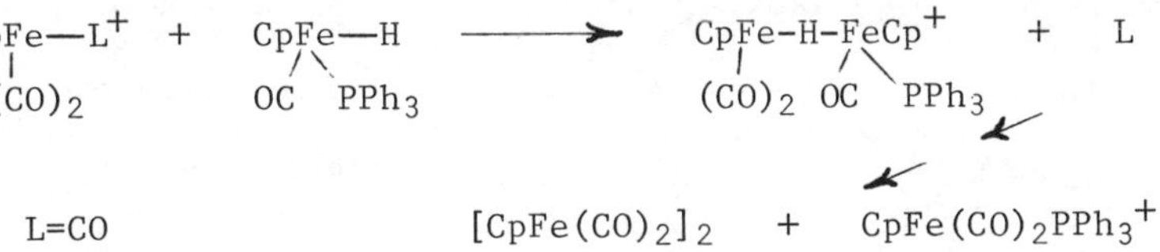

center has been reported by Bercaw et al (76). We are examining a number of first row transition organometallic hydride complexes, especially those directly derived from $H_2$, as potential reducing agents for several cationic metal carbonyl compounds.

Transition organometallic hydride complexes $CpFe(CO)PPh_3(H)$, $CpFe(Ph_2PCH_2CH_2PPh_2)H$, and $HFe(CO)_4^{-1}$ also reacted with α-alkoxyethylidene complexes 10,22 (64), but two pathways were observed.

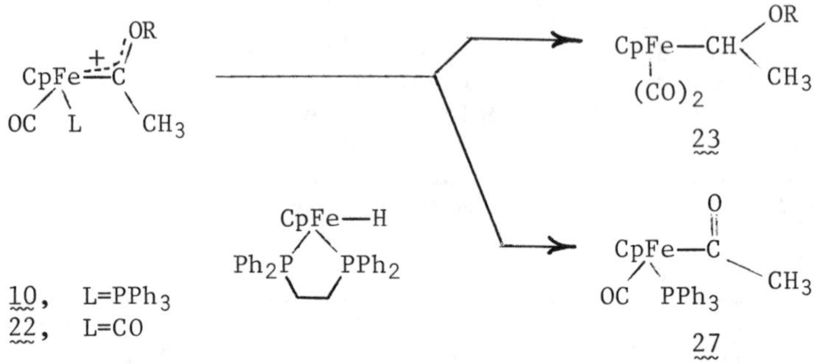


The $CpFe(CO)_2$ series of α-alkoxyethylidene compounds 22 (R=Me,Et) undergo monohydridic reduction and give only the α-alkoxyethyl complexes in good yields. In contrast, the less electrophilic $CpFe(CO)PPh_3$ series produce exclusively the parent acetyl complex. These latter dealkylation reactions, analogous to those observed with iodide, occur through nucleophilic attack of the Fe-H bond at the activating group R. Trialkylborohydride transfers one hydride to both 10 and 22, as previously noted. Therefore both the choice of auxiliary ligands and of the reducing medium controls the reaction path observed with α-alkoxyethylidene compounds and nucleophilic hydride donors.

Feasibility of transfering a hydride to bimetallic activated acetyl ligands was also investigated. The $\mu_2$-acetyl compounds 28 and 29 consumed one equivalent of $LiHBEt_3$, $LiHB(sec\text{-}butyl)_3$, $HFe(CO)_4^{-1}$, $CpFe(Ph_2PCH_2CH_2PPh_2)H$, or $CpFe(CO)PPh_3(H)$, but afforded only the starting acetyl complexes. A mechanism entailing

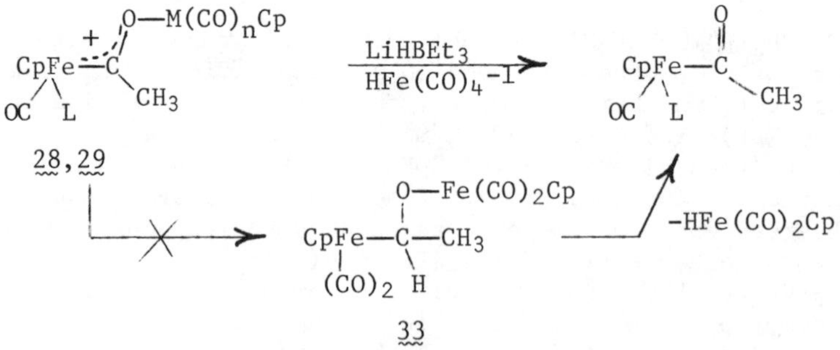


nucleophilic displacement of $CpM(CO)_n^+$ with the hydride donor (as with iodide) is tentatively favored over one involving hydride addition to the α-carbon and elimination of $HM(CO)_nCP$. We base this conclusion on the failure to trap the intermediate 33 arising from hydride addition to the α-carbon of 28. Thus reaction of 28 (L=CO) and $LiHBEt_3$ in THF at -80°, followed immediately by $HPF_6 \cdot OMe_2$ between -60° and 0°, gave no $CpFe(CO)_2(\eta^2\text{-}CH_2{=}CH_2)^+$. The intent of these experiments was to demonstrate intermediacy of 33 via O-protonation and elimination of the ethylidene salt $CpFe(CO)_2(CHMe)^+$, which would rearrange to $CpFe(CO)_2(\eta^2\text{-}CH_2{=}CH_2)^+$. We feel however that additional studies are required in order to adequately study the interaction of organometallic hydride reagents and bimetallic $\mu_2$-acyl complexes.

## Summary

We have demonstrated that $NaBH_3CN$ in alcohol reduces a CO ligand on $CpFe(CO)_3^+$ and generates an alkoxymethyl iron complex. Versatile alkoxyacetyl complexes $CpFe(CO)L(COCH_2OR)$ [L=CO,$PPh_3$,$P(OMe)_3$] derived from the alkoxyacetyl complexes then serve as a template for synthesizing other $C_2$ coordinated ligands. A significant feature of our approach to Fischer-Tropsch chemistry is that reactions utilizing alkoxyacetyl complexes take place exclusively on the coordinated $C_2$ ligand and do not involve coordinatively unsaturated iron complexes. (For a different approach in which unsaturated acyloxymethyl Mn complexes hydrogenate to glycolaldehyde, see Dombek's work (55).) Sequential electrophilic attack (or activation) of the coordinated ligands followed by reduction affords the final $\eta^1$-ethyl, carboalkoxymethyl (no reductive step), or formylmethyl ligands. Protonation then releases selectively ethane, alkyl acetate, or acetaldehyde; and heating of the ethyl complex alone or after hydride abstraction gives ethylene.

Our ultimate goal remains to devise an organometallic system in which metal reagents activate CO and subsequent ligands (including bimetallic activation of acyl ligands) to intermolecular reduction by transition metal hydrides. This essentially entails replacement of nonmetal Lewis acids and hydride donors reported in this work by transition metal analogues. Ideally the metal hydride complex would be prepared under mild conditions using $H_2$. Intermolecular transfer of hydride from the organometallic hydride complex then leaves an organometallic Lewis acid, which remains available for electrophilic activation of coordinated ligands. Work is progressing along these lines.

## Acknowledgement

We thank the Department of Energy for support of this work.

## Literature Cited


1. St-Pierre,L.E.; Brown,G.R.,Eds. "Future Sources of Organic Raw Materials-CHEMRAWN I", Pergamon, 1980.
2. Masters,C. Adv.Organometal.Chem., 1979, 17, 61-103.
3. Muetterties,E.L.; Stein,J. Chem.Rev., 1979, 79, 479-490.
4. Pruett,R.L. Ann.New York Acad.Sci., 1977, 295, 239-248.
5. Wong,K.S.; Labinger,J.A. J.Amer.Chem.Soc., 1980, 102, 3652-3653.
6. Olivé,G.H.; Olivé,S. Ang.Chem.,Internat.Ed.Engl., 1979, 18, 77-78.
7. Wong,A.; Harris,M.; Atwood,J.D. J.Amer.Chem.Soc., 1980, 102, 4529-4531.
8. Masters,C.; van der Woude,C.; van Doorn,J.A. J.Amer.Chem.Soc. 1979, 101, 1633-1644.
9. Summer,C.E.; Riley,P.E.; Davis,R.E.; Pettit,R. J.Amer.Chem. Soc., 1980, 102, 1752-1754.
10. Bradley,J.S.; Ansell,G.B.; Hill,E.W. J.Amer.Chem.Soc., 1979, 101, 7417-7419.
11. Manriquez,J.M.; McAlister,D.R.; Sanner,R.D.; Bercaw,J.E. J.Amer.Chem.Soc., 1978, 100, 2716-2724.
12. Von Gustorf,E.A.K.; Grevels,F.; Fischler,I. "The Organic Chemistry of Iron", Vol.I, Academic, 1978.
13. Ellis,J.E. J.Organometal.Chem., 1975, 86, 1-56.
14. Rosenblum,M. Acc.Chem.Res., 1974, 7, 122-128.
15. Lennon,P.; Rosan,A.M.; Rosenblum,M. J.Amer.Chem.Soc., 1977, 99, 8426-8439.
16. Lennon,P.; Madhavarao,M.; Rosan,A.; Rosenblum,M. J.Organometal Chem.,1976, 108, 93-109.
17. Cutler,A.; Ehntholt,D.; Giering,W.P.; Lennon,P.; Raghu,S.; Rosan,A.; Rosenblum,M.; Tancrede,J.; Wells,D. J.Amer.Chem. Soc., 1976, 98, 3495-3507.
18. Cutler,A.; Ehntholt,D.; Lennon,P.; Nicholas,K.; Marten,D.F.; Madhavarao,M.; Raghu,S.; Rosan,A.; Rosenblum,M. J.Amer.Chem. Soc., 1975, 97, 3149-3157.
19. Reger,D.L.; Culbertson,E.C. J.Amer.Chem.Soc., 1976, 98, 2789-2794.
20. Flood,T.C.; DiSanti,F.J.; Miles,D.L. Inorg.Chem., 1976, 15, 1910-1918.
21. Flood,T.C.; Miles,D.L. J.Organometal.Chem., 1977, 127, 33-44.
22. Attig,T.G.; Teller,R.G.; Wu,S.; Bau,R.; Wojcicki,A. J.Amer. Chem.Soc., 1979, 101, 619-628.
23. Rosenblum,M.; Waterman,P.S. J.Organometal.Chem., 1980, 187, 267-275.
24. Reger,D.L.; Coleman,C.J. Inorg.Chem., 1979, 18, 3155-3160.
25. Schilling,B.E.R.; Hoffmann,R.; Lichtenberger,D.L. J.Amer. Chem.Soc., 1979,101, 585-591.
26. Dong,D.;Slack,D.A.; Baird,M.C. Inorg.Chem., 1979,18,188-191.
27. Wojcicki,A. Adv.Organomet.Chem., 1973, 11, 87-145.
28. Calderazzo,F. Ang.Chem.,Internat.Ed.Engl., 1977,16,299-311.

29. Reger,D.L.; Coleman,C.J.; McElligott,P.J. J.Organometal.Chem., 1979, 177, 73-84, and references cited.
30. Cutler,A.R. J.Amer.Chem.Soc., 1979, 101, 604-606, and references cited.
31. Brookhart,M.; Tucker,J.R.; Flood,T.C.; Jensen,J. J.Amer.Chem. Soc., 1980, 102, 1203-1205, and references cited.
32. Johnson,M.D. Acc.Chem.Res., 1978, 11, 57-65.
33. Kochi,J.K. "Organometallic Mechanisms and Catalysis", Academic, Ch.18, 1978.
34. Rogers,W.N.; Baird,M.C. J.Organometal.Chem., 1979, 182, C65-C68.
35. Wolczanski,P.T.; Bercaw,J.E. Acc.Chem.Res., 1980, 13,121-127.
36. Jolly,P.W.; Pettit,R. J.Amer.Chem.Soc., 1966, 88, 5044-5045.
37. Green,M.L.H.; Ishaq,M.; Whiteley,R.N. J.Chem.Soc.A, 1967, 1508-1515.
38. Casey,C.P.; Andrews,M.A.; McAlister,D.R.; Rinz,J.E. J.Amer. Chem.Soc., 1980, 102, 1927-1933.
39. Sweet,J.R.; Graham,W.G. J.Organometal.Chem., 1979,173,C9-C12.
40. Saunders,A.; Bauch,T.; Magatti,C.V.; Lorenc,C.; Giering,W.P. J.Organometal.Chem., 1976, 107, 359-375.
41. Gladysz,J.A.; Selover,J.C.; Strouse,C.E. J.Amer.Chem.Soc., 1978, 100, 6766-6768.
42. Espenson,J.H.; Bakač,A. J.Amer.Chem.Soc., 1980, 102, 2488-2489, and references cited.
43. Lane,C.F. Synthesis, 1975, 135-146.
44. Bayoud,R.S.; Biehl, E.R.; Reeves,P.C. J.Organometal.Chem., 1979, 174, 297-303.
45. Davison,A.; Green,M.L.H.; Wilkinson,G. J.Chem.Soc., 1961, 3172-3177.
46. Whitesides,T.H.; Shelly,J. J.Organometal.Chem., 1975, 92, 215-226.
47. Davies,S.G.; Green,M.L.H.; Mingos,D.M.P., Tetrahedron, 1978, 34, 3047-3077.
48. Tam,W.; Wong,W.; Gladysz,J.A. J.Amer.Chem.Soc., 1979, 101, 1589-1591.
49. Treichel,P.M.; Shubkin,R.L. Inorg.Chem., 1967, 6,1328-1334.
50. Cawse,J.N.; Fiatto,R.A.; Pruett,R.L. J.Organometal.Chem., 1979, 172, 405-413.
51. King,R.B.; King,A.D.;Iqbal,M.Z.; Frazier,C.C. J.Amer.Chem.Soc. 1978, 100, 1687-1694.
52. Butts,S.B.; Strauss,S.H.; Holt,E.M.; Stimson,R.E.;Alcock,N.W.; Shriver,D.F. J.Amer.Chem.Soc., 1980, 102, 5093-5100.
53. Cutler,A.; Raghu,S.; Rosenblum,M. J.Organometal.Chem., 1974, 77, 381-391.
54. Roth,J.A.; Orchin,M. J.Organometal.Chem., 1979,172,C27-C28.
55. Dombek,B.D. J.Amer.Chem.Soc., 1979, 101, 6466-6468.
56. Su,S.; Wojcicki,A. J.Organometal.Chem., 1971, 27, 231-240.
57. Reger,D.L.; Culbertson,E.C. J.Amer.Chem.Soc., 1976, 98, 2789-2794.
58. Chow,C.; Miles,D.L.; Bau,R.; Flood,T.C. J.Amer.Chem.Soc., 1978 100, 7271-7278.

59. Casey,C.P.; Bunnell,C.A. J.Amer.Chem.Soc., 1976, 98,436-441.
60. Van Doorn,J.A.; Masters,C.; Volger,H.C. J.Organometal.Chem., 1976, 105, 245-254.
61. Gladysz,J.A.; Selover,J.C. Tetrahedron Lett.,1978,319-322.
62. Darst,K.P.; Lukehart,C.M. J.Organomet.Chem., 1979, 171,65-71.
63. Davison,A.; Reger,D. J.Amer.Chem.Soc., 1972, 94, 9237-9238.
64. Bodnar,T.; LaCroce,S.J.; Cutler,A.R. J.Amer.Chem.Soc., 1980, 102, 3292-3294.
65. Ariyaratne,J.K.P.; Bierrum,A.M.; Green,M.L.H.; Ishaq,M.; Prout,C.K. J.Chem.Soc.(A), 1969, 1309-1321.
66. Game,C.H.; Green,M.; Moss,J.R.; Stone,F.G.A. J.Chem.Soc., Dalton Trans., 1974, 351-357.
67. Marten,D.F. J.Chem.Soc.,Chem.Comm., 1980, 341-342.
68. Aris,K.R.; Brown,J.M.; Taylor,K.A. J.Chem.Soc., Dalton Trans. 1974, 2222-2228.
69. Cutler,A.; Fish,R.W.; Giering,W.P.; Rosenblum,M. J.Amer.Chem. Soc., 1972, 94, 4354-4355.
70. Cutler,A.R.; Bodnar,T., unpublished observations.
71. Wong,W.K.; Tam.W.; Gladysz,J.A. J.Amer.Chem.Soc., 1979, 101, 5440-5442.
72. Kiel,W.A.; Lin,G.; Gladysz,J.A. J.Amer.Chem.Soc., 1980, 102, 3299-3301.
73. Brookhart,M.;Nelson,G.O. J.Amer.Chem.Soc., 1977, 99, 6099-6101.
74. Casey,C.P.; Albin,L.D.; Burkhardt,T.J. J.Amer.Chem.Soc., 1977, 99, 2533-2539.
75. Berke,H.; Hoffmann,R. J.Amer.Chem.Soc., 1978, 100,7224-7236.
76. Wolcyanski,P.T.; Threlkel,R.S.; Bercaw,J.E. J.Amer.Chem.Soc., 1979, 101, 218-220.
77. Belmonte,P.; Schrock,R.R.; Churchill,M.R.; Youngs,W.J. J.Amer.Chem.Soc., 1980, 102, 2858-2860.
78. Marsella,J.A.; Caulton,K.G. J.Amer.Chem.Soc., 1980, 102, 1747-1748.
79. Gell,K.I.; Schwartz,J. J.Organometal.Chem.,1978,162,C11-C15.
80. Ginsburg,R.E.; Berg,J.M.; Rothrock,R.K.; Collman,J.P.; Hodgson,K.O.; Dahl,L.F. J.Amer.Chem.Soc., 1979, 101, 7218-7231, and references cited.
81. Reger,D.L.; Coleman,C.J. J.Organometal.Chem.,1977,131,153-162.
82. Beck,W.; Schloter,K. Z.Naturforsch.,B, 1978, 33B, 1214-1222.
83. Foxman,B.; Klemarczyk,P.T.; Liptrot,R.E.; Rosenblum,M. J.Organometal.Chem., 1980, 187, 253-265.
84. LaCroce,S.J.; Cutler,A.R. unpublished observations.
85. Johnson,E.C.; Meyer,T.J.; Winterton,N. Inorg.Chem., 1971, 10, 1673-1675.
86. Williams,W.E.; Lalor,F.J. J.Chem.Soc.,Dalton Trans., 1973, 1329-1332.
87. LaCroce,S.J.; Menard,K.P.; Cutler,A.R. J.Organometal.Chem., 1980, 190, C79-C83.
88. Menard,K.P.; Cutler,A.R., unpublished results.

RECEIVED December 8, 1980.

# 20

# Aromatic Gasoline From Hydrogen/Carbon Monoxide Over Ruthenium/Zeolite Catalysts

T. J. HUANG and W. O. HAAG

Mobil Research and Development Corporation, P.O. Box 1025, Princeton, NJ 08540

A new class of synthesis gas conversion catalysts comprising a carbon monoxide reduction catalyst combined with a ZSM-5 class zeolite has been recently reported by Chang, Lang and Silvestri (1). In elaborating on this finding, Caesar, et.al., have demonstrated that gasoline can be produced in a yield of over 60% of total hydrocarbon, constituting essentially 100% of the liquid product, by combining an iron Fischer-Tropsch catalyst with an excess volume of a ZSM-5 class zeolite (2). These zeolites are members of the group of Mobil shape selective medium pore zeolites which are active for the conversion of methanol and other oxygenates to hydrocarbons (1,3,4) or Fischer-Tropsch reaction intermediates to aromatics (1).

Ruthenium has been used as a Fischer-Tropsch catalyst to convert synthesis gas into paraffin wax under high pressure and at low temperature (5). However, at higher temperature and lower pressure, only methane is formed (6). Supported ruthenium such as Ru/alumina and Ru/silica has also been used for syngas conversion to produce gaseous, liquid and solid hydrocarbons (7-13); but, it gave a poor selectivity for liquid hydrocarbon and, again, methane becomes the major product at temperatures higher than 250°C. Futheremore, no aromatics were produced using both ruthenium dioxide and supported ruthenium catalyst.

In the Fischer-Tropsch synthesis with ruthenium as catalyst, normal paraffins are the major products. In Mobil's Distillate Dewaxing process, ZSM-5 class catalysts convert selectively high molecular weight n-paraffins into gasoline range materials (14). Thus, ruthenium-ZSM-5 class zeolites appear to be good combination for gasoline production from synthesis gas. In addition, these combination catalysts may provide a "nontrivial polystep" reaction (15) in which Fischer-Tropsch intermediates could be trapped and converted into aromatics by the zeolite component, thus producing high octane aromatic gasoline directly from synthesis gas. The successful use of the zeolites mentioned is a result of the unique properties of this class of intermediate pore zeolites, of which ZSM-5 is a prominent menber. It was chosen as a representative of this class in the present study.

0097-6156/81/0152-0307$05.00/0
© 1981 American Chemical Society

## Experimental

5% Ru(as $RuO_2$)/ZSM-5 was prepared by grinding together the appropriate amounts of $RuO_2$ and ZSM-5 zeolite, followed by pelleting and screening to 30-60 mesh. Impregnated Ru/ZSM-5 was prepared by vacuum impregnation of ZSM-5 zeolite (in $NH_4^+$ form) with $RuCl_3 \cdot 3H_2O$ in aqueous solution. After drying in vacuum, the catalyst was calcined in an oven at 538°C for two hours. Ru/$Al_2O_3$//ZSM-5 was a physical mixture of equal amounts of supported Ru - on - alumina and ZSM-5 zeolite. Supported Ru - on - alumina was prepared by vacuum impregnation of $\gamma$ - alumina with $RuCl_3 \cdot 3H_2O$ in aqueous solution, followed by drying in a rotary evaporator at about 100°C and in a vacuum oven at 102°C for two hours. Prior to syngas conversion, all ruthenium containing catalysts were reduced with hydrogen.

Syngas conversion was conducted in a down flow fashion in a fixed-bed continuous flow micro-reactor. The preheater and reaction zone were made of 1.42 cm i.d. type 321 stainless-steel tubing enclosed in a three-zone electrical resistance block heater. Gas flow was controlled using a Brooks Instrument flow controller. Liquid product was collected directly in a pressured Jerguson sight glass at ambient temperature. The exit gas passed through a condenser and a Grove back-pressure regulator to a wet test meter where the exit gas flow rate was measured. The condensed hydrocarbon in the high pressure Jerguson sight glass was further weathered to atmospheric pressure.

Product analyses were carries out by gas chromatography.

## Results and Duscussions

Generally speaking, hydrogenation of Co on ruthenium is similar to synthesis reactions on cobalt and nickel in so far as the oxygen of CO is rejected essentially as water. However, support materials may induce a shift reaction and may lead to production of some $CO_2$. As shown in Table III, the mole ratios of $H_2O$ to $CO_2$ in the reactor effluent were 29,54,and 5 for $RuO_2$, impregnated Ru/ZSM-5, and supported Ru/$Al_2O_3$, respecitvely. This is contrary to iron Fischer-Tropsch catalyst which gives $CO_2$ as the major oxygen containing product. For example, the mole ratio of $H_2O$ to $CO_2$ in the product from syngas conversion over iron catalysts (1,5,16) is generally less than 0.1. This difference arises from the fact that iron is active for water-gas shift reaction (Equation II) while ruthenium is not.

$$2H_2 + CO \longrightarrow (-CH_2-) + H_2O \quad (I)$$

$$H_2O + CO \longrightarrow CO_2 + H_2 \quad (II)$$

For ruthenium catalysts without shift activity, the stoichiometric requirement for syngas conversion is two moles of $H_2$ per mole of CO, according to Equation (I). However, the $H_2$/CO usage ratio can be less than 2 when the catalyst has shift activity (Equation II).

Effect of Zeolite Results with composite catalysts consisting of a supported $Ru/Al_2O_3$ and a zeolite are given in Table I. Although the zeolites themself have no effect on the syngas conversion, the hydrocarbon product distribution is affected by the presence of a zeolite, particularly of the ZSM-5 class. The incorporation of ZSM-5 in the catalyst not only promoted aromatics formation, but also significantly reduced the end point of the hydrocarbons. The total hydrocarbon fraction contained 66 wt% of $C_5^+$, which was essentially an aromatic gasoline (34% aromatics, 204°C boiling point at 90% overhead). It must be noted that, with ruthenium alone (Ex. 1A), no aromatics were produced and the boiling point of $C_5^+$ at 90% overhead was 322°C.

The presence of a large pore zeolite, H-mordenite, reduced the end point of $C_5^+$ only slightly. Mordenite initially gave aromatics with substantial amount of $C_{10}^+$ aromatics, but it deactivated very rapidly.

Effect of Ruthenium Loading Two ruthenium concentrations (0.5% and 1.5%) were used to study the effect of ruthenium loading on syngas conversion over physically mixed $Ru/Al_2O_3$//ZSM-5 catalysts. The results are shown in Table II. The formation of $C_1+C_2$ was greatly reduced from 40% w th 1.5% Ru to 25% with 0.5% Ru. On the other hand, the higher ruthenium loading gave a $C_5^+$ product of reduced end point (Ex. 2A and 2B). As expected, no difference in aromatics production was observed.

The same effect was seen with impregnated Ru/ZSM-5 catalysts of 1% and 5% Ru-content (Ex.2C and 2D).

Effect of Method Of Catalyst Preparation Three catalysts with different methods of preparation were used in this study and the results are given in Table III. Although they have the same ruthenium loading (5%), the degree of intimacy between ruthenium sites and active sites of ZSM-5 increased with the following order: Impregnated Ru/ZSM-5 > $RuO_2$/ZSM-5 (ground together) > Physical Mixture of $Ru/Al_2O_3$ and ZSM-5. The most striking feature was that the formation of $C_{11}^+$ heavy aromatics increased with increasing degree of intimacy between Ru and ZSM-5, as shown in Table IV. This may indicate that if ruthenium sites and acid sites of ZSM-5 are located closely together as in the case of the impregnated catalyst, the aromatics formed as zeolite sites may be further alkylated with the reaction intermediates produced at the neighboring ruthenium sites, consequently making heavy aromatics.

The variations in syngas conversion and $C_1+C_2$ selectively could be due to the difference in ruthenium surface areas as a result of different preparations.

A finely ground physical mixture of 5% Ru (as $RuO_2$)/ZSM-5 which was subsequently pelletized was used in the study of the effects of process variables on synthesis gas conversion.

Table I

SYNGAS CONVERSION OVER RUTHENIUM/ZEOLITE CATALYSTS AT 51 atm, 294°C, GHSV = 480, AND $H_2/CO$ = 2/1.

| Experiment No. | 1A | 1B | 1C |
|---|---|---|---|
| Catalyst | 0.5% $Ru/Al_2O_3$// Quartz Chips "Mixed" | 0.5% $Ru/Al_2O_3$// ZSM-5 "Mixed" | 0.5% $Ru/Al_2O_3$// H-Mordenite "Mixed" |
| Syngas Conversion, mole % | 94 | 98 | 95 |
| Reactor Effluent, wt% | | | |
| Hydrocarbons | 37 | 38 | 36 |
| $H_2$ | 0 | 0 | 1 |
| CO | 11 | 1 | 6 |
| $CO_2$ | 6 | 12 | 8 |
| $H_2O$ | 46 | 49 | 49 |
| Hydrocarbon Composition, wt% | | | |
| $C_1$+$C_2$ | 33 | 25 | 29 |
| $C_3$+$C_4$ | 8 | 9 | 7 |
| $C_5^+$ | 59 | 66 | 63 |
| Aromatics in $C_5^+$, wt% | 0 | 34 | < 5 |
| Boiling Range of $C_5^+$, °C | | | |
| 90% Overhead | 322 | 204 | 275 |
| 95% Overhead | 377 | 224 | 322 |

Table II

EFFECT OF RUTHENIUM LOADING ON SYNGAS CONVERSION
(51 atm, GHSV = 480, AND $H_2/CO$ = 2/1)

| Experiment No. | 2A | 2B | 2C | 2D |
|---|---|---|---|---|
| Catalyst | 1.5% Ru/$Al_2O_3$// ZSM-5 "Mixed" | 0.5% Ru/$Al_2O_3$// ZSM-5 "Mixed" | 5% Ru/ ZSM-5 "Impreg- nated" | 1% Ru/ ZSM-5 "Impreg- nated" |
| Temp., °C | 294 | 294 | 304 | 304 |
| Syngas Conversion, mole % | 99 | 98 | 86 | 83 |
| Reactor Effluent, wt% | | | | |
| Hydrocarbons | 39 | 38 | 38 | 32 |
| $H_2$ | 0 | 0 | 1 | 2 |
| CO | 0 | 1 | 15 | 18 |
| $CO_2$ | 17 | 12 | 2 | 3 |
| $H_2O$ | 44 | 49 | 44 | 45 |
| Hydrocarbon Composition, wt% | | | | |
| $C_1+C_2$ | 40 | 25 | 38 | 20 |
| $C_3+C_4$ | 12 | 9 | 16 | 29 |
| $C_5^+$ | 48 | 66 | 46 | 51 |
| Aromatics in $C_5^+$, wt% | 32 | 43 | 25 | 27 |
| Boiling Range of $C_5^+$, °C | | | | |
| 90% Overhead | 174 | 204 | - | - |
| 95% Overhead | 186 | 224 | - | - |

Table III

EFFECT OF METHOD OF CATALYST PREPARATION ON SYNGAS CONVERSION OVER Ru/ZSM-5 CLASS ZEOLITE. (51 atm, GHSV = 480, AND $H_2/CO$ = 2/1)

| Experiment No. | 3A | 3B | 3C |
|---|---|---|---|
| Catalyst | Ru/ZSM-5 | $RuO_2$ Plus ZSM-5 | $Ru/Al_2O_3$ Plus ZSM-5 |
| Method of Preparation | Impregnation | Ground Together | Physical Mixture |
| Particle Size, mesh | 30 - 60 | 30 - 60 | 30 - 60 |
| Ruthenium Loading, wt% | 5% | 5% | 5% |
| (Based on Total Solid) | | | |
| Temp., °C | 304 | 294 | 294 |
| Syngas Conversion, mole % | 86 | 93 | 99 |
| Reactor Effluent, wt% | | | |
| Hydrocarbons | 38 | 36 | 40 |
| $H_2$ | 1 | 0 | 0 |
| CO | 15 | 12 | 0 |
| $CO_2$ | 2 | 4 | 20 |
| $H_2O$ | 44 | 48 | 40 |
| Hydrocarbon Composition, wt% | | | |
| $C_1+C_2$ | 38 | 30 | 43 |
| $C_3+C_4$ | 16 | 17 | 14 |
| $C_5^+$ | 46 | 53 | 43 |
| Aromatics in $C_5^+$, wt% | 25 | 28 | 24 |
| Aromatics Distribution wt% | | | |
| $A_6$-$A_{10}$ | 79 | 90 | 97 |
| $A_{11}$ | 21 | 10 | 3 |

Table IV

EFFECT OF PRESSURE ON SYNGAS CONVERSION OVER 5% Ru (AS $RuO_2$)/ ZSM-5 AT 294°C, $H_2$/CO = 2/1, AND GHSV = 480.

| | | | | |
|---|---|---|---|---|
| Pressure, atm. | 13.6 | 27.2 | 51 | 75 |
| Conversion, wt% | | | | |
| CO | 63 | 78 | 86 | 90 |
| $H_2$ | 77 | 92 | 96 | 98 |
| Total Product, wt% | | | | |
| Hydrocarbon | 29.8 | 35.2 | 35.5 | 37.4 |
| $H_2$ | 2.9 | 1.4 | 0.5 | 0.2 |
| CO | 32.6 | 20.8 | 11.8 | 8.8 |
| $CO_2$ | 1.8 | 3.1 | 3.6 | 3.8 |
| $H_2O$ | 32.9 | 39.5 | 48.6 | 49.8 |
| Hydrocarbon Composition, wt% | | | | |
| $C_1$ | 52.8 | 44.5 | 26.0 | 26.1 |
| $C_2°$ | 5.9 | 5.0 | 4.3 | 3.4 |
| $C_2=$ | - | - | - | - |
| $C_3°$ | 7.7 | 7.1 | 5.1 | 3.0 |
| $C_3=$ | - | - | 0.7 | 0.2 |
| i-$C_4$ | 10.3 | 10.5 | 5.6 | 3.5 |
| n-$C_4$ | 5.0 | 6.1 | 4.5 | 3.6 |
| $C_4=$ | - | - | 0.9 | - |
| i-$C_5$ | 6.3 | 6.1 | 5.4 | 4.1 |
| n-$C_5$ | 1.5 | 2.3 | 3.5 | 4.0 |
| $C_6$+ non-aromatics | 2.0 | 7.8 | 29.5 | 40.8 |
| Aromatics | 8.5 | 10.6 | 14.7 | 11.5 |
| $C_5$+ in Total H.C., wt% | 18.3 | 26.8 | 53.1 | 60.4 |
| Aromatics in $C_5$+, wt% | 46.2 | 39.4 | 27.7 | 19.1 |
| Hydrocarbon Selectivity, wt% | 98.0 | 96.3 | 97.4 | 97.0 |

Effect of Pressure The results are listed in Table IV. The effect of pressure on conversions and selectivities are shown in Figure 1. The CO conversion increased from 63% at 13.6 atm (200psig) to 90% at 75 atm (1100psig). The hydrocarbon selectivity, defined as (total carbon converted - total carbon in $CO_2$) ÷ total carbon converted, remained steady at 97%, the rest of 3% being converted to $CO_2$. The selectivity of $C_5^+$ increased with increasing pressure while the aromatics in $C_6^+$ decreased. The $C_1+C_2$ make was substantially reduced by higher pressure, for example, from 59% at 13.6 atm to 29% at 74 atm.

Effect of Temperature Three temperatures in the range of 264 to 328°C (507 to 613°F) were used for the study of the effect of temperature. The detailed conditions and results are included in Table V. The plots of conversion and selectivities versus temperature are shown in Figure 2. Both hydrocarbon selectivity and $H_2$ and CO conversions were high in this range. The slightly lower CO conversion at higher temperature could be due to the greater yield of $C_1+C_2$. The two key features emerging from this study are the sharp increase in $C_5^+$ and the sharp decrease in $C_1+C_2$ as a function of decreasing temperature. At 264°C, the $C_1+C_2$ make was reduced to 11%. No temperature lower than 264 °C was investigated in this study although $C_3^+$ yield could possibly be increased above the 89% obtained at 264°C. The aromatics in total hydrocarbon went through a maximum at 294°C. The lower aromatics selectivity at 264°C was probably due to the poor aromatization activity of ZSM-5 at such a low temperature, while at higher temperature methane formation competes.

Effect of Space Velocity The data are given in Table VI, and the conversion and selectivities are plotted against 1/WHSV in Figure 3. Clearly, at longer contact time (or lower space velocity), $C_5^+$ decreased and $C_1+C_2$ increased. Thus the latter appear to be formed as a sequentail reaction product.

Effect of $H_2$/CO Ratio Three different $H_2$/CO ratio (1/2, 1/1, and 2/1) were employed in this study. The detailed results are listed in Table VII. As shown in Figure 4, the CO conversion increased (from 20 to 78%) with increasing $H_2$/CO ratio. Since CO conversion is stoichiometrically limited by the amount of hydrogen available in the feed in view of the absence of water-gas shift reaction, the lower the $H_2$/CO ratio, the lower the maximum attainable CO conversion. For example, with the $H_2$/CO ratio of 1/2, the maximum attainable CO conversion, based on the stoichiometry of syngas conversion over ruthenium catalysts of $2H_2$/1CO (Equation I), would be 25%. Therefore, the 20% apparent CO conversion at the $H_2$/CO ratio of 1/2 reflected 80% of the maximum attainable CO conversion. In the range of $H_2$/CO ratio employed here, the possible CO conversion was all high, amounting to about 80% of the maximum attainable CO conversion, as represented by the dotted line in Figure 4.

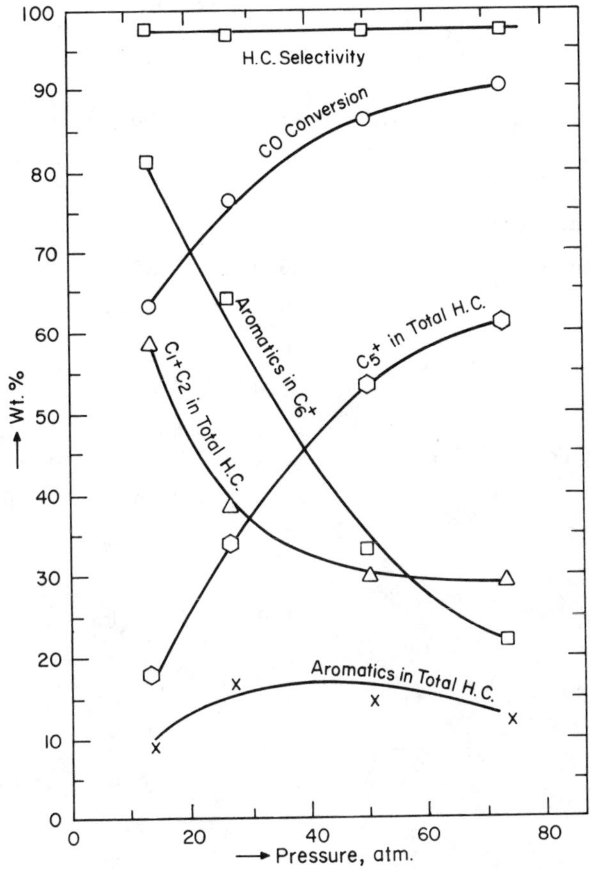


*Figure 1. Effect of pressure on syngas conversion over 5% Ru(as $RuO_2$)/ZSM-5 (294°C, GHSV = 480, and $H_2/CO$ = 2/1)*

Table V

EFFECT OF TEMPERATURE ON SYNGAS CONVERSION OVER 5% Ru (AS $RuO_2$)/ ZSM-5 AT 51 atm, GHSV = 480, AND $H_2/CO$ = 2/1.

| Temp., °C | 264 | 294 | 328 |
|---|---|---|---|
| Conversion, wt% | | | |
| CO | 93 | 86 | 78 |
| $H_2$ | 98 | 96 | 97 |
| Total Product, wt% | | | |
| Hydrocarbon | 40.0 | 35.5 | 35.8 |
| $H_2$ | 0.2 | 0.5 | 0.4 |
| CO | 6.3 | 11.8 | 18.9 |
| $CO_2$ | 1.0 | 3.6 | 4.5 |
| $H_2O$ | 52.6 | 48.6 | 40.0 |
| Hydrocarbon Composition, wt% | | | |
| $C_1$ | 10.3 | 26.0 | 61.3 |
| $C_2°$ | 1.0 | 4.3 | 6.5 |
| $C_2=$ | - | - | - |
| $C_3°$ | 1.2 | 5.1 | 4.5 |
| $C_3=$ | 0.1 | 0.7 | - |
| i-$C_4$ | 1.9 | 5.6 | 5.3 |
| n-$C_4$ | 4.2 | 4.5 | 3.4 |
| $C_4=$ | 0.2 | 0.9 | - |
| i-$C_5$ | 3.1 | 5.4 | 3.9 |
| n-$C_5$ | 4.7 | 3.5 | 1.4 |
| $C_6^+$ non-aromatics | 65.8 | 29.5 | 8.2 |
| Aromatics | 7.7 | 14.7 | 8.2 |
| $C_5^+$ in Total H.C., wt% | 81.3 | 53.1 | 19.1 |
| Aromatics in $C_5^+$, wt% | 9.4 | 27.7 | 43.1 |
| Hydrocarbon Selectivity, % | 99.2 | 97.4 | 95.8 |
| Bromine No. of Liq. Product | 90 | | |

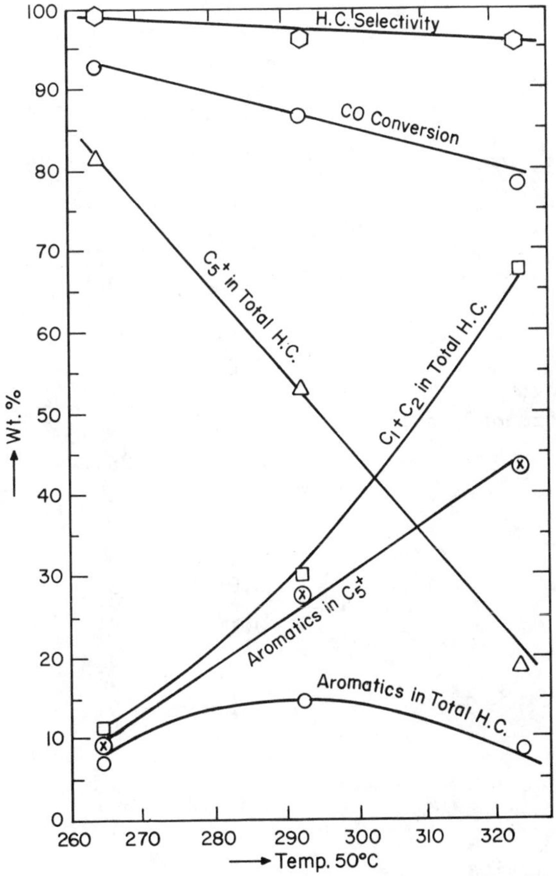


*Figure 2. Effect of Temperature on syngas conversion over 5% Ru(as $RuO_2$)/ZSM-5 (294°C, GHSV = 480, and $H_2/CO = 2/1$)*

Table VI

EFFECT OF SPACE VELOCITY ON SYNGAS CONVERSION OVER 5% Ru(AS $RuO_2$) ZSM-5 AT 294°C, 75 atm, AND $H_2/CO$ = 2/1.

| GHSV | 180 | 480 | 1428 |
|---|---|---|---|
| Conversion, wt% | | | |
| CO | 91 | 90 | 93 |
| $H_2$ | 98 | 98 | 98 |
| Total Product, wt% | | | |
| Hydrocarbon | 37.0 | 37.4 | 34.9 |
| $H_2$ | 0.2 | 0.2 | 0.2 |
| CO | 7.6 | 8.8 | 5.8 |
| $CO_2$ | 9.2 | 3.8 | 0.9 |
| $H_2O$ | 46.0 | 49.8 | 58.1 |
| Hydrocarbon Composition, wt% | | | |
| $C_1$ | 34.4 | 26.1 | 13.4 |
| $C_2°$ | 5.1 | 3.4 | 1.6 |
| $C_2=$ | - | - | - |
| $C_3°$ | 5.0 | 3.0 | 1.8 |
| $C_3=$ | 0.1 | 0.2 | 0.1 |
| i-$C_4$ | 5.9 | 3.5 | 1.8 |
| n-$C_4$ | 4.1 | 3.6 | 2.6 |
| $C_4=$ | - | - | 0.1 |
| i-$C_5$ | 4.6 | 4.1 | 3.0 |
| n-$C_5$ | 3.3 | 4.0 | 3.7 |
| $C_6^+$ non-aromatics | 23.7 | 40.8 | 58.7 |
| Aromatics | 14.0 | 11.5 | 13.3 |
| $C_5^+$ in Total H.C., wt% | 45.6 | 60.4 | 78.7 |
| Aromatics in $C_5^+$, wt% | 30.7 | 19.1 | 16.9 |
| Hydrocarbon Selectivity, wt% | 92.7 | 97.0 | 99.3 |

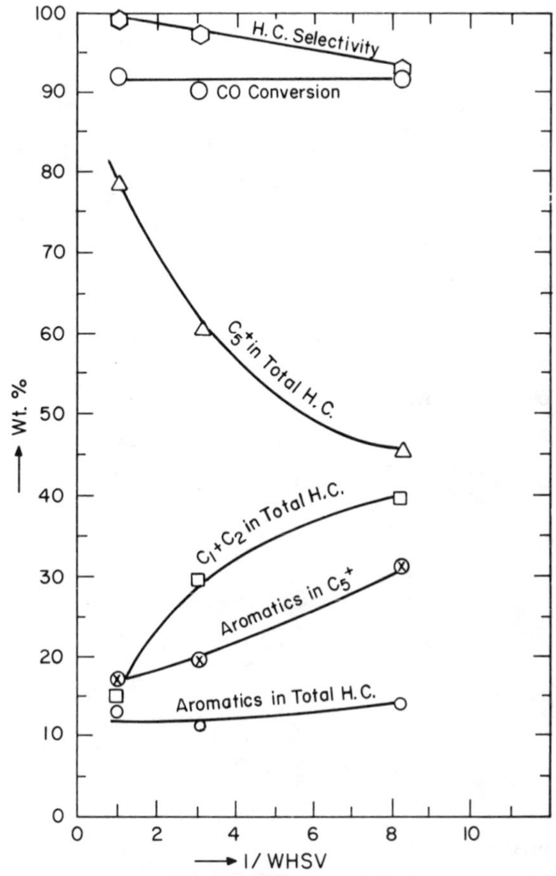


*Figure 3. Effect of space velocity on syngas conversion over 5% Ru(as $RuO_2$)/ ZSM-5 (294°C, 75 atm, and $H_2CO = 2/1$)*

Table VII

EFFECT OF $H_2$/CO RATIO ON SYNGAS CONVERSION OVER 5% Ru (AS $RuO_2$)/ ZSM-5 AT 294°C, 27 atm, AND GHSV = 480.

| $H_2$/CO, mole ratio | 1/2 | 1/1 | 2/1 |
|---|---|---|---|
| Conversion, wt% | | | |
| CO | 20 | 38 | 78 |
| $H_2$ | 74 | 85 | 92 |
| Total Product, wt% | | | |
| Hydrocarbon | 10.9 | 19.5 | 35.2 |
| $H_2$ | 0.9 | 1.0 | 1.4 |
| CO | 77.7 | 58.3 | 20.8 |
| $CO_2$ | 2.4 | 3.6 | 3.1 |
| $H_2O$ | 8.3 | 17.6 | 39.5 |
| Hydrocarbon Composition, wt% | | | |
| $C_1$ | 17.7 | 24.4 | 44.5 |
| $C_2°$ | 3.3 | 2.6 | 5.0 |
| $C_2$= | - | - | - |
| $C_3°$ | 15.7 | 12.4 | 7.1 |
| $C_3$= | 1.5 | - | - |
| i-$C_4$ | 20.5 | 19.9 | 10.5 |
| n-$C_4$ | 10.5 | 10.1 | 6.1 |
| $C_4$= | - | - | - |
| i-$C_5$ | 8.9 | 9.9 | 6.1 |
| n-$C_5$ | 2.8 | 1.4 | 2.3 |
| $C_6^+$ non-aromatics | 1.2 | 3.5 | 7.8 |
| Aromatics | 18.1 | 15.8 | 10.6 |
| $C_5^+$ in Total H.C., wt% | 30.9 | 30.6 | 26.8 |
| Aromatics in $C_5^+$, wt% | 58.4 | 51.5 | 39.4 |
| Hydrocarbon Selectivity, % | 92 | 93.5 | 86.3 |
| Octane No. (R+O) of Liquid Product | 104 | 102 | 94 |
| Boiling Range of $C_5^+$, °C 90% Overhead | 212 | 201 | 182 |

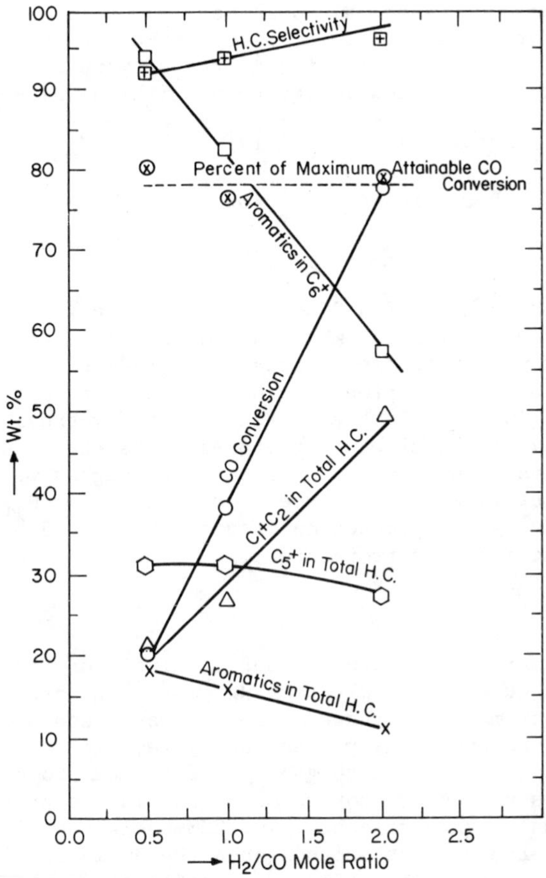


*Figure 4. Effect of $H_2/CO$ ratio on syngas conversion over 5% Ru(as $RuO_2$)/ ZSM-5 (294°C, 27 atm, and GHSV = 480)*

The yield was rather insensitive to the $H_2/CO$ ratio in this range. However, the aromatics in $C_6^+$ increased sharply with decreasing $H_2/CO$ ratio, reaching 94% at the $H_2/CO$ ratio of 1/2; the aromatics in total hydrocarbon increased from 11% at 2/1 ratio to 18% at 1/2 ratio. More importantly, $C_1$+$C_2$ formation decreased sharply with decreasing $H_2/CO$ ratio. The boiling range of $C_5^+$ was shifted toward higher boiling point as the $H_2/CO$ ratio was decreased, although more aromatics and less methane were produced with lower $H_2/CO$ ratio. Furthermore, the octane number of the $C_5^+$ fraction increased with decreasing $H_2/CO$ ratio as shown below in agreement with the higher aromatics content.

| $H_2/CO$ Ratio | 1/2 | 1/1 | 2/1 |
|---|---|---|---|
| Octane No. (R+O) | 104 | 102 | 94 |

## Conclusion

The incorporation of a ZSM-5 class zeolite into a ruthenium Fischer-Tropsch catalyst promotes aromatics formation and reduces the molecular weight of the hydrocarbons produced. These composite catalysts can produce a high octane aromatic gasoline in good yield in a single step directly from synthesis gas.

The study of the effects of process parameters reveals that (1) methane can be substantially reduced by higher pressure, shorter contact time, lower temperature, and lower $H_2/CO$ ratio; and (2) the aromatics production is greatly favored by lower $H_2/CO$ ratio at moderate temperature.

## Abstract

Ruthenium is known to be a good catalyst for producing high molecular weight paraffin wax from $H_2/CO$ at high pressure and low temperature, or making methane at low pressure and moderate temperature. However, the present study reveals that aromatic gasoline of high quality with good yield can be produced directly from synthesis gas under proper conditions in the presence of dual-functional ruthenium-containing ZSM-5 class zeolite catalysts. The nature of the product depends upon the dual-functionality of the catalyst system. The effects of method of catalyst preparation and ruthenium loading, as well as process variables such as temperature, pressure, space velocity and $H_2/CO$ ratio on the product distribution are discussed.

## Acknowledgement

We wish to thank Mr. C. L. Tatsch and Mr. R. M. Wallace for their excellent technical assistance.

## Literature Cited


1. Chang, C. D.; Lang, W. H.; Silvestri, A. J. J. Catal., 1979, 56, 268.
2. Caesar, P. D.; Brennan, J. A.; Garwood, W. E.; Ciric, J. J. Catal., 1979, 56, 274.
3. Meisel, S. L.; McCullough, J. P.; Lechthaler, C. H.: Weisz, P. B. Chemtech 1976, 6, 86.
4. Chang, C. D.; Silvestri, A. J. J. Catal., 1977, 47,249.
5. Storch, H. H.; Golumbic, N.; Anderson, R. B. "The Fischer-Tropsch and Related Syntheses", Wiley, New York, 1951, p. 309.
6. Pichler, H. Adv. Cat., Vol. IV, 1952, 271.
7. Karn, F. S.; Schultz, J. F.; Anderson, R. B. I & E C Product Research and Development, 1956, 4, 265.
8. Ekerdt, J. G.; Bell, A. T. J. Catal, 1979, 58, 170.
9. Ollis, D. F.; Vannice, M. A. J. Catal., 1975, 37, 449.
10. King, D. L. J. Catal., 1978, 51, 386.
11. Everson, R. C.; Woodburn, E. T.; Kirk, A. R. J. Catal., 1978, 53, 186.
12. Dalla Betta, R. A.; Shelef, M. J. Catal., 1977, 49, 383.
13. Jacobs, P. A.; Nijs, H. H.; Verdonck, J. J.; Uytterhoven, J. B. Preprint. Petroleum Chem. Div., Am. Chem. Soc., March 12-17, 1978, p. 469.
14. Meisel, S. L.; McCullough, J. P.; Lechthaler, C. H.; Weisz, P. B. Leo Friend Symposium, Am. Chem. Soc., Chicago, Illinois, Aug. 30, 1977.
15. Weisz, P. B. Adv. Catal., 1962, 13, 137.
16. Huang, T. J.; Haag, W. O. Unpublished data.


RECEIVED December 8, 1980.

# 21

# Mechanistic Aspects of the Homogeneous Water Gas Shift Reaction

W. A. R. SLEGEIR, R. S. SAPIENZA, and B. EASTERLING

Catalysis Group, Department of Energy and Environment, Brookhaven National Laboratory, Upton, NY 11973

The homogeneous water gas shift reaction (WGSR),

$$CO + H_2O \rightleftharpoons H_2 + CO_2 \qquad (1)$$

has been the subject of a number of reports in the recent literature (1-19). The reaction is used to provide hydrogen for processes such as ammonia synthesis, as well as to match synthesis feedgas ratios to consumption ratios in a number of downstream processes, such as hydroformylation ($H_2$:CO = 1), methanol synthesis ($H_2$:CO = 2) and methanation ($H_2$:CO = 3). Operating temperature plays a significant role in the effectiveness of a particular catalyst system, since the thermodynamically limiting conversion decreases as temperature increases ($K_{P,400^oK}$ = 1550; $K_{P,700^oK}$ = 9.5 (20)).

Currently heterogeneous catalysts are employed in industry for the WGSR, but these operate at high temperature (21). In this chapter, some homogeneous catalyst systems with high activity at modest temperatures will be described. Whether a homogeneous system will supplant currently used heterogeneous systems is dependent on a number of engineering process considerations. However, interest in this reaction centers about its deceptive simplicity, a better understanding of which should enhance the knowledge of catalytic reactions of carbon oxides. Furthermore, this reaction is related to a number of other catalytic reactions with potential synfuels importance including the Reppe hydrohydroxymethylation reaction (4,22),

$$RCH{=}CH_2 + 3CO + 2H_2O \longrightarrow RCH_2CH_2CH_2OH + 2CO_2 \qquad (2)$$

the Kölbel-Engelhardt reaction (23),

$$(3n+1)CO + (n+1)H_2O \longrightarrow H(CH_2)_nH + (2n+1)CO_2, \qquad (3)$$

0097-6156/81/0152-0325$05.00/0
© 1981 American Chemical Society

and other variations of the Fischer-Tropsch reaction, methanol synthesis from CO and $H_2O$ (1),

$$3CO + 2H_2O \longrightarrow CH_3OH + 2CO_2 \quad (4)$$

as well as reductions of organic substrates using CO and $H_2O$ (15,24),

$$ArNO_2 + 3CO + H_2O \longrightarrow ArNH_2 + 3CO_2. \quad (5)$$

This discussion will treat two particular catalytic systems affording rapid rates in basic media. The ruthenium carbonyl-trimethylamine system, exceedingly selective for the WGSR, is the most effective known homogeneous system. Cluster dissociation and formation complicate this system, but evidence strongly indicates an associative-type reaction mechanism. On the other hand, the Group VI metal carbonyls exhibit a rather straightforward dissociative-type mechanism. The application of this understanding may be helpful in selecting catalytic systems for $CO/H_2O$ reactions.

## The Ruthenium Carbonyl System

The reactions with ruthenium carbonyl catalysts were carried out in pressurized stainless steel reactors; glass liners had little effect on the activity. When trimethylamine is used as base, $Ru_3(CO)_{12}$, $H_4Ru_4(CO)_{12}$ and $H_2Ru_4(CO)_{13}$ lead to nearly identical activities if the rate is normalized to the solution concentration of ruthenium. These results suggest that the same active species is formed under operating conditions from each of these catalyst precursors. The ambient pressure infrared spectrum of a typical catalyst solution (prepared from $Ru_3(CO)_{12}$, trimethylamine, water, and tetrahydrofuran and sampled from the reactor) is relatively simple ($\nu_{C-O}$: 2080(w), 2020(s), 1997(s), 1965(sh) and 1958(m) $cm^{-1}$). However, the spectrum depends on the concentration of ruthenium in solution. The use of $Na_2CO_3$ as base leads to comparable spectra.

Although this spectrum does not correspond to any particular ruthenium carbonyl complex, it is consistent with the presence of one or more anionic ruthenium carbonyl complexes, perhaps along with neutral species. Work is in progress with a variable path-length, high pressure infrared cell designed by Prof. A. King, to provide better characterization of species actually present under reaction conditions.

After reaction, evaporation of the solvent from the $Ru_3(CO)_{12}/NMe_3$ solution, followed by protonation with $H_3PO_4$ yields principally $H_4Ru_4(CO)_{12}$, with some $Ru_3(CO)_{12}$ and $H_2Ru_4(CO)_{13}$. Isolation of the active anionic species has not been successful.

All evidence is consistent with the reaction being homogeneous. Upon cooling of the reaction to 0°C, the originally homogeneous solution normally consists of two phases. The phase separation results from the formation of ionic species (formates and carbonates) during the course of the reaction which "salt-out" the organic phase. The lower (aqueous) phase is clear and colorless. The upper (organic) phase is pale yellow to pinkish red, depending on ruthenium concentration. Visual inspection and infrared spectra indicate that the ruthenium carbonyl species are present in the upper phase. Neither phase is turbid. Filtration leaves no observable residue and the filtrate affords WGSR rates comparable to those of fresh solutions. Neither phase is paramagnetic, and neither contains appreciable suspended material as evidenced by sharp NMR signals.

The $Ru_3(CO)_{12}/NMe_3$ catalyst system is very specific for the WGSR. Although heterogeneous ruthenium is a very effective catalyst for methanation (20), no methane or higher hydrocarbons could be detected. Homogeneous ruthenium has been shown to catalyze methanol synthesis at very high temperatures and pressures (25), but only traces of methanol are detected under WGSR conditions. Small amounts of the formate ion are formed, but this seems unavoidable in reactions involving base and CO,

$$CO + OH^- \longrightarrow HCO_2^- . \qquad (6)$$

The role of formate in the WGSR will be discussed below. Generally more $H_2$ than $CO_2$ is observed at the end of the reaction. Experiments (26) suggest that this is due to the solubility of $CO_2$ in the solvent system.

Effect of Concentration and CO Pressures on the Ruthenium Carbonyl-Trimethylamine WGSR System. As shown in Figure 1, the $Ru_3(CO)_{12}/NMe_3$ WGSR system demonstrates a nearly first-order rate dependence on CO pressure at 0.5 mM $Ru_3(CO)_{12}$ concentration. (Throughout this discussion, the total ruthenium carbonyl concentration is expressed as moles $Ru_3(CO)_{12}$ added per liter of solution; this should not be construed to be the actual solution concentration of the trimer under operating conditions.) Here the initial rates of $H_2$ production are 14.6 mmol $H_2$/hr at 415 psi CO and 46.0 mmol $H_2$/hr at 1200 psi. Thus, within experimental uncertainty, a threefold increase in CO pressure leads to a threefold increase in rate.

Ford and co-workers (7) have reported a first-order rate dependence on CO pressure in the $Ru_3(CO)_{12}$/KOH system and ascribed this effect to CO participation in a rate-limiting elimination of hydrogen from a cluster species. This explanation does not fit our observations, because if loss of $H_2$ were rate-limiting, the use of KOH and $NMe_3$ as bases would be expected to lead to comparable rates for the WGSR. A comparison of activities (Laine (9): 2.3 mol $H_2$ per mol $Ru_3(CO)_{12}$ per hr using KOH/MeOH at 10 mM $Ru_3(CO)_{12}$, 800 psi CO, 135°; Slegeir (26):

5000 mol $H_2$ per mol $Ru_3(CO)_{12}$ per hr using $NMe_3$/THF at 0.02 mM, 765 psi CO, 125°) shows that the activity is greater by more than three orders of magnitude when trimethylamine is used as base. We believe there is a better explanation for this activity dependence on CO pressure in the $Ru_3(CO)_{12}/NMe_3$ system.

TABLE I

Rate Dependence on CO Pressure at 0.10 mM $Ru_3(CO)_{12}$

| $P_{CO}$, psi | mmol $H_2$ | % conversion | $t_{H_2}$ |
|---|---|---|---|
| 350 | 86 | 43 | 8600 |
| 690 | 90 | 24 | 9000 |
| 760 | 82 | 18 | 8200 |

Conditions: 0.010 mmol $Ru_3(CO)_{12}$, 20 g 25% aq. $NMe_3$, diluted to 100 mL with THF, 100°, 10 hr, 0.31 L reactor, $t_{H_2}$ = mol $H_2$/mol $Ru_3(CO)_{12}$.

As shown in Table I, at 0.1 mM $Ru_3(CO)_{12}$ concentration, CO pressure has little if any effect on activity. On the other hand, at fixed pressure, the concentration of ruthenium carbonyl has a dramatic effect on activity (see Figure 2). At 0.1 mM $Ru_3(CO)_{12}$, ruthenium carbonyl is very active for the WGSR, small decreases in catalyst concentration lead to substantial increases in activity, and no activity dependenee on CO pressure is observed. At concentrations of 0.5 mM or more, less activity is observed, changes in concentration cause smaller effects in activity and rate dependence on pressure is manifested. Diffusion effects have been shown to be unimportant (26).

It is proposed that the rate dependence on concentration and pressure involves cluster dissociation and that the monomeric species, $Ru(CO)_5$, is responsible for the high activity of this system. Dissociation is well known for ruthenium carbonyl clusters (25,27-31). Piacenti and co-workers (31) have demonstrated that at temperatures above 80° and CO pressures greater than 150 psi, monomeric ruthenium carbonyl is observed in significant quantities due to the equilibrium,

$$\underset{\mathbf{1}}{Ru_3(CO)_{12}} + 3CO \rightleftharpoons \underset{\mathbf{2}}{3Ru(CO)_5}. \quad (7)$$

Thus, increases in CO pressure favor cluster dissociation and the formation of larger quantities of **2**. High dilution should also favor the formation of **2**, resulting in greater $Ru(CO)_5$ to cluster ratios and greater WGSR activity.

Assuming the WGSR has a first-order dependence on $Ru(CO)_5$ concentration and that only trimeric and monomeric species are present, it can be shown that rate $\propto P_{CO}$ (26), in accord with

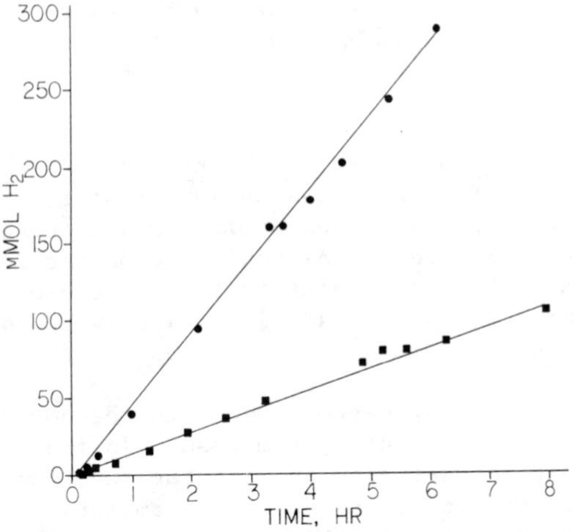


*Figure 1. Rate of $H_2$ production as a function of CO pressure at 0.50mM added $Ru_3(CO)_{12}$ concentration; 0.0507 mmol $Ru_3(CO)_{12}$, 5 g $NMe_3$, 15 g $H_2O$, solution diluted to 100 mL with THF, 100°C; (●) 1200 psi CO, (■) 415 psi CO*

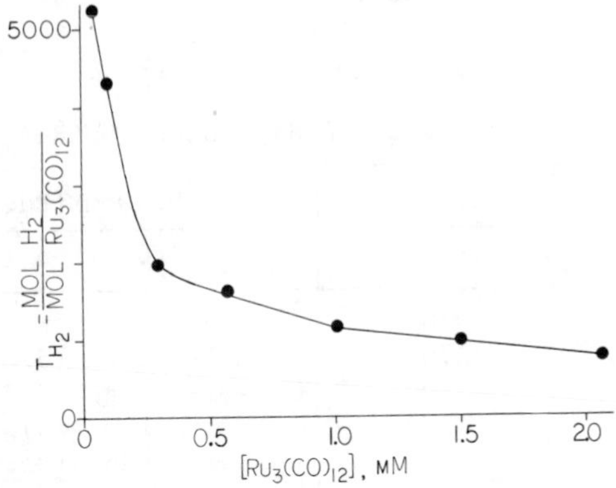


*Figure 2. $H_2$ production as a function of added $Ru_3(CO)_{12}$ concentration; 5 g $NMe_3$, 15 g $H_2O$, solution diluted to 100 mL with THF, 415 psi CO, 100°C, 5 h. The abscissa reflects the Ru carbonyl concentration based on initial catalyst loadings; clearly it does not reflect true $[Ru_3(CO)_{12}]$.*

experiments at 0.5 mM $Ru_3(CO)_{12}$. However, if tetrameric species are in equilibrium with 2, then rate $\propto P_{CO}^{1.5}$. The isolation of $H_4Ru_4(CO)_{12}$ from acidified reaction mixtures supports the existence of tetrameric species at higher ruthenium concentrations. Experiments at 2.07 mM $Ru_3(CO)_{12}$ concentration indicate that rate $\propto P_{CO}^{1.3}$, suggesting that clusters larger than the trimer may exist.

Effect of Solvent and Base on the Ruthenium Carbonyl/Trimethylamine System. Solvent plays an important role in the rate of hydrogen production. The ideal solvents are tetrahydrofuran, diglyme, and dimethoxyethane. Alcohols are only slightly less effective. Apparently the solvent must be miscible with water, promote ion formation, and be capable of weakly coordinating with the coordinately unsaturated species formed in the course of the reaction.

Small amounts of hydrocarbons added to the normal tetrahydrofuran or diglyme solvent system result in improved WGSR activity, but larger quantities inhibit the reaction (Table II). When 1-butene or 1-hexene is used, hydroformylation competes with the WGSR (4), but the rate of this process is small compared with the rate of $H_2$ production. With pentane, no olefin or aldehyde products could be detected. Calderazzo (29) has reported that $Ru(CO)_5$ is the principal product when the acetylacetonate of ruthenium is treated with synthesis gas in heptane,

$$\underset{(200\ \text{atm})}{Ru(acac)_3 + CO + H_2} \xrightarrow[180^o]{\text{heptane}} \underset{51\%}{Ru(CO)_5}, \quad (8)$$

Table II

Effect of Hydrocarbon Additions on WGSR Activity

| 0.10 mM[$Ru_3(CO)_{12}$] | | 0.50 mM[$Ru_3(CO)_{12}$] | |
|---|---|---|---|
| Hydrocarbon (g) | $t_{H_2}$ | Hydrocarbon (g) | $t_{H_2}$ |
| none | 9300 | none | 4900 |
| 1-butene (5.2) | 11700 | 1-butene (5.2) | 5280 |
| pentane (10) | 10700 | 1-hexene (7.7) | 6800 |
| pentane (18) | 7300 | pentane (13) | 6440 |
| | | 1-butene (15.5) | 5080 |
| | | pentane (65) | <25 |

Conditions: solution prepared from 20 g 25% aq. $NMe_3$, hydrocarbon and sufficient THF or diglyme to bring solution volume to 100 mL, 100°, 10 hr; experiments employing 0.10 mmol $Ru_3(CO)_{12}$ charged with 700 psi CO and those at 0.050 mmol charged with 750 psi. The experiment with 65 g pentane had no ether solvent present.

while James (32) has shown that the trimer is the principal product when essentially the same reaction is carried out in methanol,

$$Ru(acac)_3 + CO + H_2 \xrightarrow[140\text{-}160^o]{MeOH} Ru_3(CO)_{12}. \qquad (9)$$
$$(300\ atm) \qquad\qquad 76\%$$

It is thought that small additions of hydrocarbon solvents tend to enhance the formation of $Ru(CO)_5$, whereas larger concentrations seriously decrease the dielectric constant of the solvent so that the formation of ionic species in solution is suppressed.

The base has a very important effect on the efficiency of ruthenium carbonyl for the WGSR (see Table III). Amines were found to provide much better activity than Bronsted bases, and trimethylamine appears to be the base of choice, affording rates more than two orders of magnitude greater than those of Bronsted bases.

Table III

Effect of Base on the Ruthenium Carbonyl-Catalyzed Water Gas Shift Reaction

| Base | Amt Base,g | Amt $H_2O$,g | $t_{H_2}$ |
|---|---|---|---|
| $NMe_3$ | 5 | 15 | 5740 |
| $NEt_3$ | 5 | 15 | 860 |
| $NBu_3$ | 5 | 15 | 540 |
| N-Me Pyrrolidine | 5 | 15 | ∿2400 |
| $NHMe_2$ | 5 | 15 | 2200 |
| Pyridine | 4 | 16 | ∿ 300 |
| $NH_3$ | 6 | 15 | 420 |
| $Na_2CO_3$ | 3 | 20 | < 50 |
| $Li_2CO_3$ | 2 | 20 | < 50 |
| $Me_4NOH$ | 0.2 | 20 | < 50 |
| $Bu_4NOH$ | 0.03 | 20 | < 50 |
| None | 0.0 | 20 | < 50 |

Conditions: 0.05 mmol $Ru_3(CO)_{12}$, 92 mmol 1-butene, base and water diluted to 100 mL with diglyme, 750 psi CO, 100°, 10 hr, 0.31 L reactor.

The nucleophilic reaction of hydroxide with carbonyl ligands of transition metal complexes,

$$M{=}C{=}O + OH^- \longrightarrow \overline{M}{-}C\begin{matrix}\nearrow O \\ \searrow OH\end{matrix} \xrightarrow{-CO_2} \overline{M}{-}H \qquad (10)$$

is a well-known reaction (33), frequently employed in the preparation of metal hydride complexes (34). However, Bronsted bases are involved in a number of reactions which lower the nucleophile concentration:

$$CO + OH^- \longrightarrow HCO_2^- \tag{6}$$

$$CO_2 + OH^- \longrightarrow HCO_3^- \tag{11}$$

$$HCO_3^- + OH^- \longrightarrow CO_3^{2-} + H_2O \tag{12}$$

Because of these side reactions, we have not been able to maintain the apparent pH above 10.0 for any significant period of time; therefore the maximum sustainable hydroxide nucleophile concentration in an experiment of several hours is about $1.0 \times 10^{-4}$ M.

Amines hydrolyze,

$$NR_3 + H_2O \rightleftharpoons HNR_3^+ + OH^-, \tag{13}$$

to provide hydroxide concentrations comparable to those of weaker Bronsted bases (0.5 M $NMe_3$ is pH 12.2; 0.5 M $Na_2CO_3$ is pH 12.1; at room temperature in water). Thus amines may participate in the WGSR via hydroxide attack on carbonyl ligands, as is evident in the rhodium carbonyl system (26). However, direct amine attack on carbonyl ligands is known. Edgell (35,36) has reported that primary and secondary amines react with iron pentacarbonyl to form zwitterionic metallocarboxamides, **3**, which in the presence of traces of water are rapidly hydrolyzed (presumably by nucleophilic attack by water on the activated carbonyl carbon) to the hydride anion:

$$Fe(CO)_5 + R_2NH \rightleftharpoons R_2\overset{+}{N}(H)\text{-}\overset{O}{\overset{\|}{C}}{}^{-}\text{-}Fe(CO)_4 \xrightarrow{H_2O} HFe(CO)_4^- \tag{14}$$

**3**

Nesmeyanov has provided interesting examples of apparent intramolecular nucleophilic attack by amine on carbonyl ligands (37). Angelici (38,39) has demonstrated that amine attack on cationic metal carbonyl complexes is a general reaction resulting in the formation of carbamoyl complexes:

$$MCO^+ + NHR_2 \rightleftharpoons M\text{-}\overset{O}{\overset{\|}{C}}\text{-}\overset{+}{N}HR_2 \underset{}{\overset{-H^+}{\rightleftharpoons}} M\text{-}\overset{O}{\overset{\|}{C}}\text{-}NR_2 \tag{15}$$

Ammonia, primary and secondary amines are known to undergo side reactions under WGSR conditions:

$$CO + HNR_2 \longrightarrow HCONR_2 \quad (16)$$

$$CO_2 + HNR_2 \longrightarrow HOCONR_2 \quad (17)$$

$$CO_2 + 2HNR_2 \longrightarrow R_2NCONR_2 \quad (18)$$

These reactions serve to deplete the available nucleophile concentration. Furthermore, the formation of a relatively stable carbamoyl complex may serve to lower both the metal carbonyl and the nucleophile concentrations.

Angelici and Brink (40) have found that in the reactions of amine with trans-$M(CO)_4(PPhMe_2)_2^+$ (M = Mn or Re), the rate of carbamoyl formation follows the order, n-butylamine > cyclohexylamine $\gtrsim$ isopropylamine > sec-butylamine >> tert-butylamine, implying a strong steric effect in carbamoyl formation. A similar order has been observed in the rate of reaction of organic esters with amines to form amides (41). The data in Table III indicate that a steric effect may be operative in the $Ru_3(CO)_{12}/NR_3$-catalyzed WGSR, since with tertiary amines the rate follows the order, $NMe_3$ > $MeNC_4H_8$ > $NEt_3$ > $NBu_3$, which does not reflect the basicity of these amines.

Angelici (38) has correlated the reactivity of metal carbonyl complexes toward amines with C-O stretching force constants. Application of his empirical rule to the WGSR indicates that on electronic grounds, $Ru(CO)_5$ and $Ru_3(CO)_{12}$ should be comparably subject to carbonyl nucleophilic attack. In light of the steric effect observed in nucleophilic attack by amines, mononuclear species are thought to be more effective than cluster species in the ruthenium carbonyl-amine catalyzed WGSR. When the nucleophile is the much more sterically compact hydroxide group, there may be little steric bias in the base attack step.

The use of amines allows much higher nucleophile concentrations than those achievable with Bronsted bases. We have used solutions as concentrated as 6 M $Me_3N$. This vast difference in available nucleophile concentration partially explains the huge increase in rate afforded by $NMe_3$ over the rate with Bronsted bases. Very large concentrations of hydroxide may promote the base attack step but can decrease the rate of the WGSR due to inhibition of the protonation of the metal hydride species.

Additional Comments Regarding the Ruthenium Carbonyl-Trimethylamine WGSR System. A potential mechanistic pathway for a WGSR system involves the production of formate, followed by its catalytic decomposition:

$$CO + OH^- \longrightarrow HCO_2^-, \quad (6)$$

$$HCO_2^- + H_2O \longrightarrow H_2 + CO_2 + OH^- . \qquad (19)$$

Ruthenium carbonyl decomposes the formate ion in basic media, but at a rate slower than the rate of the WGSR. At 100° and 0.10 mM $Ru_3(CO)_{12}$, under 3 atm $N_2$, the rate of decomposition of trimethyl ammonium formate to $H_2$ and $CO_2$ is 0.6 mmol/hr. Under 5 atm CO the rate is slower (<0.1 mmol/hr), but the overall rate of $H_2$ production is >0.4 mmol/hr. At this low CO pressure, the rate of $H_2$ production directly from CO and $H_2O$ is more than three times that from formate decomposition. Furthermore, since increases in CO pressure result in improved $H_2$ production rates (10 mmol/hr at 50 atm CO), while apparently inhibiting the rate of formate decomposition, it may be concluded that formate decomposition has little mechanistic significance in the WGSR activity of $Ru_3(CO)_{12}/NMe_3$.

On the basis of this discussion, we propose that the $Ru_3(CO)_{12}/NMe_3$-catalyzed WGSR follows the mechanism shown in Figure 3. A similar mechanism, involving nucleophilic attack by hydroxide instead of amine, has been proposed by Pettit and coworkers (4) for the $Fe(CO)_5$/base system.

The dihydride 6, once formed should eliminate $H_2$ readily; it decomposes rapidly at 0° (42) or above 20° under 300 atm $H_2$ (27). This should be contrasted with the stability and, presumably, the catalytic activity of the cluster hydride, $H_4Ru_4(CO)_{12}$. Kaesz (43) has indicated the cluster species 1 and 8 are in

$$\underset{\mathbf{1}}{Ru_3(CO)_{12}} \underset{CO}{\overset{H_2}{\rightleftharpoons}} \underset{\mathbf{8}}{H_4Ru_4(CO)_{12}} \qquad (20)$$

equilibrium and may be interconverted. Conditions as mild as 1 atm $H_2$ at 80° allow the quantitative conversion of 1 to 8 (44). Furthermore, $H_4Ru_4(CO)_{12}$ is reported to be stable under a 1:1 mixture of CO and $H_2$ at 100° (45). If an important step in the ruthenium carbonyl-catalyzed WGSR involved elimination of $H_2$ from 8, hydrogen pressure should tend to inhibit the WGSR; if elimination of $H_2$ occurred with the monomer 6, little inhibition of the WGSR should be observed. The effect of hydrogen pressure on the rate of $H_2$ production was tested by carrying out a WGSR at 0.10 mM $Ru_3(CO)_{12}$ with an initial pressure of 310 psi $H_2$ ($CO:H_2 = 2$); the rate of $H_2$ production was identical to that in the absence of added $H_2$.

By judicious adjustment of conditions, the rate of the $Ru_3(CO)_{12}/NMe_3$-catalyzed WGSR could be significantly improved. As mentioned earlier, those factors which favor formation of $Ru(CO)_5$ -- decreases in concentration and increases in CO pressure -- favor higher turnover numbers in the WGSR. Increases in amine concentration and in temperature also improve the rates of $H_2$ production. Thus, at 155°, 0.0082 mM $Ru_3(CO)_{12}$ and 1080 psi initial CO pressure, the net hydrogen turnover number is 270,000 over a 10 hr period. This corresponds to a rate of 7.5 mol $H_2$ per mol $Ru_3(CO)_{12}$ per sec, an improvement of nearly three orders of magnitude over the rate in any other reported homogeneous system.

## The Group VI Metal Carbonyl System

The Group VI metal carbonyls demonstrate good activity in the WGSR, but differ significantly from ruthenium carbonyl in several ways. Tables IV and V summarize some WGSR experiments with chromium and tungsten carbonyls in a tetrahydrofuran-water solvent system.

TABLE IV

$Cr(CO)_6$ as WGSR Catalyst in $THF/H_2O$

| Base (mmol) | g $H_2O$ | mmol $H_2$ | $t_{H_2}$ |
|---|---|---|---|
| $K_2(CO)_3$, (12) | 7.5 | 9.5 | 0.95 |
| $NMe_3$, (85) | 15 | 2.4 | 0.24 |
| None | 20 | <0.01 | <0.001 |

Conditions: 10.0 mmol $Cr(CO)_6$, solution phase diluted to 100 mL with tetrahydrofuran, 600 psi CO charge at room temperature, 150° for 20 hr, 0.31 L reactor, $t_{H_2}$ = mol $H_2$/mol cat.

TABLE V

$W(CO)_6$ as WGSR Catalyst in $THF/H_2O$

| Base (mmol) | $P_{CO}$, psi | mmol $H_2$ | $t_{H_2}$ |
|---|---|---|---|
| $Na_2CO_3$ (2.4) | 800 | 6.0 | 4.2 |
| $Na_2CO_3$ (2.4) | 110 | 8.0 | 5.4 |
| $NMe_3$ (114) | 790 | 4.0 | 3.0 |

Conditions: 1.45 mmol $W(CO)_6$, 20 g $H_2O$ diluted to 100 mL with tetrahydrofuran, 150° for 20 hr, 0.31 L reactor.

With $Cr(CO)_6$, base clearly promotes the WGSR. However, unlike ruthenium carbonyl, chromium and tungsten carbonyls demonstrate less activity with trimethylamine than with carbonate as base.

Darensbourg and Darensbourg (46) have associated the reactivity of carbonyl ligands toward nucleophiles with C-O stretching force constants. This is reasonable since the higher the force constant, the greater the positive charge on the carbonyl carbon. The C-O stretching force constants of a variety of metal carbonyl complexes have been compiled (26). Those of the Group VI metal carbonyls (16.4 to 16.5 mdyn/Å, depending on investigator) do not differ significantly (47). The force constants of the axial carbonyls of $Ru(CO)_5$ and $Ru_3(CO)_{12}$ (17.1 (ax), 16.6 (eq) and

17.0 (ax), 16.3 (eq), respectively) are considerably higher, indicating a greater tendency for nucleophilic attack.

King and King (11) reported considerably higher WGSR activity for the Group VI metal carbonyls in methanol-water than we observe in tetrahydrofuran-water. Table VI summarizes some experiments with tungsten carbonyl in methanol-water, and shows results with $W(CO)_6$/KOH comparable with those reported by the King group. Methanol-water as solvent is preferable to THF-water, apparently because of the pronounced tendency of sodium formate solutions to "salt-out" tetrahydrofuran, and thus lead to a two-phase reaction solution.

Table VI

$W(CO)_6$ as WGSR Catalyst in $MeOH/H_2O$

| Base (mmol) | $P_{CO}$, psi | mmol $H_2$ | $t_{H_2}$ |
|---|---|---|---|
| KOH (24) | 200 | 30 | 170 |
| $NaO_2CH$ (48) | 200 | 40 | 220 |
| $K_2CO_3$ (24) | 130 | 44 | 250 |
| $K_2CO_3$ (24) | 200 | 31 | 170 |
| $K_2CO_3$ (24) | 800 | 18 | 95 |

Conditions: 0.18 mmol $W(CO)_6$, 2.5 mL $H_2O$ diluted to 100 mL with methanol, 20 hr, 0.31 L reactor.

We found little difference between the activities of this catalyst with $K_2CO_3$ and with KOH. However, a pronounced dependence on pressure was seen: for a six-fold decrease in CO pressure, the activity increased by a factor of 2.5. This tendency is in marked contrast to the activity increase with increasing CO pressure observed with ruthenium carbonyl.

This activity dependence on CO pressure could be attributed to a dissociative-type mechanism, i.e., one that necessitates loss of a carbonyl ligand as a kinetically limiting step:

$$M(CO)_6 \longrightarrow M(CO)_5 + CO. \quad (21)$$

Coordinately unsaturated species are highly reactive (48) and are usually present as solvates in the presence of ligating solvent molecules.

The possible intermediacy of the formate ion (eqs. 6 and 18) in the WGSR has been considered (2,6,10), but its involvement has not been clearly demonstrated. The Group VI metal carbonyl complexes are effective in the decomposition of formic acid (as sodium formate), as shown in Table VII. Some heterogeneity is observed in those reactions carried out under nitrogen pressure, but in no case was CO detected. The similarity in rates for WGSR

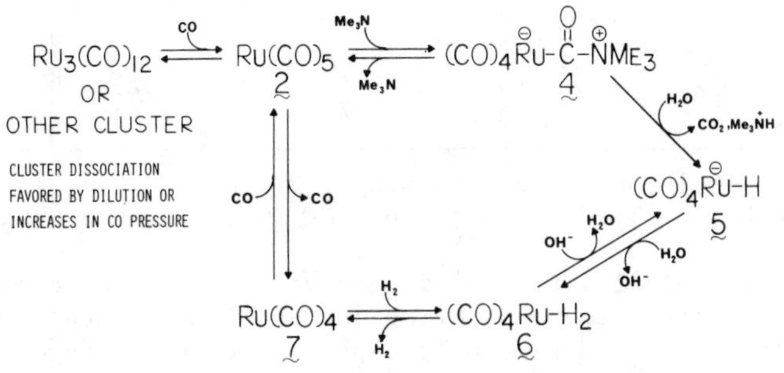


*Figure 3. Proposed mechanism for the $Ru_3(CO)_{12}/NMe_3$-catalyzed homogeneous WGSR*

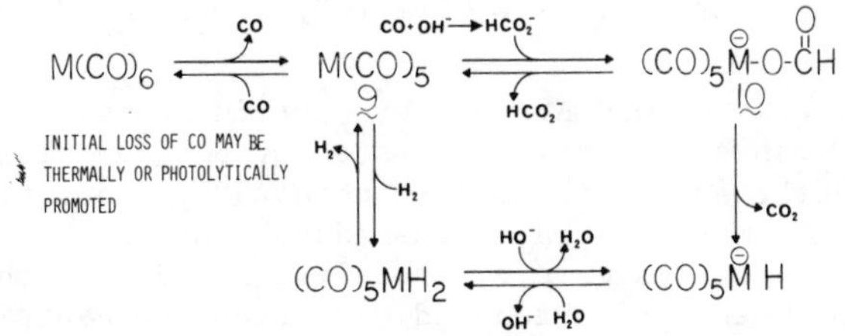


*Figure 4. Proposed mechanism for the group VI metal carbonyl-catalyzed homogeneous WGSR*

and formate decomposition and the inverse rate dependence on CO pressure have prompted us to propose the mechanism shown in Figure 4 for the Group VI metal carbonyls.

Table VII

The Activity of the Group VI Metal Carbonyls in Decomposition of Formic Acid as Formate

| Metal Carbonyl | $P_{CO}$, psi | mmol $H_2$ | mmol $CO_2$ | $t_{H_2}$ |
|---|---|---|---|---|
| $W(CO)_6$ | 200 | 40 | 13 | 220 |
| $W(CO)_6$ | 0 | 43 | 3 | 240 |
| $Mo(CO)_6$ | 0 | 41 | 2.5 | 220 |
| $Cr(CO)_6$ | 0 | 43 | <0.5 | 240 |

Conditions: 0.18 mmol metal carbonyl, 2.5 mL $H_2O$, 97.5 mL MeOH, $155^o$, 20 hr, 0.31 L reactor; 0 psi CO implies 250 psi $N_2$.

To test the validity of this mechanism, chromium carbonyl (1.0 g) was photolyzed under Ar at ambient temperature in a solution of methanol and hexamethylphosphoramide in the apparatus shown in Figure 5. The lamp was turned off periodically to check for the disappearance of slightly soluble $Cr(CO)_6$. Several photolyzing cycles were necessary to effect nearly complete conversion to the solvent-stabilized coordinately unsaturated species (equivalent to 9 in Figure 4),

$$Cr(CO)_6 \xrightarrow[S]{h\nu} CO + (CO)_5Cr\text{-}S. \qquad (22)$$

At this point, only CO and Ar were present in the gas phase. Most of the CO was swept from the system by purging with argon. After allowing the system to cool to about $30^o$, a solution of $[Et_4N]^+[O_2CH]^-$ in methanol-water was added. After about a minute, small streams of bubbles were vigorously evolved from the solution. The gases were collected and were found to contain principally Ar and $H_2$ with a small quantity of $CO_2$.

## Conclusions

It now appears that at least two mechanisms exist for the base-promoted homogeneous water gas shift reaction, differing in the method of hydride formation. The "associative mechanism", first proposed by Pettit and co-workers (1,4), involves nucleophilic attack on a carbonyl ligand and it has two variations. One involves hydroxide attack, leading to the formation of a metallocarboxylic acid (species 11 in Figure 6), and is evident in the $Fe(CO)_5$/base-catalyzed system (1). The other involves the formation of a readily hydrolyzable, zwitterionic metallocarboxamide, 12, in accord with the work of Edgell (35,36) and is evident in the $Ru_3(CO)_{12}/NMe_3$ system.

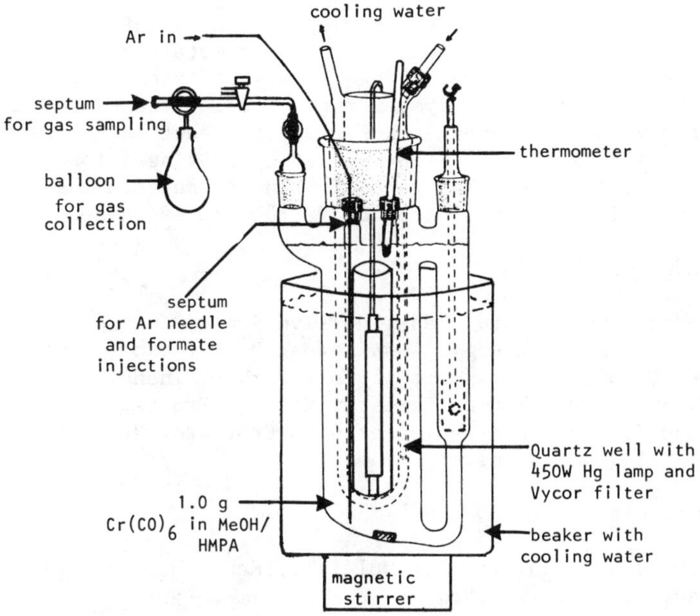


*Figure 5. Schematic of photolysis apparatus*

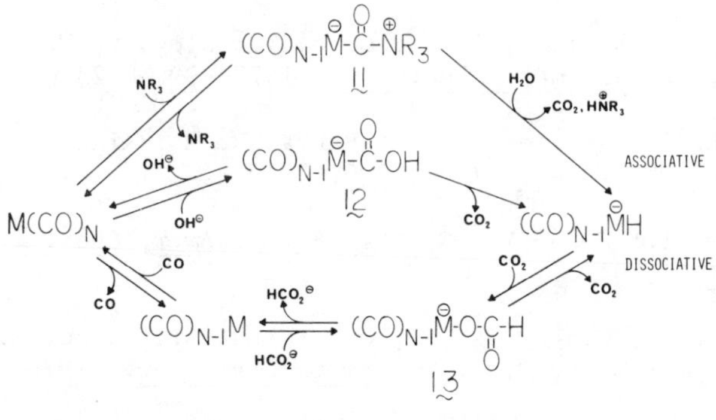


*Figure 6. Associative and dissociative mechanisms for hydride formation*

The group VI metals preferentially follow a "dissociative mechanism" in which loss of CO precedes formation of the catalytically active coordinately unsaturated species. The key intermediate, the formato complex, **13**, is very similar to surface formate intermediates observed during the course of the heterogeneous WGSR (49,50) and thus the Group VI metal carbonyl WGSR may be one of the best homogeneous models of heterogeneous reactions. Furthermore, this system further demonstrates the importance of oxygen-bound species in catalytic reactions of carbon monoxide (51); intermediate **13** may be regarded as a formyl complex of a metal oxide. The rôle of bound formate in the WGSR and other catalytic reactions is being further investigated.

## Acknowledgments

The authors wish to thank the Division of Chemical Sciences, U. S. Department of Energy, Washington, D. C., under Contract No. DE-AC02-76CH00016 for support of this work. Financial support of B. A. E. by the Brookhaven Semester Student Program is gratefully acknowledged. We are also indebted to Professor R. Pettit for many helpful discussions.

## Literature Cited

1. Pettit, R.; Mauldin, C.; Cole, T.; Kang, H. Ann. New York Acad. Sci., 1977, 295, 151.

2. Laine, R. M.; Rinker, R. G.; and Ford, P. C. J. Amer. Chem. Soc., 1977, 99, 252.

3. Cheng, D-H.; Hendriksen, D. E.; Eisenberg, R. J. Amer. Chem. Soc., 1977, 99, 2791.

4. Kang, H-C.; Mauldin, C. H.; Cole, T.; Slegeir, W.; Cann, K.; Pettit, R. J. Amer. Chem. Soc., 1977, 99, 8323.

5. King, R. B.; Frazier, C. C.; Hanes, R. M.; King, A. D. J. Amer. Chem. Soc., 1978, 100, 2925.

6. Yoshida, T.; Ueda, Y.; Otsuka, S. J. Amer. Chem. Soc., 1978, 100, 3941.

7. Ford, P. C.; Rinker, R. G.; Ungermann, C.; Laine, R. M.; Landis, V.; Moya, S. A. J. Amer. Chem. Soc., 1978, 100, 4595.

8. Cheng, C-H.; Eisenberg, R. J. Amer. Chem. Soc., 1978, 100, 5968.

9. Laine, R. M. J. Amer. Chem. Soc., 1978, 100, 6451.

10. Ford, P. C.; Rinker, R. G.; Laine, R. M.; Ungermann, C.; Landis, V.; Moya, S. A. Adv. Chem. Ser., 1979, 173, 81.

11. Frazier, C. C.; Hanes, R.; King, A. D.; King, R. B. Adv. Chem. Ser., 1979, 173, 94.

12. Darensbourg, D. J.; Darensbourg, M. Y.; Burch, R. R.; Froelich, J. A.; Incorvia, M. J. Adv. Chem. Ser., 1979, 173, 106.

13. Pettit, R.; Cann, K.; Cole, T.; Mauldin, C. H.; Slegeir, W. Adv. Chem. Ser., 1979, 173, 121.

14. Ungermann, C.; Landis, V.; Moya, S. A.; Cohen, H.; Walker, H.; Pearson, R. G.; Rinker, R. G.; Ford, P. C. J. Amer. Chem. Soc., 1979, 101, 5922.

15. Pettit, R.; Cann, K.; Cole, T.; Mauldin, C. H.; Slegeir, W. Ann. New York Acad. Sci., 1980, 333, 101.

16. Baker, E. C.; Hendriksen, D. E.; Eisenberg, R. J. Amer. Chem. Soc., 1980, 102, 1020.

17. King, A. D.; King, R. B.; Yang, D. B. J. Amer. Chem. Soc., 1980, 102, 1028.

18. Darensbourg, D. J.; Baldwin, B. J.; Froelich, J. A. J. Amer. Chem. Soc., 1980, 102, 4688.

19. King, A. D.; King, R. B.; Yang, D. B. J. Chem. Soc., Chem. Commun., 1980, 529.

20. Mills, G. A., Steffgen, F. W. Catal. Rev., 1973, 8, 159.

21. "Catalyst Handbook"; Springer-Verlag: London, 1970; Chapters 5, 6, 7.

22. Reppe, W.; Vetter, H. Ann. Chem., 1953, 582, 133.

23. Kölbel, H.; Engelhardt, F. Erdol u. Kohle, 1949, 2, 52.

24. Cann, K.; Cole, T.; Slegeir, W.; Pettit, R. J. Amer. Chem. Soc., 1978, 100, 3969.

25. Bradley, J. S. "Abstracts of Papers"; 34th Southwest Regional Meeting, ACS, Corpus Christi, Tx., November 1978; IHSC54.

26. Slegeir, W. A. "Ph.D. Dissertation", The University of Texas at Austin, December 1979.

27. Whyman, R. J. Organometal. Chem., 1973, 56, 339.

28. Moss, J. R.; Graham, W. A. G. J. Chem. Soc. Dalton, 1977, 95.

29. Calderazzo, F.; L'Eplattenier, F. Inorg. Chem., 1967, 6, 1220.

30. Dombek, B. D. "Abstracts of Papers"; Second Chemical Congress of the North American Continent, Las Vegas, Nv., Aug. 24-29, 1980; INOR 208.

31. Piacenti, F.; Bianchi, M.; Frediani, P.; Benedetti, E. Inorg. Chem., 1971, 10, 2759.

32. James, B. R.; Rempel, G. L.; Teo, W. K. Inorg. Synth., 1976, 16, 45.

33. Darensbourg, D. J. Israel J. Chem., 1977, 15, 247.

34. Kaesz, H. D.; Saillant, R. B. Chem. Rev., 1972, 72, 231.

35. Edgell, W. F.; Yang, M. T.; Bulkin, B. J.; Bayer, R.; Koizumi, N. J. Amer. Chem. Soc., 1965, 87, 3080.

36. Edgell, W. F.; Bulkin, B. J. J. Amer. Chem. Soc., 1966, 88, 4839.

37. Nesmeyanov, A. N.; Salńikova, T. N.; Struchkov, Y. T.; Andrianov, V. G.; Progrebnyak, A. A.; Rybin, L. V.; Rybenskaya, M. I. J. Organometal. Chem., 1976, 117, C16.

38. Angelici, R. J. Accounts Chem. Res., 1972, 5, 335.

39. Angelici, R. J.; Blacik, L. Inorg. Chem., 1972, 11, 1754.

40. Angelici, R. J.; Brink, R. W. Inorg. Chem., 1973, 12, 1067.

41. Arnett, E. M.; Miller, J. G.; Day, A. R. J. Amer. Chem. Soc., 1950, 72, 5635.

42. Cotton, J. D.; Bruce, M. I.; Stone, F. G. A. J. Chem. Soc. A., 1968, 2162.

43. Kaesz, H. D. Chem. Brit., 1973, 344.

44. Knox, S. A. R.; Koepke, J. W.; Andrews, M. A.; Kaesz, H. D. J. Amer. Chem. Soc., 1975, 97, 3942.

45. Johnson, B. F. G.; Johnston, R. D.; Lewis, J.; Robinson, B. H.; Wilkinson, G. J. Chem. Soc. A., 1968, 2856.

46. Darensbourg, D. J.; Darensbourg, M. Y. Inorg. Chem., 1970, 9, 1691.

47. Cotton, F. A.; Kraihanzel, C. S. J. Amer. Chem. Soc., 1962, 84, 4432.

48. Perutz, R. N.; Turner, J. J. J. Amer. Chem. Soc., 1975, 97, 4791.

49. Ueno, A.; Onishi, T.; Tamaru, K. Trans. Faraday Soc., 1970, 66, 756.

50. Aenenomiya, Y. J. Catal., 1979, 57, 64.

51. Sapienza, R. S.; Sansone, M. J.; Spaulding, L. D.; Lynch, J. F. in M. Tsutsui, Ed., "Fundamental Research in Homogeneous Catalysis", Vol. 3; Plenum: New York, 1979; pp. 179-197.

RECEIVED December 8, 1980.

# INDEX

# INDEX

## D

## N